Basic Mechanics with Engineering Applications

T0187641

Basic Mechanics with Engineering Applications

J. N. Fawcett and J. S. Burdess

Routledge
Taylor & Francis Group

LONDON AND NEW YORK

First published by Butterworth-Heinemann

First published 1988

This edition published 2011 by Routledge
2 Park Square, Milton Park, Abingdon, Oxon OX14 4RN
711 Third Avenue, New York, NY 10017, USA

Routledge is an imprint of the Taylor & Francis Group, an informa business

British Library Cataloguing in Publication Data
A catalogue record for this book is available from the British Library

ISBN 978-0-415-50317-4

Preface

Modern societies depend upon machines to provide standards of living far beyond the level that man alone can provide. The human body is very limited in its ability to generate large, or very small forces and cannot work at high speeds or with high accuracy for long periods. To overcome these limitations man has striven to develop machines which augment his own strength and reduce the drudgery of repetitive tasks. We all now depend upon machines to provide us with our food, clothing, energy, transport and defence. Most households, in addition to possessing some form of transport, have machines to reduce the effort involved in washing, cleaning and preparing food. Agriculture and medicine depend upon machines to increase their efficiency. Less familiar are the machines used on an industrial scale for knitting and weaving, for food processing and packaging and for the manufacture of metal goods of all sorts.

All engineers will at some time have a professional interest in machines, perhaps as a designer, but more often as a user of machines. A sound understanding of the physical principles which govern the behaviour of machines is therefore an essential part of an engineer's education.

A machine is an assembly of fixed and moving bodies connected together in some manner. Dynamics is the study of bodies in motion and provides the basic tools for the analysis and synthesis of machines. The subject is usually split into two parts. The first, kinematics, is the study of the geometric properties of motion, and the second, kinetics, is concerned with the forces which cause, or maintain, motion.

The book is intended as a first year undergraduate course for Mechanical Engineering students, and as a course in basic mechanics suitable for students in other engineering subjects.

The authors have taught Machine Dynamics courses at undergraduate and postgraduate level for a number of years. They have found that the good text books available generally cover a full undergraduate course and use a formal vector approach. For first year students, or for those engineering students who follow only a limited course in Mechanics, they feel that a course restricted to consider only plane motion is adequate. The use of formal vector algebra is not required for a basic course of this type and an informal vector approach can give a greater understanding of the basic principles. It is essential however that derivations of formulae should be rigorous and that any assumptions made in their derivation are fully explained.

In each section of this book the basic theory is presented in a rigorous manner and then applied to typical engineering problems. The relevance of the mathematical model used to represent each real system is considered and the meaning of the results is discussed. The importance of understanding the underlying theory, and the use of free body diagrams or system boundaries is emphasised throughout the text.

It is expected that students following this text will develop a thorough understanding of basic mechanics which will enable them to tackle many engineering problems. It will also provide a valuable foundation for more advanced courses in Mechanics.

The authors have carefully checked the text and the examples for student solution. They would welcome notification of any errors which remain, and suggestions for improving the layout and content of the book would also be appreciated.

Finally the authors would like to thank Sheila Stone and Joyce McLean for their efforts well beyond the call of duty in the preparation of the manuscript and figures for the book.

J. N. Fawcett
J. S. Burdess
1988

Contents

1

Kinematics

Basic Theory

1.1 Vector quantities

Postion, velocity and acceleration cannot be described solely in terms of magnitude; their directions must also be specified. Quantities which require a definition of both magnitude and direction are called *vector* quantities; those defined by magnitude only are referred to as *scalar* quantities. Heavy type will be used to denote a vector, e.g. *r* is a vector with magnitude *r*. (A straight line above the symbol, or a wavy line beneath are alternative methods of notation.) A vector quantity is represented graphically by the line and arrow head shown in Fig. 1.1. The length of the line is proportional to the magnitude of the vector and the direction is specified by the orientation of the line and the direction of the arrowhead.

The addition and multiplication of vectors must be carried out according to the rules of vector algebra. For the purposes of this elementary text, which is restricted to the study of motion in a plane, we need only consider the rules of vector addition.

Vector addition

The sum *c* of two vectors *a* and *b* is defined by the diagonal OC of the parallelogram OABC of Fig. 1.2. The lines OA, OB represent the vectors *a* and *b* in both magnitude and direction. Upon completion of the parallelogram, its diagonal OC represents the sum, or resultant, of *a* and *b*. This method of addition is referred to as the *parallelogram rule* of vector addition. The length OC and its direction may be obtained by drawing the parallelogram to some convenient scale, or from simple geometry.

We represent the addition of two vectors by the vector equation

$$c = a + b$$

where the heavy type indicates that we require a *vector sum,* as given by the parallelogram rule, rather than the usual scalar sum involving magnitude alone.

An equivalent and more convenient method of adding two vectors is to draw the first vector *a*, and then the second vector *b* starting from the *tip* of vector *a*, as in Fig. 1.3. The sum, or resultant, *c* will be the vector *from* the *tail* of *a* to the *tip* of *b* and will close the triangle. This is identical to △OAC of Fig. 1.2. We could of course have started with the vector *b* and then added *a*. This would be equivalent to △OBC of Fig. 1.2. The order of the addition is therefore seen to be unimportant, i.e. $a + b = b + a$.

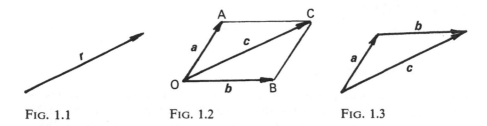

FIG. 1.1   FIG. 1.2   FIG. 1.3

If we wish to add more than two vectors we may extend the above method. For example, to add the four vectors *a, b, c* and *d* the sum or resultant *e* can be obtained by placing them tip to tail as in Fig. 1.4(a), i.e. $e = a + b + c + d$.

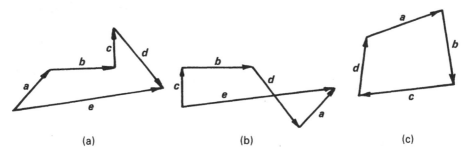

(a)   (b)   (c)

FIG. 1.4

The order can be seen to be unimportant if we recombine them as in Fig. 1.4(b) where $e = c + b + d + a$.

If the vectors being added form a *closed* polygon the length of the resultant will be zero, as in Fig. 1.4(c), and their vector sum will be zero, i.e. $a + b + c + d = 0$.

Components of a vector

It is often convenient to consider a vector in terms of two mutually perpendicular vectors whose sum is equal to the original vector. In Fig. 1.5 the vector *c* is the sum of the two vectors *a* and *b* which are *perpendicular* to one another. Vectors *a* and *b* are referred to as the *components* of *c*. From Fig. 1.5 it can be seen that the magnitude of *c*, and the angle, ϕ, between *a* and *c*, are given by

$$c = \sqrt{a^2 + b^2}, \quad \text{and} \quad \tan \phi = b/a.$$

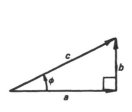

FIG. 1.5

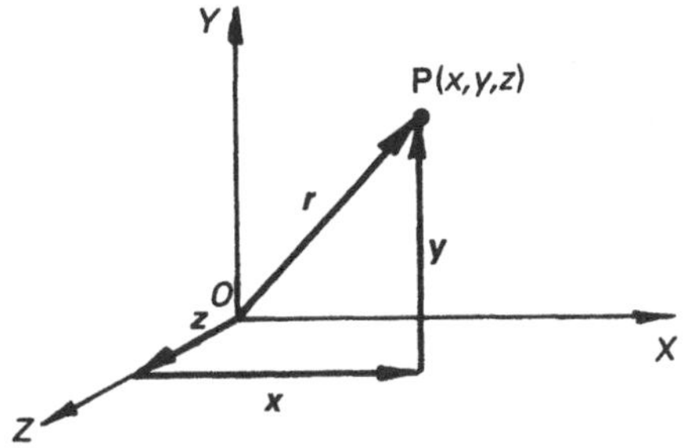

FIG. 1.6

1.2 Motion of a point

Frames of reference

Before we can study the motion of a point we must first define its position in space. We therefore need to choose a fixed point (or origin), and some fixed directions in space to which the position of the point can be referred. The set formed by the origin and the fixed directions is called a *frame of reference*. The position of the origin, and the reference directions, may be chosen arbitrarily.

A reference frame having three reference directions is required to specify the position of a point in space and in most cases it is convenient to arrange these directions to be mutually perpendicular, for example *OX*, *OY*, *OZ* as shown in Fig. 1.6. Using this reference frame, a point at position P is specified by the position vector *r*. This vector has components *x*, *y*, *z*, of magnitude *x*, *y*, *z*, along the reference directions *OX*, *OY* and *OZ*.

The reference frame *OXYZ* is assumed to be fixed in space, and motion measured relative to this frame is called *absolute* motion. If *OXYZ* were moving, we would have to take into account the motion of the frame itself in order to obtain absolute values. For most engineering purposes, a reference frame attached to the earth may be considered fixed since the errors introduced by neglecting the earth's motion are usually negligible. Also, many engineering components are constrained to move in a plane, e.g. *OXY* in Fig. 1.6. In this plane the position vector can be defined using two components, *x* and *y*.

In this text we will limit ourselves to considerations of motion in a plane and will assume that reference frames attached to the earth may be considered fixed in space.

1.3   Motion of a point along a straight line

This is a special case of plane motion. Since we may choose the position of our reference frame arbitrarily, let us arrange that the axis OX lies along the straight line. The OY axis is superfluous.

Position and displacement

The *position* of a point at P may be described by its distance x along the line OX, as shown in Fig. 1.7. In this case the position vector of the point is $r = x$. If the point now moves to P′ such that its new position vector is $x + \delta x$, the change in position δx is called the *displacement* of the point.

FIG. 1.7

The number of degrees of freedom of a point is always equal to the number of *independent* co-ordinates needed to specify its displacement. In this case the displacement can be expressed in terms of a change in the single coordinate x, so that a point moving on a straight line has *one degree of freedom*. Since the direction of the displacement vector is specified by the straight line, all vectors must lie along OX. Hence when adding these vectors, the vector triangle of Fig. 1.3 will reduce to a straight line. This is equivalent to simple scalar addition, taking account of sign.

The recommended SI units for position and displacement are metres, m, or millimetres, mm.

Velocity

Consider the point P with position x at time t. Let the point be moving such that a small time δt later it has moved to a new position $x + \delta x$. The displacement δx is the change in position, and is assumed to be positive in the same sense as x, i.e. along OX. The velocity v of the point is defined as its rate of change of position and is given by

$$v = \lim_{\delta t \to 0} \frac{(x + \delta x) - x}{\delta t} = \frac{dx}{dt}. \tag{1.1}$$

This limit defines the velocity of the point at position x. The derivative dx/dt is often written as $\dot{x}$, where the dot denotes differentiation with respect to time.

Velocity, like position, is a vector quantity and in the case of straight line motion its direction is defined by its sign, which is the same as that of the incremental displacement δx. The recommended SI unit of time is the second, s, so that the SI unit of velocity will be m s^{-1}.

Acceleration

If the point has velocity v at time t, and time δt later has velocity $v + \delta v$, then the acceleration a of the point is defined as its rate of change of velocity, i.e.

$$a = \lim_{\delta t \to 0} \frac{(v + \delta v) - v}{\delta t} = \frac{\mathrm{d}v}{\mathrm{d}t} = \frac{\mathrm{d}^2 x}{\mathrm{d}t^2} = \ddot{x}. \tag{1.2}$$

The recommended SI unit of acceleration is m s^{-2}.

Again the direction is defined by its sign and is positive if δv is positive. Thus, the acceleration of the point is in the same direction as the incremental *change* in velocity, but not necessarily in the same direction as the displacement. For example, a positive displacement would give a positive velocity, but during the displacement the magnitude of the velocity could decrease. In this case δv would be negative, and the acceleration would therefore be negative. This situation arises when applying the brakes in a vehicle.

It is important to note that once we have defined the positive direction of the position x, we have also defined $\dot{x}$ and $\ddot{x}$ positive in the same direction. In the above case x, $\dot{x}$ and $\ddot{x}$ are *all* positive along OX.

Given an expression for the position of a point as a function of time, i.e. $x = f(t)$, it is now possible, using eqns (1.1) and (1.2), to calculate its velocity and acceleration by successive differentiation of x with respect to time.

Example 1.1 A point moves on a straight line with its position, measured from an origin O, given by $x = A \sin \omega t$. Obtain expressions for its velocity and acceleration, if A and ω are constants.

Its velocity

$$v = \frac{\mathrm{d}x}{\mathrm{d}t} = \frac{\mathrm{d}}{\mathrm{d}t} (A \sin \omega t) = A\omega \cos \omega t, \tag{i}$$

and its acceleration

$$a = \frac{\mathrm{d}v}{\mathrm{d}t} = \frac{\mathrm{d}}{\mathrm{d}t} (A\omega \cos \omega t) = -A\omega^2 \sin \omega t = -\omega^2 x. \tag{ii}$$

The acceleration of the point is therefore proportional to the position of the particle, and the negative sign shows that the acceleration is in the opposite sense from the position, and will always be directed towards O.

Graphs showing the variation of x, v and a with time are shown in Fig. 1.8. The position is zero when $\sin \omega t = 0$, i.e. when $\omega t = 0$, $\pi, 2\pi, \ldots$ and the motion repeats itself after a time $2\pi/\omega$. The point has extreme positions of $\pm A$. The velocity of the point depends upon $\cos \omega t$ and at $t = 0$ has a value

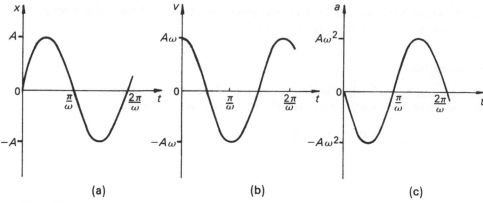

FIG. 1.8

of $+A\omega$. As time increases v reduces and is zero when $\omega t = \pi/2$. At this point the position is at its maximum value. When $t = \pi/\omega$, $\cos \omega t = -1$ and the velocity has its maximum negative value of $-A\omega$. The cycle continues until, at $\omega t = 2\pi$, $v = A\omega$, and the cycle repeats itself. The acceleration of the point varies with $-\sin \omega t$ between $\pm A\omega^2$ as shown in Fig. 1.8(c).

Note that, at any instant, the velocity is equal to the slope of the x-t curve (i.e. $\mathrm{d}x/\mathrm{d}t$), and the acceleration is equal to the slope of the v-t curve, (i.e. $\mathrm{d}v/\mathrm{d}t$).

This type of motion occurs in vibrating systems, and is called *simple harmonic motion*. The motion is characterised by its *amplitude, A,* and *frequency, ω.* It can be seen to repeat itself after a time, $2\pi/\omega$, which is called the *periodic time* of the motion.

Other variables

Position is not always known as a function of time, but may be given as a function of another variable u, i.e. $x = f(u)$. In this case we must apply the chain rule of differentiation when obtaining the velocity, i.e.

$$v = \frac{\mathrm{d}x}{\mathrm{d}t} = \frac{\mathrm{d}x}{\mathrm{d}u}\frac{\mathrm{d}u}{\mathrm{d}t}. \tag{1.3}$$

We can also use the chain rule when finding the acceleration. For example, if velocity were given as a function of, say, position, i.e. $v = f(x)$, then

$$a = \frac{\mathrm{d}v}{\mathrm{d}t} = \frac{\mathrm{d}v}{\mathrm{d}x}\frac{\mathrm{d}x}{\mathrm{d}t}.$$

But $\dfrac{\mathrm{d}x}{\mathrm{d}t} = v$, so that $a = v\dfrac{\mathrm{d}v}{\mathrm{d}x}$.   (1.4)

Integration to give velocity and position

We have shown that expressions for velocity may be obtained from position, and acceleration from velocity, by differentiation with respect to time. By performing the reverse process, i.e. by integration with respect to time, we can obtain velocity from acceleration, and position from velocity.
From eqn (1.2),

$$\mathrm{d}v = a\,\mathrm{d}t. \tag{1.5}$$

If the acceleration, a, can be expressed as a known function of time [i.e. $a = f(t)$], then by integrating eqn (1.5) the velocity at some time t is given by

$$v = \int a\,\mathrm{d}t + A \tag{1.6}$$

where A is an arbitrary constant of integration.

Similarly we can find the position of the point by integrating, with respect to time, an expression for its velocity.

From eqn (1.1)

$$\mathrm{d}x = v\,\mathrm{d}t,$$

and if v is a known function of time,

$$x = \int v\,\mathrm{d}t + B, \tag{1.7}$$

where B is again a constant of integration.

The constants A and B can be determined if the values of x and v are known at some time during the motion. A and B can also be calculated if the position x is known at *two* times during the motion. Usually the constants follow from eqns (1.6) and (1.7) by knowing v at time t_1 and x at time t_2. In many cases $t_1 = t_2 = 0$ and it is the initial velocity and position of the particle which are often used to determine the constants of integration.

Example 1.2 A point accelerates along a straight line with constant acceleration a. Let us obtain expressions for its velocity and position as functions of time, if at $t = 0$, $\dot{x} = v_0$ and $x = 0$.

From eqn (1.2)

$$a = \frac{\mathrm{d}v}{\mathrm{d}t},$$

so that $v = \displaystyle\int \mathrm{d}v = \int a\,\mathrm{d}t + A,$ $\qquad\qquad$ (i)

where A is a constant of integration.

Since a is a constant $\int a \, dt = a \int dt$, so that eqn (i) can be integrated to give

$$v = at + A. \tag{ii}$$

A can be evaluated from the condition that $v = v_0$ at $t = 0$. Substituting this value into eqn (ii) gives $A = v_0$ so that the velocity of the particle at any time t is given by

$$v = at + v_0. \tag{iii}$$

Similarly, from eqn (1.1),

$$v = \frac{dx}{dt},$$

so that eqn (iii) may be expressed as

$$\frac{dx}{dt} = at + v_0.$$

Thus $$x = \int (at + v_0) \, dt + B$$

which, since a and v_0 are constants, can be integrated to give

$$x = \frac{at^2}{2} + v_0 t + B, \tag{iv}$$

where B is a constant of integration.

Using the condition that at $t = 0$, $x = 0$ eqn (iv) gives $B = 0$, so that the position of the particle at any time t is given by

$$x = \tfrac{1}{2}at^2 + v_0 t. \tag{v}$$

Expressions (iii) and (v) are well known, but it must be remembered that they apply *only* in cases where the acceleration is *constant*. A typical example is that of a particle falling freely under gravity near the earth's surface, when its acceleration would be constant at g (9.81 m s^{-2}) vertically downwards.

Graphical solution

In the above example it was possible to evaluate the integrals $\int a \, dt$ and $\int v \, dt$ to obtain closed form analytical solutions. If, however, the functions a and v are such that the integrals cannot be evaluated analytically the results may be expressed as

$$\int_{v_0}^{v} dv = \int_{t_0}^{t} a \, dt,$$

so that $$v - v_0 = \int_{t_0}^{t} a \, dt, \tag{1.8}$$

where v_0 and v are the velocities of the point at t_0 and t respectively.

Similarly

$$\int_{x_0}^{x} dx = \int_{t_0}^{t} v \, dt,$$

so that $x - x_0 = \int_{t_0}^{t} v \, dt.$ (1.9)

The integral on the right-hand side of eqn (1.8) represents the area under the acceleration–time curve over the interval (t_0, t).

Equations (1.8) and (1.9) may therefore be expressed as

$$v - v_0 = \text{area under acceleration–time curve between}$$
$$t_0 \text{ and } t,   \text{(1.10)}$$

and $x - x_0 = $ area under velocity–time curve between t_0 and t,
as shown in Fig. 1.9.

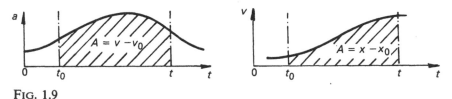

FIG. 1.9

These areas can be evaluated graphically or numerically (e.g. using the trapezoidal rule) to give the change in velocity or the displacement between the two times.

It is instructive to re-examine the curves of Fig. 1.8 and verify that eqns (1.10) apply.

Change of variable

When acceleration is given as a function of time the variables can be separated, as in eqn (1.6), so that the equation can be integrated directly. However, if the acceleration is given as a function of displacement, i.e. $a = f(x)$, the variables cannot be separated.

In this case

$$a = \frac{dv}{dt} = f(x)   \text{(1.11)}$$

and direct integration with respect to time is *not* possible. However, we can change the variable from t to x by writing

$$\frac{dv}{dt} = \frac{dv}{dx} \cdot \frac{dx}{dt} = v \frac{dv}{dx},$$

as in eqn (1.4).

Thus

$$a = v\frac{dv}{dx} = f(x),$$

so that $v\, dv = f(x)\, dx.$

Defining the definite integral from an initial state in which $v = v_0$ and $x = x_0$ gives

$$\int_{v_0}^{v} v\, dv = \int_{x_0}^{x} f(x)\, dx, \tag{1.12}$$

so that $\frac{1}{2}(v^2 - v_0^2) = \int_{x_0}^{x} f(x)\, dx,$

$$= \int_{x_0}^{x} a\, dx. \tag{1.13}$$

Thus the area under the acceleration–displacement curve is equal to $\frac{1}{2}(v^2 - v_0^2)$.

Example 1.3   A drag racing car reaches a maximum speed of $90\ \mathrm{m\ s^{-1}}$. After crossing the finish line the driver releases a braking parachute to reduce the speed of the car. When the speed has been reduced to $18\ \mathrm{m\ s^{-1}}$ he applies the wheel brakes to bring the car to rest. The acceleration between 90 and $18\ \mathrm{m\ s^{-1}}$ is proportional to the square of the velocity and is given by $a = -k_1 v^2$ where k_1 is a constant. Between $18\ \mathrm{m\ s^{-1}}$ and rest the acceleration is constant so that $a = -k_2$.

Calculate the distance travelled between the opening of the braking parachute and the application of the brakes, and the total time and distance taken for the car to be brought to rest, if it is estimated that

$$k_1 = \tfrac{1}{450}\ \mathrm{m^{-1}} \quad \text{and} \quad k_2 = 3\ \mathrm{m\ s^{-2}}.$$

The dimensions of the car are unimportant to our solution so the car may be considered as a point travelling along a straight line.

Between 90 and $18\ \mathrm{m\ s^{-1}}$,

$$a = \frac{dv}{dt} = -k_1 v^2. \tag{i}$$

Separating the variables and integrating we have

$$\int \frac{dv}{v^2} = -k_1 \int dt + C_1,$$

giving $-\dfrac{1}{v} = -k_1 t + C_1.$ \hfill (ii)

If we measure time from the instant at which the parachute opens, then at $t = 0$, $v = 90\ \mathrm{m\ s^{-1}}$ and from eqn (ii), $C_1 = -\tfrac{1}{90}\ \mathrm{s\ m^{-1}}$.

Thus from eqn (ii),

$$t = \frac{C_1 + 1/v}{k_1}.$$ (iii)

When the speed has reduced to 18 m s^{-1},

$$t = (\tfrac{1}{18} - \tfrac{1}{90})450 = 20 \text{ s}.$$

To obtain the distance travelled as the speed reduces we can express eqn (i) in terms of the position x, i.e.

$$a = \frac{dv}{dt} = \frac{dv}{dx} \cdot \frac{dx}{dt} = v\frac{dv}{dx}.$$

Thus $\quad v\dfrac{dv}{dx} = -k_1 v^2.$ (iv)

Separating the variables and integrating gives

$$\int \frac{dv}{v} = -k_1 \int dx + D_1,$$

i.e. $\quad \ln v = -k_1 x + D_1.$

When $x = 0$, $v = 90$ m s^{-1} so that $D_1 = \ln 90$ and

$$k_1 x = \ln 90 - \ln v = \ln\left(\frac{90}{v}\right).$$

Thus when $v = 18$ m s^{-1},

$$x = \frac{1}{k_1} \ln 5 = 725 \text{ m}.$$

The car therefore travels 725 m between the opening of the parachute and the application of the brakes. The reader should verify that the same answer can be obtained from eqn (iii) by integrating v with respect to time over the interval 0 to 20 s.

At speeds less than 18 m s^{-1}, and for $t > 20$ s,

$$a = \frac{dv}{dt} = -k_2.$$ (v)

Since k_2 is a constant, eqn (v) can be integrated directly to give

$$\int dv = -k_2 \int dt + C_2$$ (vi)

so that $\quad v = -k_2 t + C_2.$

When $t = 20$ s, $v = 18$ m s^{-1} so that $C_2 = 78$ m s^{-1}.

Hence when $v = 0$, i.e. the car is brought to rest,

$$0 = -3t + 78$$

and therefore $t = 26$ s.

To find the distance travelled we can again express the acceleration in terms of the position so that eqn (v) becomes

$$v \frac{dv}{dx} = -k_2.$$

Integrating gives

$$\int v \, dv = -k_2 \int dx + D_2$$

or
$$\frac{v^2}{2} = -k_2 x + D_2.$$

When $x = 725$ m, i.e. when the brakes are applied, $v = 18 \text{ m s}^{-1}$, so that $D_2 = 2332 \text{ m}^2 \text{ s}^{-2}$.

Therefore

$$x = \frac{1}{k_2}\left(2332 - \frac{v^2}{2}\right).$$

The distance travelled when the car comes to rest is given by setting $v = 0$,

i.e. $x = \frac{2332}{3} = 777$ m.

The reader should again verify that this result could have been obtained from eqn (vi) by integrating with respect to time over the interval 20 to 26 s.

The total time required to stop the car from its maximum speed is therefore 26 s, and the minimum length of track for a safe stop is 777 m.

1.4   Angular motion

The angular position of a line can be defined by the angle made by the line with some reference direction, e.g. *OX*, in Fig. 1.10. Since only *one* coordinate is required, e.g. θ, the angular motion of a line is directly analogous to the straight line motion of a point.

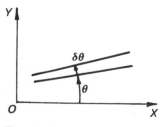

FIG. 1.10

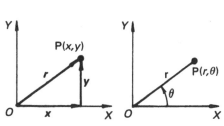

FIG. 1.11

If θ is defined positive in an anticlockwise direction from OX, and the line makes an angle θ with OX at time t, and an angle $\theta + \delta\theta$ at $t + \delta t$, then the angular velocity of the line is defined as

$$\omega = \lim_{\delta t \to 0} \frac{\delta\theta}{\delta t} = \frac{\mathrm{d}\theta}{\mathrm{d}t} = \dot{\theta}. \tag{1.14}$$

The angular velocity, ω, of the line is the rate of change of the angular position of the line, and mathematically, eqn (1.14) is the same as eqn (1.1).

Similarly the angular acceleration, α, of the line is defined as

$$\alpha = \lim_{\delta t \to 0} \frac{\delta\omega}{\delta t} = \frac{\mathrm{d}\omega}{\mathrm{d}t} = \dot{\omega} = \ddot{\theta}. \tag{1.15}$$

Angular displacement is measured in radians (rad) or degrees (°), which are dimensionless. Therefore angular velocity and acceleration have units of rad s^{-1} and rad s^{-2} respectively. As in the case of the motion of a point in a straight line, the sign of the quantities is sufficient to indicate their directions. It is usual to consider the anticlockwise direction as positive.

The equations for angular motion may be manipulated in the same way as those for straight line motion to give expressions such as

$$\omega = \int_{t_0}^{t} \alpha \, \mathrm{d}t + \omega_0, \tag{1.16}$$

which is the equivalent of eqn (1.8), or if $\alpha = f(\theta)$,

$$\tfrac{1}{2}\omega^2 = \int_{\theta_0}^{\theta} f(\theta) \, \mathrm{d}\theta + \tfrac{1}{2}\omega_0^2,$$

the equivalent of eqn (1.13).

It is important to understand that, since a point is of zero size, it is impossible to draw a line in a point. We cannot therefore define its angular orientation so that it is not meaningful to refer to its angular velocity or angular acceleration. *Points* therefore have only *linear* velocities and accelerations.

1.5   Motion of a point in a plane

Two coordinates are necessary to specify the position of a point in the OXY plane, e.g. the Cartesian co-ordinates x, y of Fig. 1.11(a), or the polar coordinates r, θ of Fig. 1.11(b). If the point is moving freely in the plane, changes in both of the co-ordinates used, i.e. δx, δy or δr, $\delta\theta$ are necessary to specify its displacement.

A point moving *freely* in a plane therefore has *two* degrees of freedom.

Cartesian co-ordinates

If we consider first the motion of the point in terms of Cartesian co-ordinates, the position vector, r, may be expressed in terms of its components x along

OX and y along OY, i.e.

$$r = x + y. \tag{1.17}$$

Let us now give the point displacements δx, δy in some small time δt. By considerations similar to those of Section 1.3 the velocity and acceleration components of the point in the directions OX, OY are

$$v_x = \dot{x}, \quad v_y = \dot{y} \quad \text{and} \quad a_x = \ddot{x}, \quad a_y = \ddot{y}.$$

The actual velocity v and acceleration a of the point are obtained by vector addition of the components, i.e.

$$v = v_x + v_y, \quad a = a_x + a_y$$

as in Figs. 1.12(a) and 1.12(b).

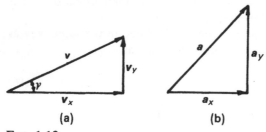

(a)   (b)

FIG. 1.12

The magnitude v of the velocity vector is given in terms of the component magnitudes by

$$v = \sqrt{v_x^2 + v_y^2}. \tag{1.18}$$

The inclination, γ, of v to the axis OX is given by

$$\gamma = \tan^{-1}\left(\frac{v_y}{v_x}\right). \tag{1.19}$$

Similar results may be obtained for acceleration.

Polar co-ordinates

Corresponding expressions for the velocity and acceleration of a point may be obtained in polar co-ordinates by considering the point at P(r, θ) to have moved to P'($r + \delta r$, $\theta + \delta\theta$) in time δt, as shown in Fig. 1.13. Note that θ defines the angular position of the *line* OP.

The radial component of the displacement vector PP' along the direction OP

$$\delta R_r = (r + \delta r) \cos \delta\theta - r.$$

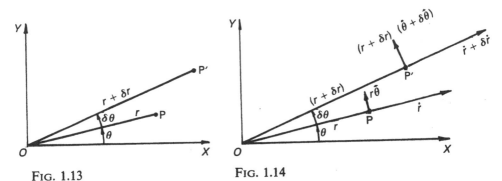

Fɪɢ. 1.13   Fɪɢ. 1.14

The radial component of the velocity is therefore

$$v_r = \lim_{\delta t \to 0} \frac{\delta R_r}{\delta t} = \lim_{\delta t \to 0} \frac{(r + \delta r)\cos \delta\theta - r}{\delta t} = \frac{dr}{dt} = \dot{r}, \tag{1.20}$$

and is positive along OP.

Similarly the velocity of the point in the tangential direction, i.e. perpendicular to OP, is given by

$$v_t = \lim_{\delta t \to 0} \frac{\delta R_\theta}{\delta t} = \lim_{\delta t \to 0} \frac{(r + \delta r)\sin \delta\theta}{\delta t} = r\frac{d\theta}{dt} = r\dot{\theta}. \tag{1.21}$$

The component v_t is positive in the direction the point P would move when the line OP is given a positive (anticlockwise) change in θ.

Vector addition of the radial and tangential components will give the resultant velocity of the point, i.e.

$$v = v_r + v_t.$$

Let the velocity of the point have radial and tangential components $\dot{r}$ and $r\dot{\theta}$ at time t, and $(\dot{r} + \delta\dot{r})$, $(r + \delta r)(\dot{\theta} + \delta\dot{\theta})$ at time δt later as shown in Fig. 1.14.

The absolute change in velocity of the point in the radial direction OP is obtained by resolving the velocity vector at P′ along OP and subtracting $\dot{r}$,

i.e. $\delta v_r = (\dot{r} + \delta\dot{r})\cos \delta\theta - (r + \delta r)(\dot{\theta} + \delta\dot{\theta})\sin \delta\theta - \dot{r}.$

The radial component of the acceleration of the point is therefore

$$a_r = \lim_{\delta t \to 0} \frac{\delta v_r}{\delta t} = \ddot{r} - r\dot{\theta}^2, \tag{1.22}$$

since as $\delta t \to 0$, $\sin \delta\theta \to \delta\theta$ and $\cos \delta\theta \to 1$.

Similarly, the absolute change in the tangential component of velocity (perpendicular to OP) is given by

$$\delta v_t = (r + \delta r)(\dot{\theta} + \delta\dot{\theta})\cos \delta\theta + (\dot{r} + \delta\dot{r})\sin \delta\theta - r\dot{\theta},$$

and

$$a_t = \lim_{\delta t \to 0} \frac{\delta v_t}{\delta t} = r\ddot{\theta} + 2\dot{r}\dot{\theta}. \tag{1.23}$$

The acceleration of the point is then given by adding the two components a_r and a_t vectorially, i.e.

$$\boldsymbol{a} = \boldsymbol{a}_r + \boldsymbol{a}_t.$$

These results can also be obtained by a co-ordinate transformation from Cartesian to polar co-ordinates. From eqn (1.11),

$$x = r \cos \theta, y = r \sin \theta \quad \text{and} \quad r = \sqrt{x^2 + y^2}.$$

Consider, for example, the radial component of the velocity of the point. If we resolve $\dot{x}$ and $\dot{y}$ into components along the radial and tangential directions, we obtain, from Fig. 1.15,

$$v_r = \dot{x} \cos \theta + \dot{y} \sin \theta$$

$$= \left[\frac{d}{dt}(r \cos \theta) \right] \cos \theta + \left[\frac{d}{dt}(r \sin \theta) \right] \sin \theta$$

$$= (\dot{r} \cos \theta - r\dot{\theta} \sin \theta) \cos \theta + (\dot{r} \sin \theta + r\dot{\theta} \cos \theta) \sin \theta$$

$$= \dot{r}.$$

Corresponding expressions may be obtained for v_t, a_r and a_t. The reader is recommended to check that this is so.

It is useful to remember these expressions for v_r, v_t, and a_r, a_t, as they are used frequently in kinematics and dynamics.

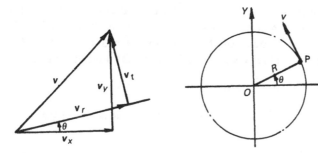

FIG. 1.15   FIG. 1.16

Example 1.5 *Motion in a circle* A point P moves on a circular path of radius R such that the magnitude of its velocity, v, remains constant. Let us find the acceleration of the point.

In polar co-ordinates (eqn 1.20) $v_r = \mathrm{d}r/\mathrm{d}t$. For this case $r = R$ and is constant so that $v_r = 0$. From eqn (1.21), $v_t = R\dot{\theta}$, where $\dot{\theta}$ is the angular velocity of the line OP in Fig. 1.16. The velocity of the particle has a constant magnitude, v, in the tangential direction so that $\dot{\theta} = v/R$, and is constant.

From eqns (1.22) and (1.23),

$$a_r = \ddot{r} - r\dot{\theta}^2 = -R\dot{\theta}^2 = -\frac{v^2}{R} \qquad \text{since } \ddot{r} = 0,$$

and $\quad a_t = r\ddot{\theta} + 2\dot{r}\dot{\theta} = 0, \qquad$ since $\ddot{\theta} = \dot{r} = 0.$

Hence we can see that the acceleration of a point moving on a circular path at constant speed is given by $a_r = -R\dot{\theta}^2 = -v^2/R$ and is radially inwards, as shown by the minus sign.

We should note that even though the *magnitude* of its velocity is constant, the point has an acceleration. This acceleration arises because the *direction* of the velocity vector is constantly changing. The change in velocity is $\delta v = v\,\delta\theta$, as shown in Fig. 1.17 and is radially inwards. Hence the acceleration, a of the point has magnitude

$$a = \lim_{\delta t \to 0} \frac{\delta v}{\delta t} = \lim_{\delta t \to 0} \frac{v\,\delta\theta}{\delta t} = v\dot{\theta} = \frac{v^2}{R} = R\dot{\theta}^2$$

and is in the direction of δv which is radially inwards.

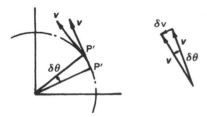

FIG. 1.17

It is also important to note that whilst a point moving *freely* in a plane has two degrees of freedom, a point moving on a circular path has only one degree of freedom. Since R is not a variable, the displacement of the point can be expressed in terms of the single coordinate θ. The displacement of a point moving along *any* prescribed path can always be expressed in terms of a single variable. A point moving along a prescribed path has, therefore, only one degree of freedom.

Example 1.6 *Motion under gravity* An approximation to the free motion of a projectile may be obtained by assuming it to behave as a point which has a constant acceleration towards the earth. This neglects the effect of air resistance, which would cause the vertical acceleration to vary with the speed of the projectile and which would also produce an acceleration component in the horizontal direction.

Consider a shell fired from a ship's gun at a floating target 7.5 km from the ship, both ship and target being stationary. If the velocity of the shell as it leaves the gun is 400 m s^{-1}, calculate the gun elevation α necessary for the shot to hit the target.

Let the gun be at the origin of axes OXY, with OX horizontal, and let the initial velocity of the shell be v. The target T will lie on OX 7.5 km from O as shown in Fig. 1.18.

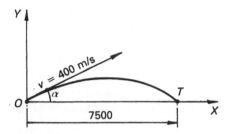

FIG. 1.18

The acceleration of the shell will be constant and due to gravity only. It will be negative since we have chosen OY to be positive in the upward direction. Hence the acceleration of the shell has components

$$\ddot{x} = 0 \tag{i}$$

and $\quad \ddot{y} = -g.$ $\hfill$ (ii)

Consider the horizontal motion. Integrating eqn (i) gives $\dot{x} = A$, which is a constant. We know that at $t = 0$, the velocity component of the shell along OX is $v \cos \alpha$. Hence,

$$\dot{x} = \frac{dx}{dt} = v \cos \alpha. \tag{iii}$$

Integrating eqn (iii) gives

$$x = (v \cos \alpha)t + B.$$

The constant B can be found from the initial conditions. At $t = 0$, $x = 0$, so that $B = 0$.

Thus $\quad x = (v \cos \alpha)t,$ $\hfill$ (iv)

and the displacement along OX increases linearly with time.

For motion in the vertical direction the integration of eqn (ii) gives

$$\dot{y} = -gt + C.$$

Again, from the initial conditions, i.e. at $t = 0$, $\dot{y} = v \sin \alpha$, we find that the constant of integration $C = v \sin \alpha$, so that

$$\dot{y} = v \sin \alpha - gt. \tag{v}$$

Integrating eqn (v) gives

$$y = (v \sin \alpha)t - \tfrac{1}{2}gt^2 + D.$$

The condition $y = 0$ at $t = 0$ gives $D = 0$, so that

$$y = (v \sin \alpha)t - \tfrac{1}{2}gt^2. \tag{vi}$$

When the target is hit, the shell will have co-ordinates $x = 7500$ m, $y = 0$. Hence, from eqn (vi),

$$0 = (400 \sin \alpha)t - \tfrac{1}{2}gt^2.$$

There áre two values of t which satisfy this equation, i.e. $t = 0$ and $t = 800/g \sin \alpha$. These correspond to the times at which the shell leaves the gun and hits the target.

We need the second value to substitute into eqn (iv) for the horizontal component. Hence

$$7500 = (400 \cos \alpha)\left(\frac{800}{g} \sin \alpha\right) = \frac{320\,000}{g} \sin \alpha \cos \alpha.$$

Now $\sin \alpha \cos \alpha = \tfrac{1}{2} \sin 2\alpha$, so that

$$\sin 2\alpha = \frac{75g}{1600} = 0.46,$$

and $2\alpha = 28°$ or $(180 - 28) = 152°$.

Thus there are two possible values of α, i.e. 14° and 76°. These correspond to a low or high trajectory for the shell.

Example 1.7   A missile is launched from a stationary ship and follows some trajectory in a vertical plane. The path of the missile is observed by a radar scanner which is located at the launching point. The radar gives the position of the missile in polar coordinates r, θ, where r is the distance of the missile from its launcher and θ is the angle made by r from the horizontal as shown in Fig. 1.19(a). Test firings show that r and θ may be expressed as functions of time in the form

$$r = 2t - \frac{t^2}{40}; \qquad \theta^2 = 1600 - t^2,$$

where the units of r and θ are km and degrees, and t is the time, in seconds, after firing. What is the velocity and acceleration of the missile 30 s after launching?

(Note that the constants in these equations are *not* dimensionless, e.g. the constant 2 will have units of km s^{-1} and -1 will have units of deg^2 s^{-2}.) To calculate the velocity and acceleration of the missile at $t = 30$ s we have to calculate the values of r, $\dot{r}$, $\ddot{r}$, θ, $\dot{\theta}$ and $\ddot{\theta}$ for use in eqns (1.20), (1.21), (1.22) and (1.23).

Now $r = 2t - \dfrac{t^2}{40}$,

$\hspace{10cm}$ (i)

therefore at $t = 30$ s

$\hspace{2cm} r_{30} = 37.5$ km.

By differentiating eqn (i) with respect to time, the radial component of velocity $\dot{r} = 2 - t/20$

so that $\dot{r}_{30} = 0.5$ km s^{-1}

and $\ddot{r} = -\frac{1}{20}$ so that $\ddot{r}_{30} = -0.05$ km s^{-1}.

Also, $\theta^2 = 1600 - t^2$ giving $\theta_{30} = 26.5°$.
Differentiating with respect to time gives

$\hspace{2cm} 2\theta\dot{\theta} = -2t,$ so that $\dot{\theta}_{30} = -\frac{30}{26.5}$ deg s$^{-1} = -0.02$ rad s^{-1}.

Differentiating again we obtain

$\hspace{2cm} \dot{\theta}^2 + \theta\ddot{\theta} = -1,$

so that $\ddot{\theta}_{30} = -\dfrac{(1 + \dot{\theta}_{30}^2)}{\theta_{30}} = -0.086$ deg s^{-2}

$\hspace{4cm} = -0.0015$ rad s^{-2}.

The velocity of the missile is the vector sum of its radial and tangential components, $v_r = \dot{r}$ and $v_t = r\dot{\theta}$.

Thus $v_r = 0.5$ km s^{-1}, and $v_t = 37.5 \times (-0.02) = -0.750$ km s^{-1}.

The radial velocity v_r is positive and is therefore in the direction of increasing r. The tangential component v_t is negative and is therefore in the direction the shell would move if r were constant and θ were decreasing.

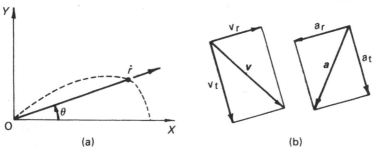

FIG. 1.19

The velocity of the shell is determined by vector addition as shown in Fig. 1.19(b), and its magnitude

$$v = \sqrt{v_r^2 + v_t^2} = 0.9 \text{ km s}^{-1}.$$

Similarly the acceleration of the missile is the sum of the components

$$a_r = \ddot{r} - r\dot{\theta}^2 = -0.5 - 37.5(-0.02)^2 = -0.065 \text{ km s}^{-2},$$

and

$$a_t = r\ddot{\theta} + 2\dot{r}\dot{\theta} = 37.5(-0.0015) + 2 \times 0.5(-0.02) = -0.076 \text{ km s}^{-2}.$$

The minus signs indicate that a_r is in the direction of r decreasing, and a_t in the direction of θ decreasing. The magnitude of a is

$$a = \sqrt{0.065^2 + 0,076^2}$$
$$= 0.1 \text{ km s}^{-2}.$$

The directions of the velocity and acceleration vectors are as shown in Fig. 1.19(b).

Relative motion of two points

Let us consider two points, A and B, moving in the OXY plane, and let their position vectors be r_A and r_B. From Fig. 1.20 these vectors may be related by the vector equation

$$r_B = r_A + r_{BA}. \tag{1.24}$$

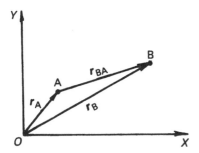

FIG. 1.20

The vector r_{BA} is the position vector of B *relative to* A. Equation (1.24) may also be expressed in terms of components,

$$x_B = x_A + x_{BA}, \quad \text{and} \quad y_B = y_A + y_{BA}.$$

By differentiating eqn (1.24) with respect to time, we obtain

$$v_B = v_A + v_{BA}, \tag{1.25}$$

where v_{BA} is the velocity of B *relative* to A. If A and B are moving on parallel paths with velocities of equal magnitude, $v_{BA} = 0$.

Similarly, by differentiating eqn (1.25) with respect to time we obtain

$$a_B = a_A + a_{BA}. \tag{1.26}$$

Example 1.8   A ship steams due north at a speed of $12\ \text{m s}^{-1}$. A second ship, 1000 m due east of the first ship, has a speed of $16\ \text{m s}^{-1}$ in a north easterly direction. What is the velocity of the second ship relative to the first?

Let us choose axes OXY such that the axis OY points due north and the origin coincides with the position of the first ship, A, at the instant considered. The second ship B will be 1000 m along OX from O as shown in Fig. 1.21.
Now

$$v_B = v_A + v_{BA}. \tag{i}$$

Equation (i) can be solved graphically since v_A and v_B are known. v_B is first drawn to some suitable scale as shown in Fig. 1.22. The vectors on the right-hand side of eqn (i) must add together to equal v_B so we now draw v_A starting from the tail of v_B. The line joining the tips of the two vectors represents v_{BA} and for eqn (i) to be satisfied its direction must be as shown. From the diagram $v_{BA} = 11\ \text{m s}^{-1}$, 3° south of east.

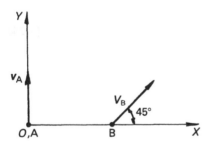

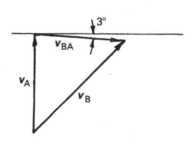

FIG. 1.21   FIG. 1.22

Alternatively, in components,

$$\dot{x}_B = \dot{x}_A + \dot{x}_{BA}, \qquad \dot{y}_B = \dot{y}_A + \dot{y}_{BA}.$$

From Fig. 1.21, $\dot{x}_A = 0$, so that $\dot{x}_{BA} = \dot{x}_B = 16 \cos 45 = 11.4\ \text{m s}^{-1}$.
Similarly,

$$\dot{y}_{BA} = \dot{y}_B - \dot{y}_A = 11.4 - 12 = -0.6\ \text{m s}^{-1}.$$

Hence

$$v_{BA} = \sqrt{\dot{x}_{BA}^2 + \dot{y}_{BA}^2} = \sqrt{11.4^2 + 0.6^2} = 11.4\ \text{m s}^{-1},$$

at an angle which is $\tan^{-1}(0.6/11.4) = 3°$ south of east.

1.6   Plane motion of a rigid body

When studying the gross motion of bodies it is generally permissible to assume that the bodies themselves are rigid, i.e. that the distance between any two points on a body remains constant. Engineering materials do deform under load, but this deformation is usually only a very small percentage of the body dimensions and in service a body would normally break before significant deformation occurred. For a great many engineering situations the rigid body assumption is adequate but nevertheless cases do arise where the deflection of the body itself can be significant and must be taken into account, e.g. in vibrating systems and high precision machinery.

To define the position of a rigid body in a plane we need to specify both the location of a point in the body and the orientation of a line fixed in the body. The position of the point is specified by two co-ordinates. These, together with a co-ordinate associated with the orientation of the chosen line, uniquely define the position of the body.

The displacement of the body can be expressed in terms of changes in these three co-ordinates so that a body moving freely in a plane has three degrees of freedom. In Fig. 1.23 the position of the body is specified by the coordinates x_A, y_A, of point A, and the angle, θ, made with OX, by a line joining point A with a second point B in the body. The anticlockwise direction is usually taken as positive for θ.

The position of any other point on the body can be expressed in terms of these co-ordinates. The point B in Fig. 1.23, distant l from A, will have co-ordinates

$$x_B = x_A + l \cos \theta, \quad y_B = y_A + l \sin \theta. \tag{1.27}$$

A point not on AB, e.g. point C on Fig. 1.23, where AC makes a constant angle α with AB, will have co-ordinates

$$x_C = x_A + l_2 \cos(\theta + \alpha), \quad \text{and} \quad y_C = y_A + l_2 \sin(\theta + \alpha).$$

Other combinations of three co-ordinates may be used to specify the position of a body in a plane. For example x_A, y_A and y_B could be used, as in Fig. 1.24. Note that x_A, y_A, x_B, y_B cannot be specified independently, since the

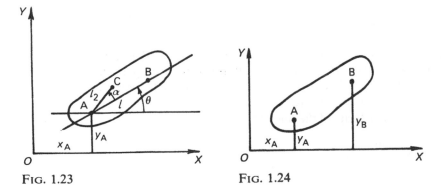

FIG. 1.23   FIG. 1.24

fourth co-ordinate can always be expressed in terms of the other three, i.e.

$$x_B = x_A \pm \sqrt{l^2 - (y_B - y_A)^2},$$

where l is the constant length AB. The alternative negative sign means that in Fig. 1.24, B could lie to the left of A for the same value of y_B.

Pure translation

A body moving with pure translation does not rotate. All points in the body will therefore move along equal parallel paths. If the points move along straight paths the motion is said to be rectilinear, Fig. 1.25(a), and if the paths are not straight the motion is said to be curvilinear, Fig. 1.25(b).

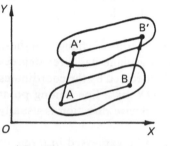

a) Rectilinear translation   b) Curvilinear translation

FIG. 1.25

Since all points in the body move through the same distance in the same time interval, they will all have identical velocities and accelerations so that $v_A = v_B$ and $a_A = a_B$.

Since θ is maintained at a fixed value, changes in only two co-ordinates, i.e. x_A and y_A, are needed to specify the displacement of the body. A body moving freely in pure translation has therefore only two degrees of freedom.

Pure rotation

A body in pure rotation rotates about an axis which passes through a fixed point. All points on the body therefore move in circular paths, as shown in Fig. 1.26.

The motion of the body may be specified by considering the motion of a straight line in the body which passes through the fixed point O. The position of the body is therefore defined by θ, the angle made by the line with some fixed direction OX, as in Fig. 1.26.

The angular velocity of the line OA will be

$$\omega = \frac{d\theta}{dt} = \dot{\theta},$$

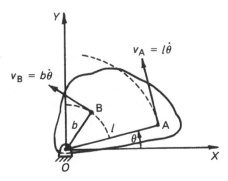

FIG. 1.26

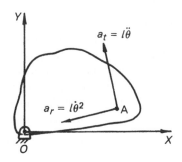

FIG. 1.27

and its angular acceleration

$$\alpha = \frac{d^2\theta}{dt^2} = \ddot{\theta}.$$

Since the body is assumed to be rigid, all lines in the body will have the same angular velocity and angular acceleration. We can therefore refer to ω and α as the angular velocity and angular acceleration of the body.

The velocity of a point in the body will depend upon its distance from O. For example in Fig. 1.26 a typical point A, distant l from O, is moving on a circular path of radius l.

From eqn (1.21), its velocity $v_A = l\dot{\theta}$, and is perpendicular to OA.

The acceleration of a point moving on a circular path has two components, one tangential to the path and the other radially inwards towards O. These components are given by eqns (1.22) and (1.23) with r given the constant value l.

Therefore

$$\boldsymbol{a}_A = \boldsymbol{a}_r + \boldsymbol{a}_t, \tag{1.28}$$

where $a_r = l\dot{\theta}^2$ along AO and $a_t = l\ddot{\theta}$ perpendicular to AO as shown in Fig. 1.27. If the body were rotating at a constant angular velocity, i.e. $\ddot{\theta} = 0$, the acceleration of the point would be $l\dot{\theta}^2$ towards O. (Remember that points on a body which is rotating at constant *angular* velocity about a fixed axis move on a circular path and do have a *linear* acceleration—see p. 16).

In pure rotation the displacement of the body can be specified in terms of changes in the single co-ordinate θ since the position of O is fixed. A body in pure rotation has therefore only one degree of freedom.

We can see that all points in the body have *different* velocities and accelerations. It is therefore *not* meaningful to refer to the velocity or acceleration of a body which is rotating about a fixed point. Ony the *angular* velocity and *angular* acceleration of the body have any meaning, and from these two quantities we can obtain the velocity and acceleration of any *point* in the body.

General motion in a plane

If a body moving in a plane has both translational and rotational motion it is said to move with general plane motion. Let us consider the body in Fig. 1.28 and assume that it is in position 1 at time t, and will reach position 2 a short time δt later. The motion of the body may be considered by choosing *any* point A and a line AB of length l in the body. Let the displacements of A and B in the time interval be given by the vectors δr_A and δr_B. The motion from 1 to 2 may be considered in two stages:

(i)   pure translation from A_1B_1 to A_2B' in which both A and B have the same displacement δr_A and

(ii) pure rotation about A_2 from A_2B' to A_2B_2 during which the line AB rotates through a small angle $\delta\theta$ and B moves along the arc of a circle of radius AB from B' to B_2.

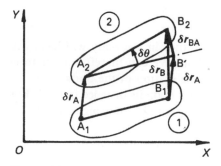

FIG. 1.28

Since point A has been chosen arbitrarily we should note that this angular motion is *not* about a fixed point. During the rotation $\delta\theta$, the point B is displaced $B'B_2$ *relative* to point A. Let this *relative* displacement be given by δr_{BA}.

Now, from the vector triangle of Fig. 1.28,

$$\delta r_B = \delta r_A + \delta r_{BA}. \tag{1.29}$$

Dividing by δt and taking the limit as $\delta t \to 0$,

$$v_B = v_A + v_{BA}. \tag{1.30}$$

The vector v_{BA} is the velocity of point B *relative to* point A and its direction is the same as that of the displacement δr_{BA}, i.e. perpendicular to AB. It is also the velocity point B would have if the body were rotating with its actual angular velocity $\dot\theta$ about A as a fixed point; i.e. *relative to* A, B always moves on a circular path. From our earlier consideration of the motion of a point on a circular path (example 1.5), we see that the direction of v_{BA} must be perpendicular to AB, and $v_{BA} = l\dot\theta$, as shown in Fig. 1.29(a). (Note that any

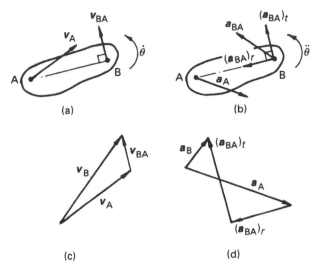

FIG. 1.29

component of v_{BA} along AB would mean that AB would change its length with time. This is not possible for a rigid body.) If the vectors v_A and v_{BA} are known, eqn (1.30) can be solved graphically, as shown in Fig. 1.29(c), to give the magnitude and direction of the velocity of point B.

If we now differentiate eqn (1.30) with respect to time we obtain an expression which relates the acceleration of point B to that of point A, i.e.

$$a_B = a_A + a_{BA}. \tag{1.31}$$

The acceleration of point B relative to A, a_{BA}, may be found by again considering point B to move on a circular path *relative to A*. By setting $r = l$ in eqns (1.22) and (1.23) it can be seen that a_{BA} will have two components,

$$a_{BA} = (a_{BA})_r + (a_{BA})_t, \tag{1.32}$$

where $(a_{BA})_r = l\dot{\theta}^2$ and is *along BA*,

and $(a_{BA})_t = l\ddot{\theta}$ and is *perpendicular to AB*, as shown in Fig. 1.29(b).

Hence $a_B = a_A + (a_{BA})_r + (a_{BA})_t. \tag{1.33}$

If a_A, $\dot{\theta}$ and $\ddot{\theta}$ are known then all of the vectors on the right-hand side of eqn (1.33) are known in both magnitude and direction. We can then solve eqn (1.33) graphically, as shown in Fig. 1.29(d), to give the magnitude and direction of the acceleration of point B.

Instantaneous centre (IC)

If the point A in eqn (1.30) were chosen such that at a particular instant $v_A = 0$, then $v_B = v_{BA}$.

This point A, which *instantaneously* has zero velocity, is called the instantaneous centre of rotation, or the velocity pole of the body. The velocity of any point in the body *at that instant* may be found by considering the body to be in *pure rotation* about the instantaneous centre.

If the direction of the velocities of two points in a body are known, then it is possible to locate the instantaneous centre. Let v_B, v_C be the velocities of points B and C on the body shown in Fig. 1.30. The direction of v_B must be perpendicular to a line joining B to the instantaneous centre (see pure rotation, p. 24).

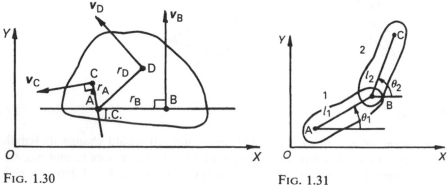

FIG. 1.30   FIG. 1.31

The same applies to point C, so that if lines are constructed through points B and C perpendicular to v_B and v_C, the instantaneous centre will be located at their intersection.

Since the body may be considered to be rotating about the instantaneous centre $v_B = r_B\omega$ and $v_C = r_C\omega$. If v_B and r_B are known, the angular velocity, ω, of the body can be found. Having found ω, the magnitude of v_C can be calculated and the magnitude and direction of the velocity of any other point in the body found directly. For example in Fig. 1.30 the velocity of point D is $v_D = r_D\omega$, and is perpendicular to r_D.

We should note that the instantaneous centre need not be within the body. If v_B in Fig. 1.30 were parallel to v_C, their perpendiculars would also be parallel. For this case the instantaneous centre would lie at infinity and the body would be in pure translation.

For bodies moving with general plane motion the instantaneous centre does *not* have zero acceleration and this method cannot be extended to find accelerations.

Connected bodies: degrees of freedom

We have seen that a single rigid body moving freely in a plane has three degrees of freedom. A system made up of two such bodies which are not connected in any way would therefore have six degrees of freedom. Now let us consider the case where the two bodies are connected together by a simple pin joint, i.e. a bearing, with negligible clearance. In Fig. 1.31 bodies 1 and 2 are connected by a pin joint with centre B. Since point B is common to both bodies, the displacement of B on body 1 is the same as the displacement of B on body 2. Let this be the only constraint on independent free motion of the bodies.

The displacement of point A on body 1 can be defined in terms of changes in the components x_A, y_A. The displacement of point B, relative to A, is a function of the change in the co-ordinate θ_1, which defines the angular position of the line AB, and the fixed length l_1. Relative to B, the displacement of a point C on body 2 can be expressed as a function of the change in θ_2, which defines the angular position of the line BC, and the fixed length l_2. Thus four variables are necessary to specify the displacement of C, i.e. if δx_A, δy_A, $\delta \theta_1$, $\delta \theta_2$ are known, the displacement of C can be found. Two bodies connected by a simple pin joint therefore have *four* degrees of freedom. Consequently a simple pin joint removes two degrees of freedom. If we now consider n bodies joined together by p pin joints, the connected system would have $(3n - 2p)$ degrees of freedom (each unconnected body has three degrees of freedom and each pin joint removes two of these).

However, connected systems of this type are seldom allowed to move without further constraint. *One* of the bodies is normally fixed to earth, e.g. the frame of a machine. This fixed body will therefore have zero degrees of freedom, so that a system of n bodies, one of which is fixed, connected by p pin joints would have $[3(n - 1) - 2p]$ degrees of freedom.

Two bodies may be connected by a sliding connection which prevents relative rotation, but allows relative translation along some line. This connection will also remove two degrees of freedom. A typical example is the piston cylinder connection in a reciprocating engine, where the displacement of the piston relative to the cylinder can be expressed in terms of the single variable, x in Fig. 1.32. In this case the guide, i.e. the cylinder, would be a part of the engine frame.

A sliding connection need not prevent relative rotation between the connected bodies. For example, consider the connection shown in Fig. 1.33. A pin fixed in body 2 at C moves in a slot in body 1. We can see that relative

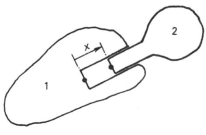

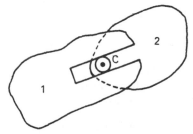

FIG. 1.32   FIG. 1.33

motion at the point C is prevented only in a direction perpendicular to the slot. Relative rotation of the two bodies can still occur, as can motion of the pin along the slot. This type of connection therefore removes only *one* degree of freedom.

We may now express the number of degrees of freedom, f, of a system of n connected bodies, one of which is fixed, as

$$f = 3(n - 1) - 2p - q, \tag{1.34}$$

where p is the number of connections which remove 2 degrees of freedom, and q the number of connections removing one degree of freedom.

It is of course possible to connect two bodies in such a way that three degrees of freedom are removed, e.g. by welding. The two bodies together would now have only three degrees of freedom since no relative motion can occur. They therefore behave as a single rigid body and would be counted as such when using eqn (1.34).

Let us find the number of degrees of freedom of the following systems.

(i)   Consider two bodies, one of which is fixed, connected by a single pin joint, as shown in Fig. 1.34. In eqn (1.34), $n = 2$, $p = 1$, and $q = 0$, so that $f = 3 - 2 = 1$. This represents a rigid body rotating about a fixed point (see p. 24). Practical examples are rotating machines such as electric motors, steam or gas turbines, in which the frame of the machine is fixed and the rotating body is the armature, or turbine shaft and discs. The connection between the two bodies is via the shaft bearings, which are located at the geometric centre of the shaft. The type of bearing, e.g. ball, roller, plain, does not affect the kinematic behaviour, and since they all allow only relative rotation (neglecting clearances and deformation) they can be considered as pin joints.

(ii)   Now let us add an extra body and two pin joints, giving three bodies connected by three pin joints, as shown in Fig. 1.35.

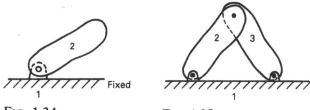

FIG. 1.34   FIG. 1.35

In this case $n = 3$, $p = 3$, $q = 0$, so that $f = (3 \times 2) - (2 \times 3) = 0$. This arrangement therefore has no degrees of freedom and represents a *structure* which cannot move.

(iii)   If a fourth body and pin joint are added, as in Fig. 1.36, $f = (3 \times 3) - (2 \times 4) = 1$. Since this arrangement has *one* degree of freedom, the changes in the angular position of the bodies, and the displacements of all points in the bodies can be expressed in terms of one variable e.g. θ_2 in Fig. 1.36. Four bodies connected in this way constitute a four bar linkage mechanism. If the pin joint between links 4 and 1 is replaced by a sliding connection which prevents rotation, as shown in Fig. 1.37, the arrangement becomes a slider crank mechanism which also has one degree of freedom.

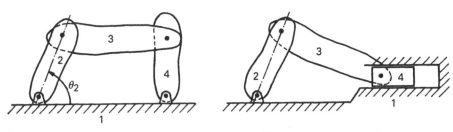

FIG. 1.36   FIG. 1.37

(iv) With five bodies, one of which is fixed, and five pin joints, as in Fig. 1.38, $f = (3 \times 4) - (2 \times 5) = 2$. A system with two degrees of freedom requires two variables, e.g. θ_2 and the angular position of any other link such as θ_5, to define the displacement of all points in the bodies.

(v) If we now add another pin joint to our five link arrangement, as in Fig. 1.39, $f = (3 \times 4) - (2 \times 6) = 0$, and we again have a structure.

Note that if the extra pin joint had been added between two of the links already connected in Fig. 1.38, e.g. between 2 and 3 as shown in Fig. 1.40, no relative rotation of the two links would now be possible. These two links would now behave as a single rigid body. The resulting system is therefore effectively, as in Fig. 1.36, a four bar linkage with one degree of freedom. Equation (1.34) would yield an incorrect result of -1 degrees of freedom in this case. This apparent inconsistency arises because the second pin joint could have been replaced by a joint which removes only one degree of freedom such as that shown in Fig. 1.41 (cf. Fig. 1.33) without altering the behaviour of the system. If we now substitute into eqn (1.34) $p = 5$ and $q = 1$, then $f = 1$, which is the correct result for this case.

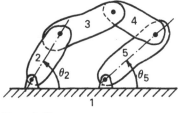

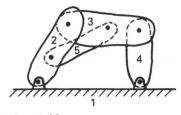

FIG. 1.38   FIG. 1.39

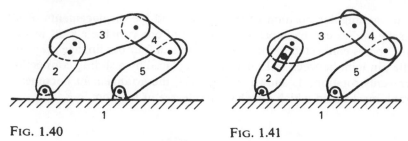

FIG. 1.40   FIG. 1.41

Thus where a joint which removes two degrees of freedom can be replaced by a joint which removes one degree of freedom without altering the constraint of the system, it must be considered as such in eqn (1.34).

(vi)   The addition of a further link with pin joints to links 3 and 4, as in Fig. 1.42, gives $f = (3 \times 5) - (2 \times 8) = -1$. The connection between links 3 and 6 is shown coincident with that between links 3 and 5, but is counted as two separate connections because, kinematically, the connection behaves as if links 5 and 6 are individually connected to link 3. This again represents a structure, but since it has fewer than zero degrees of freedom it is a *redundant* structure. Redundancy is discussed more fully on p. 69

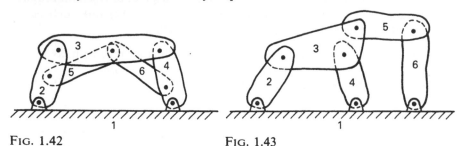

FIG. 1.42   FIG. 1.43

(vii)   Different arrangements of six links, with seven pinned connections, can produce mechanisms with one degree of freedom, e.g. as shown in Fig. 1.43. For this case, $f = (3 \times 5) - (2 \times 7) = 1$. This represents a six bar linkage mechanism.

The kinematics of some common mechanisms which have one degree of freedom will be considered in some detail later.

Engineering Applications

It is possible to determine the velocities and accelerations of machine components in connected systems by successive application of eqns (1.30) and

(1.31) to each rigid body in turn. A graphical solution of these vector equations, involving drawing to scale or sketching the vector diagrams representing these equations, is usually convenient. Alternatively, velocities may be determined using the instantaneous centre of the body.

A more general approach is to work from first principles, remembering that velocity and acceleration are the first and second derivatives, with respect to time, of position. Hence by writing down an expression for the position of a point on a body, or the angular position of a line in a body, analytical expressions for linear or angular velocities and accelerations may be derived. However, the algebra involved, even in physically simple systems, can be very complex.

As examples of the application of these general principles we will now consider the motion of a number of engineering systems. In some cases both graphical and analytical approaches will be used to show their relative advantages.

1.1 Rolling wheel

Consider a wheel of radius r rolling along the horizontal line OX in Fig. 1.44 such that its centre C has a velocity v_C. The wheel has both translational and rotational motion and may be considered as a rigid body in general plane motion. This assumes that the deformation which would normally occur at the contact point is ignored. We will also assume pure rolling, i.e. that no slip occurs at the contact point.

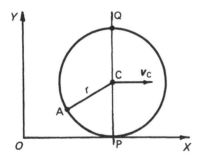

FIG. 1.44

Let P be the point on the wheel instantaneously in contact with OX. We can apply eqn (1.30) to the points C and P, i.e.

$$v_C = v_P + v_{CP}. \tag{i}$$

Now if there is no slip at P there is no relative motion between point P on the wheel and a coincident point on OX. Thus $v_P = 0$, so that

$$v_C = v_{CP}.$$

The direction of the velocity of C relative to P is therefore along OX and its magnitude can be expressed in terms of the angular velocity $\dot{\theta}$ of the wheel as

$$v_{CP} = r\dot{\theta}.$$

Hence the angular velocity of the wheel,

$$\dot{\theta} = \frac{v_C}{r}.$$

By inspection of Fig. 1.44, $\dot{\theta}$ is in a clockwise direction.

The velocity of any point on the wheel can now be found. Let us consider the point A on the rim of the wheel.

Equation (1.30) gives

$$v_A = v_C + v_{AC}. \tag{ii}$$

The vector v_{AC} has magnitude $r\dot{\theta}$ and is perpendicular to AC as shown on Fig. 1.45(a). The velocity of point A is obtained by a graphical solution of eqn (ii). The magnitudes and directions of v_C and v_{AC} are known and can be drawn to scale, as shown in Fig. 1.45(b). The vector v_A is then equal to their vector sum as shown. We should also note that v_A is perpendicular to the line AP. This is because P has zero velocity and is an instantaneous centre, so that the magnitude of v_A may also be expressed as

$$v_A = \dot{\theta}(AP).$$

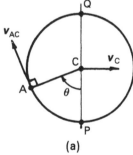

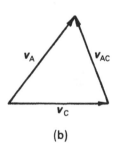

(a)   (b)

FIG. 1.45

The result could also be obtained in components by resolving v_{AC} along OX and OY, i.e.

$$(v_A)_x = v_C + (v_{AC})_x = v_C - r\dot{\theta}\cos\theta = r\dot{\theta}(1 - \cos\theta),$$

$$(v_A)_y = (v_{AC})_y = r\dot{\theta}\sin\theta.$$

If we now consider the uppermost point, Q, on the wheel,

$$v_Q = v_C + v_{QC}. \tag{iii}$$

The vector v_{QC} has magnitude $r\dot\theta$ and is perpendicular to QC as shown in Fig. 1.46(a). It is also parallel to v_C. The vector v_Q is found from eqn (iii) and its graphical solution is shown in Fig. 1.46(b). In component form,

$$(v_Q)_x = v_C + r\dot\theta = 2r\dot\theta = 2v_C,$$

and $\quad (v_Q)_y = 0.$

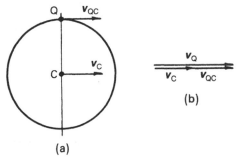

(b)

(a)

FIG. 1.46

It is of interest to note that the point Q at the top of the wheel has a velocity which is twice that of the centre, so that when driving along a motorway at $100\ \mathrm{kmh^{-1}}$ the tops of the tyres have a velocity of approximately $200\ \mathrm{kmh^{-1}}$.

The acceleration of point A may be found in a similar manner since, for a rigid body in general plane motion,

$$a_A = a_C + a_{AC}. \tag{iv}$$

If the centre of the wheel has a constant velocity, $a_C = 0$, and $\dot\theta$ is constant. a_{AC} has therefore only a radial component of magnitude $r\dot\theta^2$ along AC.

Substituting these values into eqn (iv) we see that the direction of a_A is along AC and

$$a_A = r\dot\theta^2 = \frac{v_C^2}{r}.$$

The components of a_A are

$$(a_A)_x = r\dot\theta^2 \sin\theta, \text{ and } (a_A)_y = r\dot\theta^2 \cos\theta.$$

We should note that at the point P, $\theta = 0$, so that

$$(a_P)_x = 0 \text{ and } (a_P)_y = r\dot\theta^2.$$

Hence while P has zero velocity and may be considered as an instantaneous centre of rotation, it does *not* have zero acceleration.

Since P is an instantaneous centre, the velocities v_A and v_Q, but not the accelerations, could have been obtained by considering the wheel to be in pure rotation about P.

The velocity and acceleration of point A can also be obtained from first principles by noting that the position of A has components given by

$$x_A = x_c - r \sin \theta$$

and $$y_A = r - r \cos \theta.$$ (v)

Differentiation of these expressions with respect to time yields

$$\dot{x}_A = \dot{x}_c - r\dot{\theta} \cos \theta = v_c - r\dot{\theta} \cos \theta$$

and $$\dot{y}_A = r\dot{\theta} \sin \theta.$$ (vi)

Since $(v_A)_x = \dot{x}_A$, $(v_A)_y = \dot{y}_A$ and $v_c = r\dot{\theta}$ we have

$$(v_A)_x = r\dot{\theta}(1 - \cos \theta)$$

and $$(v_A)_y = r\dot{\theta} \sin \theta,$$

which agrees with the expressions given above.

By differentiating eqn (vi) with respect to time the reader should verify the expressions given for the acceleration of A.

1.2   Belt and chain drives

A belt or chain drive may be considered kinematically as two cylinders, which can rotate about their centres, connected by an inextensible string, as shown in Fig. 1.47. In this case all points on the string must have velocities of equal magnitude.

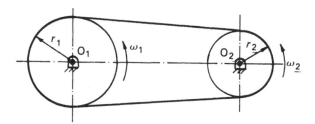

FIG. 1.47

If we also assume that no slip occurs between the string and the pulleys, then the magnitude of the velocities of points on the surfaces of the pulleys must be equal to that of the belt. The pulleys are rigid bodies rotating about fixed points, so that if their angular velocities are ω_1 and ω_2 respectively, the magnitude v of the velocity of points on their surfaces is given by

$$v = \omega_1 r_1 = \omega_2 r_2,$$

i.e. $$\omega_2 = \omega_1 r_1 / r_2,$$

and the two pulleys rotate in the same sense.

The velocity ratio, ω_2/ω_1, of a real belt drive will deviate very slightly from this value because of creep between the belt and pulleys.

This velocity error is not usually significant but can lead to long term *positional* errors. In camshaft and distributor drives in internal combustion engines such errors cannot be tolerated. Chains or toothed belts, which by eliminating slip give accurate positioning, must be used in such applications.

1.3   Toothed gearing

The geometry of gear teeth is a subject beyond the scope of this text and for our purposes it is sufficient to observe that correctly designed gear wheels behave kinematically as cylinders in *rolling* contact. Gear teeth allow the transmission of much greater loads than would be possible by relying on friction alone. The diameters of the kinematically equivalent cylinders are equal to the *pitch circle diameters* of the gears.

The simplest form of gear train consists of two gear wheels, as shown in Fig. 1.48. Let the gears have pitch circle diameters d_1, d_2, with centres O_1, O_2, and let C be their point of contact.

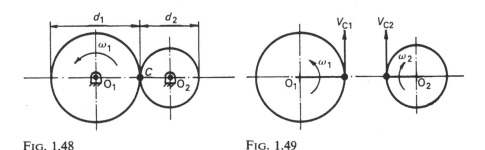

FIG. 1.48   FIG. 1.49

Each gear wheel may be considered as a rigid body rotating about a fixed point. Since there can be no slip at C, the velocity of point C on gear 1 must be equal to that of point C on gear 2.

If gear 1 is rotating with an angular velocity ω_1 in the anticlockwise direction, v_{C_1} will be in the direction shown in Fig. 1.49 and will have magnitude $\omega_1 d_1/2$.

Velocity v_{C_2} must be equal to v_{C_1} (the teeth prevent any slip), so that by inspection of Fig. 1.49, gear 2 will rotate in a clockwise direction. Thus

$$v_{C_1} = \frac{\omega_1 d_1}{2} = v_{C_2} = \frac{\omega_2 d_2}{2}, \tag{i}$$

or $$\frac{\omega_2}{\omega_1} = \frac{d_1}{d_2}.$$

ω_2 is proportional to the gear ratio d_1/d_2 and is in the *opposite* direction from ω_1.

The tangential components of the acceleration of points C must be the same on both gears, so that the ratio of their angular accelerations is also d_1/d_2, and $\dot{\omega}_2$ will be in the opposite direction from $\dot{\omega}_1$. Similarly, their angular displacements θ_1, θ_2 will be in opposite directions and in the same ratio.

A gear train may consist of more than two gears, e.g. three are shown in Fig. 1.50. Let gear 1 rotate in the anticlockwise direction with angular velocity ω_1. From eqn (i),

$$\omega_2 = \frac{d_1}{d_2}\omega_1. \tag{ii}$$

Also $$\omega_3 = \frac{d_2}{d_3}\omega_2 = \frac{d_2}{d_3}\frac{d_1}{d_2}\omega_1 = \frac{d_1}{d_3}\omega_1 \tag{iii}$$

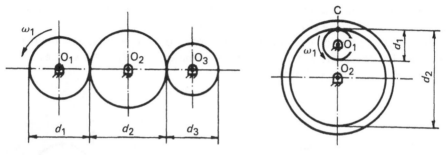

FIG. 1.50   FIG. 1.51

and is in the *same* direction as ω_1. The ratio between ω_3 and ω_1 is therefore independent of the diameter of the intermediate gear, and the direction of ω_3 is the same as that of ω_1. Any number of intermediate wheels may be introduced without altering the overall gear ratio, but the direction of rotation of the output gear will depend upon whether an odd or even number of gears is used.

The simple gears considered above have been solid cylinders with teeth on the outside, i.e. externally meshing gears. Another common type of gear wheel consists of a hollow cylinder with teeth on the inside, which is often referred to as a ring gear or an internally toothed gear.

Consider a ring gear of pitch circle diameter d_2 rotating about its centre O_2 and meshing with a simple gear of diameter d_1, centre O_1, as shown in Fig. 1.51.

As above, the point of mesh C must have the same velocity on both gears. If gear 1 rotates with an angular velocity ω_1 in the anticlockwise direction, then $v_{C_1} = \omega_1 d_1/2$ in the direction shown in Fig. 1.52(a).

v_{C_2} must be in the same direction as v_{C_1}, and therefore gear 2 will rotate in an anticlockwise direction, as shown in Fig. 1.52(b).

Now

$$v_{C_2} = \frac{\omega_2 d_2}{2} = \frac{\omega_1 d_1}{2}$$

so that $\omega_2 = \omega_1 d_1/d_2$ and is in the *same* direction as ω_1.

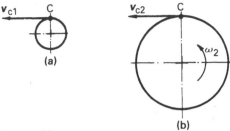

FIG. 1.52

A pair of internally meshing gears will therefore both rotate in the same direction, whereas a pair of externally meshing gears rotate in opposite directions.

1.4 Epicyclic gear trains

Epicyclic gear trains are characterised by having at least one gear wheel whose centre moves on a circular path.

A typical epicyclic gear train is shown diagrammatically in Fig. 1.53. The gear S rotates about a fixed axis through O_1, and meshes with gear P. Gear P rotates about an axis O_2 which can move around axis O_1. Gear P is also in mesh with an internally toothed gear R which has centre O_1 and is fixed so that it cannot rotate. If gear S is rotated, gear P will move around the annulus between R and S. Gear S is normally referred to as a sun gear, P as a planet gear and R as a ring gear. The planet axis O_2 is carried on a rigid body which rotates about the axis O_1. This body, referred to as a planet carrier, is shown in Fig. 1.53, and may be used as an output shaft.

Suppose that the sun is given an angular velocity ω_s in the positive anticlockwise sense. Let us find the angular velocity of the planet carrier and hence the overall gear ratio of the epicyclic gear. Assume that the pitch circle radii of the sun, planet and ring gears are r_s, r_p and r_R, respectively.

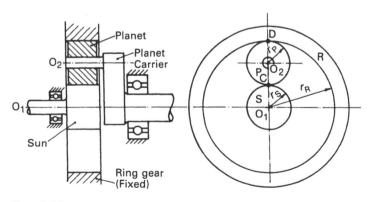

FIG. 1.53

The sun may be considered as a rigid body rotating about a fixed point O_1. The velocity of C, the point of contact between sun and planet, is

$$v_C = \omega_s r_s, \tag{i}$$

in the direction shown in Fig. 1.54(a).

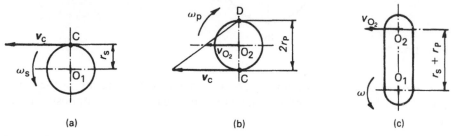

FIG: 1.54

Now the planet has both rotation and translation, and is therefore in general plane motion. However, since the ring gear is fixed, we know that point D must have zero velocity, and may therefore be considered as an instantaneous centre of rotation. The planet may therefore be considered to be *instantaneously* in pure rotation *about* D. The velocity of C is known, and from Fig. 1.54(b)

$$v_C = \omega_p 2 r_p, \tag{ii}$$

where ω_p is the angular velocity of the planet gear and will be in the clockwise direction as shown.

The velocity of point O_2 can now be found by considering the planet to rotate about its instantaneous centre D, i.e.

$$v_{O_2} = \omega_p r_p.$$

Substitution from eqns (i) and (ii) gives

$$v_{O_2} = \frac{v_C}{2} = \frac{\omega_s r_s}{2} \tag{iii}$$

in the direction shown in Fig. 1.54(b).

We may now consider O_2 as a point on the planet carrier which is a rigid body rotating about the fixed point O_1.

From Fig. 1.54(c),

$$v_{O_2} = \omega(r_s + r_p) \tag{iv}$$

where ω is the angular velocity of the planet carrier. Thus from eqns (iii) and (iv),

$$\omega = \frac{\omega_s r_s}{2(r_s + r_p)}$$

and is in the positive *anticlockwise* sense.

The gear ratio is given by

$$n = \frac{\omega}{\omega_s} = \frac{r_s}{2(r_s + r_p)}$$

or alternatively, since $r_R = r_s + 2r_p$,

$$n = \frac{r_s}{r_s + r_R}.$$

The input and output shafts will rotate in the same direction.

Other configurations are possible, e.g. the sun may be fixed, and the ring gear given an angular velocity ω_R. The output would again be taken from the planet carrier. If the planet carrier is fixed, all the gears will rotate about fixed centres so that the gear will no longer be epicyclic.

One advantage of epicyclic gear trains is that they have coaxial input and output shafts. They are widely used, a common application being in vehicle automatic gearboxes.

1.5   Eccentric cam mechanism

The simple eccentric cam mechanism, shown in Fig. 1.55, consists of two bodies, a circular cam of radius R, which rotates about a fixed point O distant e from its geometric centre G, and a follower which is constrained to move along a straight line whilst remaining in contact with the cam. The surface of the follower is normally flat and perpendicular to its direction of motion.

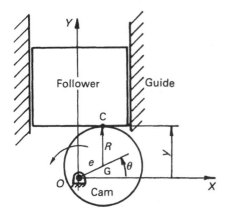

FIG. 1.55

The cam may be considered as a rigid body rotating about the fixed point O, and the follower is in pure translation along the axis OY in Fig. 1.55. Normally the cam would rotate at a constant angular velocity ω. Let us find the velocity and acceleration of the follower.

The geometry of the mechanism is very simple and an analytical method is convenient. First we write an expression for the displacement of the follower with the cam in a general position defined by θ in Fig. 1.55, i.e.

$$y = e \sin \theta + R. \tag{i}$$

Differentiating both sides of this equation with respect to time gives the velocity of the follower as

$$\dot{y} = e\dot{\theta} \cos \theta,$$

and its acceleration as

$$\ddot{y} = e\ddot{\theta} \cos \theta - e\dot{\theta}^2 \sin \theta.$$

Now if the cam rotates at constant angular velocity ω,

$$\dot{\theta} = \omega, \text{ and } \ddot{\theta} = 0.$$

Thus $\dot{y} = e\omega \cos \theta$, and $\ddot{y} = -e\omega^2 \sin \theta.$ (ii)

The follower will therefore move with simple harmonic motion (see p. 50).

The velocity and acceleration of the point of contact C across the face of the follower may be found in a similar manner. From the geometry of the system

$$x_C = e \cos \theta,$$

and by successive differentiation with respect to time,

$$\dot{x}_C = -e\omega \sin \theta \quad \text{and} \quad \ddot{x}_C = -e\omega^2 \cos \theta.$$

The eccentric cam with flat follower is the simplest of a family of cam mechanisms. Usually the cam profile will not be circular, and the follower may be of the translating type, or the pivoted type which rotates about a fixed point, as shown in Fig. 1.56(a) and (b). If a part of the cam profile is a circular arc with centre at the axis of rotation, the follower will remain stationary when in contact with this part of the profile. A well known application of cam mechanisms is in the operation of the inlet and exhaust valves in piston engines. They are also widely used in a variety of forms to produce complex motions in machinery used in the food processing, packaging and textile industries.

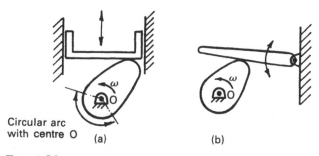

Circular arc
with centre O   (a)   (b)

FIG. 1.56

1.6 Slider crank mechanism

The best known application of the slider crank mechanism is in internal combustion engines. It consists of a crank *OR* rotating about a fixed point *O*, connected to a piston S by a connecting rod RS. The piston is constrained to slide along the cylinder axis. There are many other applications of the basic slider crank mechanism, e.g. in electrical switchgear where the slider is a moving contact driven towards, or away from, a fixed contact by rotating the crank. In this case the crank would operate only over a limited arc, whereas in the engine mechanism the crank rotates over a full revolution. Other applications arise in packaging and textile machinery.

Let us suppose that the crank in Fig. 1.57(a) rotates at some constant angular velocity ω, and that we wish to find the velocity and acceleration of the piston and the angular velocity and acceleration of the connecting rod.

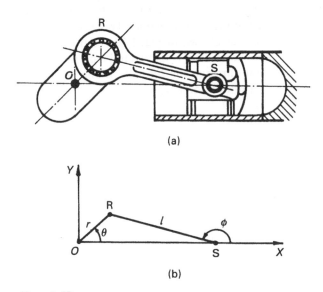

(a)

(b)

FIG. 1.57

We must first examine each rigid body in the mechanism and identify its type of motion, i.e. pure translation, pure rotation, or general plane motion. Crank *OR* is rotating about fixed point *O* and is therefore in pure rotation. Since slider S moves along a straight line and does not rotate, it is in rectilinear translation. The connecting rod RS has both translation and rotation and is therefore in general plane motion. If we assume that each body is rigid, and neglect any clearances at the bearings, the mechanism may be represented by the kinematically equivalent line diagram in Fig. 1.57(b) where the links are replaced by lines joining the bearing centres and the point S is constrained to slide along *OX*.

Graphical method

By drawing Fig. 1.57(b) to scale, for a given value of θ, x_S can be found. Next the velocity of S can be determined by considering each body in the mechanism in turn, starting with the input link. Link OR is rotating about a fixed point O. The velocity of point R therefore has magnitude

$$v_R = r\omega \qquad \qquad \text{(i)}$$

and has direction perpendicular to OR (see p. 25) as shown in Fig. 1.58.

Link RS is in general plane motion. For such a body, eqn (1.30) gives

$$v_S = v_R + v_{SR}. \qquad \qquad \text{(ii)}$$

In this equation we know v_R in both magnitude and direction. We also know that the direction of v_S must lie along the axis OX, and that v_{SR}, since RS is a rigid body, must be perpendicular to RS (see p. 26). Equation (ii) can now be solved graphically. We start with the vector v_R on the right-hand side of eqn (ii) which we know in both magnitude and direction. This vector can be drawn to scale in the correct direction as shown in Fig. 1.59. To this vector a second vector v_{SR} must be *added*. The line of v_{SR} is known to be perpendicular to RS so that we can now draw a line through the *tip* of v_R along the line of v_{SR}. At this stage there are two possibilities for v_{SR}, one towards the top and the other towards the bottom of Fig. 1.59.

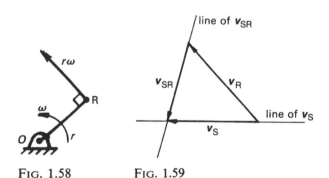

FIG. 1.58   FIG. 1.59

We now consider the right-hand side of eqn (ii) and note that the line of v_S lies along OX. The vector v_S must equal the *sum* of the other two vectors. Thus we start from the *tail* of v_R and draw a line along the line of v_S. The intersection of this line with the line representing v_{SR} locates the tips of the vectors v_{SR} and v_S. Arrows showing their directions can now be added such that v_S becomes the *sum* of v_R and v_{SR}, as in eqn (ii). The magnitudes of v_S and v_{SR} can be measured from the vector diagram, which is often referred to as a velocity diagram.

The angular velocity of the link RS can now be found. Link RS is sketched, as in Fig. 1.60, and the direction of the vector v_{SR}, is obtained from Fig. 1.59.

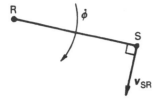

FIG. 1.60

Since

$$v_{SR} = l\dot{\phi}, \text{ we obtain } \dot{\phi} = \frac{v_{SR}}{l}. \tag{ii}$$

By inspection of Fig. 1.60, $\dot{\phi}$ is seen to be in a *clockwise* direction.

The piston is in pure translation so that all points on the piston have the same velocity. Point S on the connecting rod has the same velocity as the coincident point S on the piston. Hence the velocity of the piston is v_S. For the crank position considered, the vector diagram of Fig. 1.59 shows that the piston is moving to the left.

The accelerations of the system can be found in a similar manner. Crank OR rotates about a fixed point so that

$$a_R = (a_R)_r + (a_R)_t,$$

where $(a_R)_r = \omega^2 r$ along RO and $(a_R)_t = \dot{\omega}r$ perpendicular to RO. In this case $\dot{\omega} = 0$ (constant angular velocity) so that $(a_R)_t = 0$. The acceleration of R is therefore radially inwards along RO and has magnitude $a_R = \omega^2 r$.

The link RS is in general plane motion, and we now know the acceleration of end R of this link.

Using eqn (1.31) we have

$$a_S = a_R + a_{SR}.$$

In this vector equation we know the magnitude and direction of a_R. The line of a_S must lie along OX, and a_{SR} has two components, $(a_{SR})_r = \dot{\phi}^2 l$ along SR, and $(a_{SR})_t = \ddot{\phi}l$ perpendicular to RS. Since $\dot{\phi}$ has been found from the velocity diagram, the component $(a_{SR})_r$ is known in both magnitude and direction. Only the line of $(a_{SR})_t$ is known. Thus, in the vector equation for link RS,

$$a_S = a_R + (a_{SR})_r + (a_{SR})_t, \tag{iii}$$

only the *magnitudes* of a_S and $(a_{SR})_t$ are unknown and it is possible to obtain a graphical solution. First we draw a_R to scale as in Fig. 1.61(a) and then add the vector $(a_{SR})_r = \dot{\phi}^2 l$ along the direction SR, i.e. draw in $(a_{SR})_r$ to scale starting from the tip of a_R. The vector $(a_{SR})_t$ will be perpendicular to $(a_{SR})_r$ and its line can be drawn in as shown. The sum of these three vectors, a_S, must lie along OX so that the line of a_S can now be drawn in through the tail

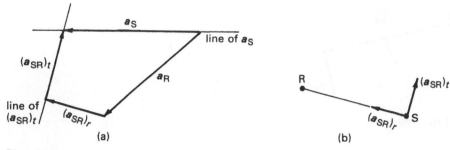

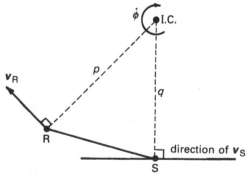

Fig. 1.61

of a_R. The intersection of the lines in the directions of $(a_{SR})_t$ and a_S defines the tips of the vectors a_S and $(a_{SR})_t$. Arrows showing the directions of the vectors can now be added such that the vector diagram satisfies eqn (iii). The magnitudes of the unknown accelerations are obtained by measurement of the vectors.

Since $(a_{SR})_t = l\ddot{\phi}$, the angular acceleration $\ddot{\phi}$ of the connecting rod can be found, and by inspection of Fig. 1.61(b) can be seen to be in an *anticlockwise* direction.

All points on the piston have acceleration a_S since it is in pure translation. For the position given in Fig. 1.57, the vector diagram of Fig. 1.61(a) shows that the acceleration of the piston is towards O.

The vector diagram of Fig. 1.61(a) is often referred to as an acceleration diagram.

A complete picture of the variation in velocity and acceleration of the components of the mechanism can be obtained by drawing a series of vector diagrams at increments of the crank angle θ.

As an alternative to the above method the *velocities* of the components could have been obtained using the instantaneous centre method. Link RS is a rigid body on which the directions of the velocities of two points, R and S, are known. By drawing perpendiculars to these directions through R and S, as in Fig. 1.62, the instantaneous centre can be located (see p. 28). Let the distances

Fig. 1.62

R(IC) and S(IC) be p and q respectively. The magnitude of v_R is known from the motion of point R on the crank OR, so that the angular velocity $\dot{\phi}$, of the link RS, is *clockwise* and is given by

$$\dot{\phi} = \frac{v_R}{p} = \frac{r\omega}{p}.$$

The velocity of point S, therefore, is $v_S = \dot{\phi}q$ along XO. The instantaneous centre approach applies only to velocities and the accelerations must now be found as before.

Analytical method

Let the crank OR make an angle θ with OX, as shown in Fig. 1.57(b). It is now possible to write an expression for the position, x_S, of point S, in terms of θ, ϕ, r and l.

From Fig. 1.57(b),

$$x_S = r \cos \theta + l \cos(\pi - \phi) = r \cos \theta - l \cos \phi. \tag{iv}$$

By differentiating both sides of eqn (iv) with respect to time, the velocity of the slider is given by

$$v_S = \frac{dx_S}{dt} = -r\dot{\theta} \sin \theta + l\dot{\phi} \sin \phi. \tag{v}$$

$$\left[\text{Note: } \frac{d}{dt}(r \cos \theta) = \frac{d\theta}{dt}\frac{d}{d\theta}(r \cos \theta) = \dot{\theta}(-r \sin \theta) \right].$$

Similarly, differentiation of eqn (v) with respect to time gives the acceleration

$$a_S = \frac{dv_S}{dt} = -r\ddot{\theta} \sin \theta - r\dot{\theta}^2 \cos \theta + l\ddot{\phi} \sin \phi + l\dot{\phi}^2 \cos \phi. \tag{vi}$$

$$\left[\text{Note: } \frac{d}{dt}(r\dot{\theta} \sin \theta) = r\left(\ddot{\theta} \sin \theta + \dot{\theta}\frac{d}{dt} \sin \theta \right) = r[\ddot{\theta} \sin \theta + \dot{\theta}(\dot{\theta} \cos \theta)] \right].$$

Now the crank speed $\dot{\theta} = \omega$ and is constant, so that $\ddot{\theta} = 0$. Therefore

$$a_S = -r\omega^2 \cos \theta + l\ddot{\phi} \sin \phi + l\dot{\phi}^2 \cos \phi.$$

From Fig. 1.57 we see that ϕ is not independent of θ since the vertical position of point S is

$$y_S = r \sin \theta - l \sin(\pi - \phi) = r \sin \theta - l \sin \phi = 0. \tag{vii}$$

Differentiating this equation with respect to time gives

$$r\omega \cos \theta - l\dot{\phi} \cos \phi = 0 \tag{viii}$$

and then

$$-r\omega^2 \sin \theta - l\ddot{\phi} \cos \phi + l\dot{\phi}^2 \sin \phi = 0. \tag{ix}$$

Also from eqn (vii)

$$\cos \phi = \sqrt{1 - \sin^2 \phi} = \sqrt{1 - \left(\frac{r}{l}\right)^2 \sin^2 \theta}. \tag{x}$$

By substituting ϕ, $\dot{\phi}$ and $\ddot{\phi}$ from eqns (vii), (ix), (x) into eqns (iv), (v) and (vi), expressions for x_S, v_S and a_S may be obtained in terms of the position, θ, of the crank and its angular velocity, ω. Even for this very simple mechanism the analytical solution is complex. A numerical solution would be most suitable. For a given mechanism geometry and speed, i.e. for known values of r, l and ω, values of x_S, v_S and a_S could be calculated for a series of values of θ, by using a computer or a calculator.

1.7   Four bar linkage

A typical four bar linkage is shown in Fig. 1.63. It consists of three moving links, O_1A, AB, O_2B, and a fixed link O_1O_2. The links O_1A and O_2B are rigid bodies which rotate about the fixed points O_1 and O_2. The link AB—called the coupler—has both translational and rotational motion and is therefore in general plane motion. It is of interest to note that if link O_2B were very long, the point B would move along an almost straight line so that the mechanism would be equivalent to the slider crank mechanism.

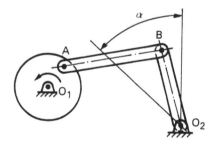

FIG. 1.63

With the geometry of Fig. 1.63 the crank O_1A can turn through a full revolution during which time the link O_2B would have oscillated through some angle α as indicated. This type of mechanism is called a crank-rocker four bar linkage and is very common in a wide variety of machinery used in the weaving, sewing, packaging, and other manufacturing industries. By changing the geometry such that the shortest link becomes the fixed link, both O_1A and O_2B can turn through complete revolutions, as shown in Fig. 1.64(a). If O_1A rotates at a constant angular velocity, O_2B will rotate at a non constant angular velocity. In Fig. 1.64(a), as A moves from A to A′, B moves from B to B′. It can be seen that B will take the same time to move from B to B′ as from B′ back to B, and therefore the angular velocity of O_2B will be non-uniform.

A further variation is obtained by fixing the link opposite the shortest link, as shown in Fig. 1.64(b), in which case neither O_1A nor O_2B can turn through a complete revolution. This variation is called a rocker-rocker four bar linkage.

The kinematic analysis of the four bar linkage, or of any other mechanism, follows the same procedure as for the slider crank mechanism and may be either analytical or graphical.

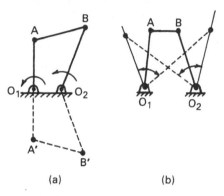

(a)   (b)

FIG. 1.64

Let us consider the mechanism of Fig. 1.65 in which the kinematically equivalent line diagram of an actual four bar linkage is shown. O_1A is the input link and rotates at a constant angular velocity $\dot{\theta}_2 = \omega$. Let us find the angular velocity and acceleration of links AB and O_2B when the angle θ_2 between the input link and the fixed link is 60°. The link lengths are $l_1 = (1 + 2\sqrt{3})a$, $l_2 = 2a$, $l_3 = 4a$, and $l_4 = (2 + \sqrt{3})a$, and points A and B are both on the same side of O_1O_2.

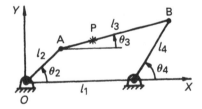

FIG. 1.65

We will obtain solutions using both the direct analytical and graphical methods.

Analytical method

The links of the mechanism must form a closed loop. Therefore the projection of links 2 and 3 along OX must equal the projections of links 1 and 4 along

OX, i.e.

$$l_2 \cos \theta_2 + l_3 \cos \theta_3 = l_1 + l_4 \cos \theta_4. \tag{i}$$

Similarly the projections along *OY* give

$$l_2 \sin \theta_2 + l_3 \sin \theta_3 = l_4 \sin \theta_4. \tag{ii}$$

The general solution of these simultaneous, non-linear, algebraic equations to yield θ_3 and θ_4 as functions of the input angle θ_2 is not easy. However, for the *special case* of $\theta = 60°$ and the link lengths given, $\theta_3 = 30°$ and $\theta_4 = 90°$, as shown in Fig. 1.66. We should note that the alternative closure, shown in broken line of Fig. 1.66, is also a possible solution to eqns (i) and (ii). For this case A and B are on *opposite* sides of O_1O_2.

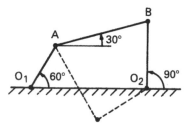

FIG. 1.66

Equations (i) and (ii) can now be differentiated with respect to time to give equations for the angular velocities $\dot\theta_3$ and $\dot\theta_4$, i.e.

$$-l_2\dot\theta_2 \sin \theta_2 - l_3\dot\theta_3 \sin \theta_3 = -l_4\dot\theta_4 \sin \theta_4, \tag{iii}$$

and $$l_2\dot\theta_2 \cos \theta_2 + l_3\dot\theta_3 \cos \theta_3 = l_4\dot\theta_4 \cos \theta_4. \tag{iv}$$

If the angles θ_2, θ_3 and θ_4 are known, the solution of these equations is straightforward since they are simple linear simultaneous equations in $\dot\theta_3$ and $\dot\theta_4$.

Substituting for θ_2, θ_3 and θ_4 and with $\dot\theta_2 = \omega$ we obtain

$$\frac{\sqrt{3}\,l_2\omega}{2} + \frac{l_3\dot\theta_3}{2} = l_4\dot\theta_4$$

and $$\frac{l_2\omega}{2} + \frac{\sqrt{3}\,l_3\dot\theta_3}{2} = 0.$$

These equations give

$$\dot\theta_3 = -\frac{\omega}{2\sqrt{3}} \quad \text{and} \quad \dot\theta_4 = \frac{2\omega}{2\sqrt{3}+3}.$$

For this position, the angular velocity $\dot\theta_3$ is negative. The minus sign means that link AB is rotating in the *clockwise* direction. Since $\dot\theta_4$ is positive, link O_2B is rotating in the anticlockwise direction.

The angular accelerations $\ddot{\theta}_2$, $\ddot{\theta}_3$ and $\ddot{\theta}_4$ are obtained by differentiating eqns (iii) and (iv) with respect to time giving

$$l_2\ddot{\theta}_2 \sin\theta_2 + l_2\dot{\theta}_2^2 \cos\theta_2 + l_3\ddot{\theta}_3 \sin\theta_3 + l_3\dot{\theta}_3^2 \cos\theta_3$$
$$= l_4\ddot{\theta}_4 \sin\theta_4 + l_4\dot{\theta}_4^2 \cos\theta_4 \qquad\qquad (v)$$

and $\quad l_2\ddot{\theta}_2 \cos\theta_2 - l_2\dot{\theta}_2^2 \sin\theta_2 + l_3\ddot{\theta}_3 \cos\theta_3 - l_3\dot{\theta}_3^2 \sin\theta_3$
$$= l_4\ddot{\theta}_4 \cos\theta_4 - l_4\dot{\theta}_4^2 \sin\theta_4. \qquad\qquad (vi)$$

These equations are linear simultaneous equations in angular acceleration and, if the angles and angular velocities are known, $\ddot{\theta}_3$ and $\ddot{\theta}_4$ can be found as functions of the input angular acceleration $\ddot{\theta}_2$. In the particular case under consideration $\dot{\theta}_2 = \omega$, which is constant, and therefore $\ddot{\theta}_2 = 0$.

Substituting values for θ_2, θ_3 and θ_4 and the corresponding angular velocities into eqns (v) and (vi) gives

$$\frac{l_2\omega^2}{2} + \frac{l_3\ddot{\theta}_3}{2} + \frac{\sqrt{3}}{24}l_3\omega^2 = l_4\ddot{\theta}_4,$$

and $\quad -\dfrac{\sqrt{3}}{2}l_2\omega^2 + \dfrac{\sqrt{3}}{2}l_3\ddot{\theta}_3 - \dfrac{l_3\omega^2}{24} = \dfrac{4l_4\omega^2}{(2\sqrt{3}+3)^2}.$

Hence $\quad \ddot{\theta}_3 = 0.45\,\omega^2$,

and $\quad \ddot{\theta}_4 = 0.58\,\omega^2$.

Since these expressions are positive, the angular acceleration of links AB and O_2B are in the anticlockwise direction.

The above results are independent of the link parameter a so that, for a given input angular velocity, the angular velocity and acceleration of each link are independent of the scale of the mechanism; i.e. if the link lengths are *all* multiplied by the same factor, there will be no change in *angular* velocity or acceleration. Also, the angular velocities of the links are directly proportional to ω, and their angular accelerations to ω^2. These statements apply to all single degree of freedom mechanisms, i.e. if the input speed is doubled, the angular velocities of the links are doubled and their angular accelerations are increased by a factor of four.

The *linear* velocity and acceleration of any point on the mechanism may now be calculated. For example, take a point P on AB where P is distant b from A.

From the geometry of Fig. 1.65, the position of P is given by the co-ordinates

$$x_P = l_2 \cos\theta_2 + b \cos\theta_3,$$
$$y_P = l_2 \sin\theta_2 + b \sin\theta_3.$$

Hence by differentiating the foregoing displacement equation with respect to time, the velocity and acceleration components of P are

$$\dot{x}_P = -l_2\omega_2 \sin\theta_2 - b\dot{\theta}_3 \sin\theta_3,$$

$$\dot{y}_P = l_2\omega_2 \cos\theta_2 + b\dot{\theta}_3 \cos\theta_3,$$

and

$$\ddot{x}_P = -l_2\omega^2 \cos\theta_2 - b\ddot{\theta}_3 \sin\theta_3 - b\dot{\theta}_3^2 \cos\theta_3,$$

$$\ddot{y}_P = -l_2\omega^2 \sin\theta_2 + b\ddot{\theta}_3 \cos\theta_3 - b\dot{\theta}_3^2 \sin\theta_3.$$

The most difficult part of the solution is calculating the angles θ_3 and θ_4. Once these are available the remainder of the calculation is concerned only with the solution of simultaneous algebraic equations.

Graphical method

As in the case of the slider crank mechanism, a graphical method may be used to solve the vector equations for the bodies which comprise the linkage. Following the steps taken in the slider crank example, we first draw to scale the kinematically equivalent diagram of the mechanism. This is shown in Fig. 1.67, with $a = 1$ unit. In this particular case the geometry is so simple that a sketch will suffice, but in general an accurate scale drawing will be necessary. We now consider the input link O_1A which is rotating at a constant angular velocity about a fixed point. The velocity of point A which is moving on a circular path is therefore known in both magnitude and direction, i.e. $v_A = \omega l_2$ perpendicular to O_1A as shown in Fig. 1.68(a).

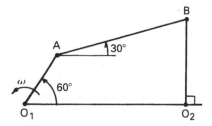

FIG. 1.67

Since link AB is a rigid body in general plane motion, the velocity of point B on AB is given by

$$v_B = v_A + v_{BA}. \tag{vii}$$

In this equation v_A is known in magnitude and direction. The line of v_{BA} must be perpendicular to AB (see p. 26) but its magnitude is unknown. If we now consider B as a point on the link O_2B, which rotates about the fixed point O_2, we see that the line of v_B must be perpendicular to O_2B, as in Fig. 1.68(c),

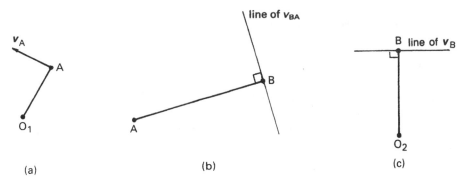

FIG. 1.68

but the magnitude is unknown. We now have enough information to solve eqn (vii).

First draw v_A to scale, as shown in the vector diagram of Fig. 1.69, where

$$v_A = \omega l_2 = 2a\omega.$$

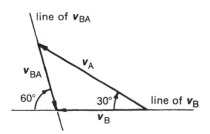

FIG. 1.69

The right hand side of eqn (vii) requires that the vector v_{BA} be *added* to v_A, so that we now draw a line through the tip of v_A perpendicular to the link AB. The sum of v_A and v_{BA} is v_B which is perpendicular to O_2B. Therefore starting from the tail of v_A we draw a line along the line of v_B. The intersection of this line with that of v_{BA} is at the tips of these two vectors and the arrows are drawn so that the resulting triangle represents the vector equation (vii).

The magnitudes of v_B and v_{BA} can be measured directly from the scale drawing. Alternatively we can calculate the values from the geometry of the vector triangle. In this case

$$v_{BA} \sin 60 = v_A \sin 30 = 2\omega a \sin 30$$

so that $v_{BA} = \dfrac{2\omega a}{\sqrt{3}}.$

Also $v_B = v_A \cos 30 - v_{BA} \cos 60 = 2\omega a \cos 30 - \dfrac{2\omega a}{\sqrt{3}} \cos 60$

so that $v_B = \dfrac{2\omega a}{\sqrt{3}}$.

Note that v_B and v_{BA} are *linear* velocities of *points* in the mechanism, and if a and ω have units of m and rad s^{-1} respectively, v_B and v_{BA} would be in m s^{-1}.

The directions of v_{BA} and v_B are obtained from Fig. 1.69 and are shown again in Fig. 1.70.

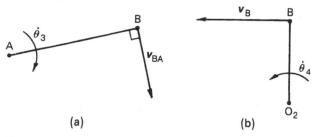

(a)   (b)

FIG. 1.70

From Fig. 1.70(a) it can be seen that the angular velocity of link AB is in the *clockwise* direction and

$$\dot{\theta}_3 = \frac{v_{BA}}{l_3} = \frac{\omega}{2\sqrt{3}}.$$

Similarly, referring to Fig. 1.70(b), the angular velocity of link O_2B is

$$\dot{\theta}_4 = \frac{v_B}{l_4} = \frac{2\omega}{2\sqrt{3}+3}.$$

and is in the *anticlockwise* direction. These values agree with values obtained from the analytical method.

The angular accelerations of the links may be found in a similar manner. Again starting with point A as a point on link O_1A, we know that its acceleration is

$$a_A = \omega^2 l_2 = 2a\omega^2$$

radially inwards along AO_1. There is no tangential component of a_A since ω is constant.

For link AB, which is in general plane motion,

$$a_B = a_A + a_{BA}. \tag{viii}$$

Consider the right-hand side of this equation. The vector a_A is known in both magnitude and direction. The acceleration a_{BA} of B relative to A has two components, $(a_{BA})_r$ along BA and $(a_{BA})_t$ perpendicular to BA. The magnitude and direction of the radial component are known, i.e.

$$(a_{BA})_r = \dot{\theta}_3^2 l_3 = 0.33\ a\omega^2,$$

and its direction is along BA.

The magnitude of $(a_{BA})_t$ is unknown.

We shall now examine the left-hand side of eqn (viii) and consider B as a point on the link O_2B. Now

$$a_B = (a_B)_r + (a_B)_t$$

where $(a_B)_r$ is the radial component along BO_2 and

$(a_B)_t$ is the tangential component perpendicular to O_2B.

The magnitude of the component $(a_B)_r$ is known and is

$$(a_B)_r = \dot{\theta}_4^2 l_4 = 0.357\ a\omega^2.$$

Thus in eqn (viii) the lines and directions of all the vectors are known, and only the magnitudes of $(a_{BA})_t$ and $(a_B)_t$ are unknown. Equation (viii) can therefore be solved graphically.

Taking the right side of the equation first, we can draw a_A to scale as shown in Fig. 1.71.

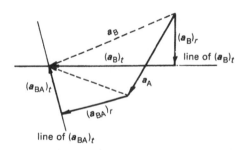

FIG. 1.71

To this vector we add $(a_{BA})_r$. Since the line of the tangential component is known, we can draw in the line of $(a_{BA})_t$ through the *tip* of $(a_{BA})_r$. Remember that the line of $(a_{BA})_t$ is perpendicular to the line of $(a_{BA})_r$. We now consider the left-hand side of eqn (viii) and draw in $(a_B)_r$ starting from the tail of a_A. The line of $(a_B)_t$, which is perpendicular to $(a_B)_r$, is now constructed through the tip of $(a_B)_r$. The intersection of $(a_B)_t$ and $(a_{BA})_t$ is at the tips of these two vectors and the arrows can be drawn in so that Fig. 1.71 represents eqn (viii) completely.

Vectors representing a_B and a_{BA} can now be drawn in if required. The magnitude of each vector can be obtained by measurement from the scale drawing or from the geometry of the vector diagram.

From Fig. 1.71,

$$(a_{BA})_t \cos 30 = a_A \cos 30 + (a_{BA})_r \cos 60 - (a_B)_r,$$

so that $(a_{BA})_t = 1.77\ a\omega^2$.

Also $(a_B)_t = a_A \sin 30 + (a_{BA})_r \sin 60 + (a_{BA})_t \sin 30$,

giving $(a_B)_t = 2.17\ a\omega^2$.

The angular accelerations of the links can now be obtained from these tangential components of linear acceleration.

Thus $\ddot{\theta}_3 = \dfrac{(a_{BA})_t}{l_3} = \dfrac{1.77\ \omega^2}{4} = 0.45\ \omega^2,$

and $\ddot{\theta}_4 = \dfrac{(a_B)_t}{l_4} = 0.58\ \omega^2.$

The directions of $\ddot{\theta}_3$ and $\ddot{\theta}_4$ can be found by considering the directions of $(a_{BA})_t$ and $(a_B)_t$. From Fig. 1.71 it can be seen that both $\ddot{\theta}_3$ and $\ddot{\theta}_4$ will be in the positive anticlockwise direction.

The velocity and acceleration of any point such as P (Fig. 1.65) can now be obtained since

$$v_P = v_A + v_{PA}, \quad \text{and} \quad a_P = a_A + a_{PA}.$$

Since both the angular velocity and acceleration of the link AB are now known, the vectors on the right-hand side of these equations are known both in magnitude and direction. Taking P as, for example, the mid point of AB, the reader is recommended to construct the vector diagrams from which v_P and a_P can be obtained.

Exercises

1   A vehicle which is travelling in a straight line has its position defined by its distance x from some origin O. If it starts from rest at a position x_O and has a constant acceleration a, obtain, from first principles, expressions for its velocity and position as functions of time. Also obtain an expression for the velocity of the vehicle as a function of its position x.

2   A particle moves along a straight line such that its displacement $x = at^2 + b\sin \omega t$, where a, b and ω are constants.
    If at $t = 0$, $\dot{x} = 8\ \mathrm{m\ s}^{-1}$ and at $t = \pi/2\omega$, $\ddot{x} = 3\ \mathrm{m\ s}^{-2}$ find a and b in terms of ω. What are the units of a, b and ω.

3   A rotor has an angular acceleration which is given by

$$\ddot{\Theta} = Ke^{-ct},$$

    where K and c are constants.

If the rotor starts from rest at $t = 0$, obtain expressions for
(i)   the maximum speed of the rotor,
(ii) the time taken to reach 90% of the maximum speed,
(iii) the angle turned through when the speed reaches 90% of its maximum value.

4   A vehicle P travelling in a horizontal plane OXY carries two accelerometers which measure accurately its component accelerations a_x and a_y in the *fixed* directions OX and OY. The vehicle starts from rest at O.

$$\text{For } 0 < t < 10 \text{ s} \quad a_x = 2 \text{ m s}^{-2}, \quad a_y = 0,$$

$$\text{for } 10 < t < 30 \text{ s} \quad a_x = 0, \quad a_y = 4 \text{ m s}^{-2}$$

$$\text{and for } 30 \text{ s} < t \quad a_x = 0, \quad a_y = 0.$$

By determining the equations for the position of the vehicle sketch its track over the plane. Find also the magnitude and direction of its final velocity.

5   Two projectiles P and Q are fired simultaneously, their paths being such that they collide during flight. P has initial velocity V, and is fired at an angle α from the horizontal. Q is fired from a point which is h vertically above the launching point of P and has velocity $V/2$ at an angle β from the horizontal.
    If $\cos \beta = 2/\sqrt{5}$, show that $\tan \alpha = 2$ and that collision will occur when $t = (2\sqrt{5}/3)(h/V)$.

6   A straight rod AB of length 2 m rotates about a fixed pivot at end A.
    (a)   Find the magnitude and direction of the acceleration of end B when the rod rotates at a constant angular velocity of 4 rad/s.
    (b)   If the rod has an instantaneous angular velocity of 4 rad/s and the acceleration vector of end B makes an angle of 30° with the rod, what is the angular acceleration of the rod?
    (c)   If the rod were at rest but had an angular acceleration of 20 rad/s^2 what would be the acceleration of end B?

7   A cylindrical roller of radius R is in rolling contact with two parallel plates A and B as shown in Fig. 1.72. If the plates have velocities v_A and v_B whose vectors are parallel as shown, find the velocity of the centre O of the roller, and the angular velocity of the roller. Assume that there is no slip between the roller and the plates.

8   The point A on link AB in which AB = 2 m moves with constant acceleration of 10 ms^{-2} in the direction shown in Fig. 1.73. The link is initially at rest with AB horizontal as shown but it has a constant angular

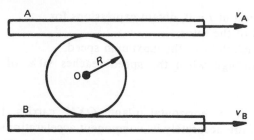

FIG. 1.72

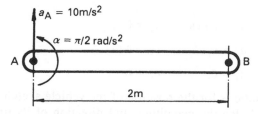

FIG. 1.73

acceleration of $\pi/2$ rad s^{-2} in the anticlockwise sense. Find the magnitude and direction of the acceleration of end B after 2 s.

9 A toothed belt drive has pulleys of pitch circle diameter 100 mm and 225 mm. If the smaller pulley rotates at 400 rev/min in an anticlockwise direction, what is the angular velocity of the larger pulley?

10 A simple gear train consists of two gears having 28 and 48 teeth respectively. If the smaller gear rotates at 1200 rev/min in a clockwise direction, what is the angular velocity of the larger gear? How would the output angular velocity be affected if an idler gear with 52 teeth were interposed between the two gears?

11 A simple tracked vehicle is shown in Fig. 1.74. If the vehicle speed is v and the radius of the driving sprocket is r, what is the angular velocity of the sprocket and the velocity of the top run of the track?

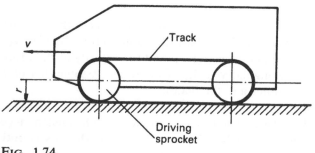

FIG. 1.74

12 A geared electrical driven winch is shown diagrammatically in Fig. 1.75. The pitch circle diameter of the motor pinion is 150 mm, that of the gear fixed to the winch barrel is 900 mm and the barrel itself has a diameter of 400 mm. Find the velocity of the load when the motor runs at 500 rev/min.

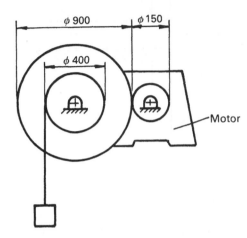

FIG. 1.75

13 An epicyclic gear train is shown in Fig. 1.76 in which the arm OA rotates in a clockwise direction at angular velocity ω as shown. The ring gear G_1 is fixed. What is the angular velocity of the sun gear G_2?

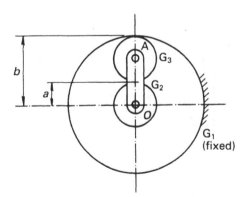

FIG. 1.76

14 A bevel epicyclic gear arrangement is shown in Fig. 1.77. Determine the angular velocities of gear G_2 and of the arm AB if gear G_3 is fixed and G_1 rotates with angular velocity ω_1.

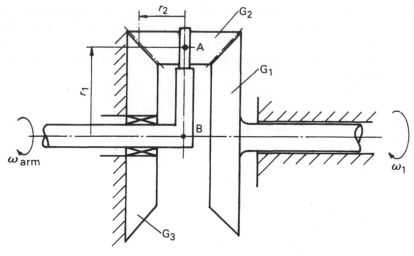

FIG. 1.77

15 A compound epicyclic gear arrangement is shown in Fig. 1.78. Arm A is driven by the input shaft and carries a compound gear BC. C meshes with a fixed gear E and B meshes with a gear D. Gear D is attached to the output shaft which passes through the centre of the fixed gear E. Find the ratio of the speed of the output shaft to that of the input shaft. Gears B, C and D have pitch circle diameters 2*b*, 2*c*, 2*d*.

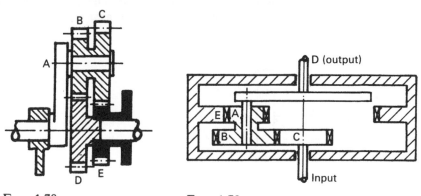

FIG. 1.78   FIG. 1.79

16 An epicyclic gearbox is shown in Fig. 1.79. Gear C (16 teeth) meshes with wheel B (32 teeth) of the compound planet wheel. Wheel A of the compound planet wheel has 12 teeth and meshes with the fixed ring gear E and runs on a shaft fixed to the output shaft D as shown. The teeth on A and E have the same pitch as those on B and C. Calculate the speed of the output shaft D when the input shaft C has a speed of 1000 rev/min.

17 The cam mechanism shown in Fig. 1.80 consists of an eccentrically mounted disc of radius R and a flat faced follower. The eccentricity of the disc is ε and the face of the follower is inclined to the horizontal at a constant angle α.

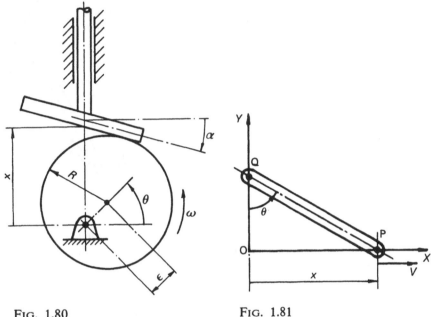

FIG. 1.80   FIG. 1.81

Obtain expressions for the velocity and acceleration of the follower as functions of the cam angle θ if the cam rotates at a constant angular velocity ω. Assume that the cam and follower are always in contract.

18 The pins P and Q in the link shown in Fig. 1.81 are constrained to move along OX, OY respectively. Obtain analytical expressions for the angular velocity $\dot{\theta}$ and angular acceleration $\ddot{\theta}$ of the link as functions of x when P moves along OX with a constant velocity v.

19 Fig. 1.82 shows a simple robot arm. Write down the equations from which you could calculate the velocity and acceleration of point P in terms of θ_1, θ_2 and their derivatives.

20 Fig. 1.83 shows a line diagram of a simple robot arm ABC. Point A is fixed and the links AB and BC are driven by torque motors at A and B.
      When the arm is in the position shown the velocity and acceleration of C are specified as shown.
      Determine the required angular velocities and angular accelerations of the links AB and BC.

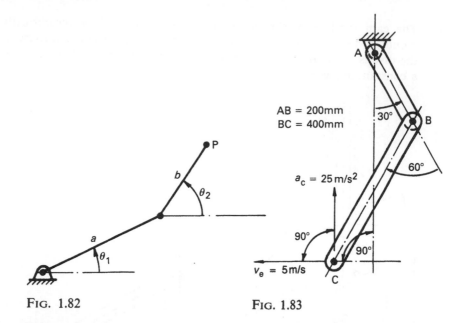

AB = 200mm
BC = 400mm

a_c = 25 m/s²

v_e = 5m/s

Fig. 1.82   Fig. 1.83

21 A four-bar linkage is shown in Fig. 1.84. For the position shown determine the angular velocity and angular acceleration of the output link O_2B when the input link O_1A has a constant angular velocity of 100 rad/s anticlockwise.

22 A reciprocating roller-mechanism (Fig. 1.85) consists of a cylindrical roller of centre A which rolls without slip on a horizontal plane. It is driven by a slider B which moves in a vertical straight line and which is connected to the roller by a straight rod AB freely pinned at its ends to the roller and to the slider respectively. The radius of the roller is 25 mm and the

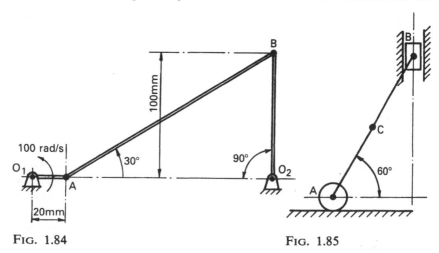

Fig. 1.84   Fig. 1.85

length AB is 200 mm. In the position shown, the slider has a downward velocity of 600 mm/s and zero acceleration. Find for the given position (i) the angular acceleration of the roller, and (ii) the resultant linear acceleration of C, the mid-point of AB.

23 A four bar linkage ABCD is shown in Fig. 1.86, in which the crank AB rotates at a constant angular velocity of 10 rad/s.

Find the angular velocities of links BC and CD when the crank AB is at 60°.

Also determine the angular accelerations of BC and CD.

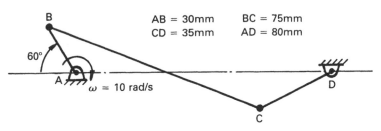

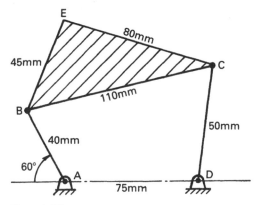

FIG. 1.86

24 A four bar linkage ABCD is shown in Fig. 1.87 in which the coupler is the triangular link BCE.

Find the velocity and acceleration of point E when the input crank AB makes an angle of 60° with the horizontal. At this instant AB has angular velocity and acceleration of 50 rad/s and 1600 rad/s² respectively in the clockwise direction.

FIG. 1.87

2
Newton's laws of motion

2.1 Laws of Motion

Having studied the geometric properties of motion we now need to consider the forces necessary to produce motion, and the behaviour of bodies when forces are applied to them. The relationship between force and motion was formulated by Sir Isaac Newton (1642-1727). His results were first published in his *Principia* (1687).

Newton's laws of motion may be expressed as follows.

Law 1 The absolute velocity *v* of a particle* remains constant if there is no net external force applied to the particle. Note that *v* is a vector so that both *magnitude and direction* will be constant.

Law 2 When a force *F* acts on a particle the absolute acceleration of the particle is directly proportional to the magnitude of the force and is in the direction of the force.

Law 3 When two particles A and B are in contact the force applied to particle A by particle B, at the contact point, is equal in magnitude but opposite in direction from the force applied to particle B by particle A.

It should be noted that the First Law is a special case of the Second Law since it represents the condition when the acceleration of the particle is zero.

The Second Law may be formulated mathematically as

$$F = kma,$$ (2.1)

where *F* is the force applied to the particle, *m* is the mass of the particle, and *a* is the acceleration of the particle. *k* is a constant of proportionality. If more than one force acts on the particle, *F* is equal to the vector sum of all of the individual forces. The acceleration vector will always be in the same direction as the resultant force vector.

Laws 1 and 2 apply only to motion relative to frames of reference which are either stationary or translating with constant velocity. Such frames are called *inertial frames of reference* and the acceleration measured in such a frame is always the *absolute* acceleration of the particle. However, the errors

---

* A body with mass but negligible dimensions.

introduced by neglecting the accelerations due to rotation of the Earth are insignificant in most engineering applications, and the accelerations measured relative to the Earth may be considered as absolute.

From the Second Law we also see that when the acceleration of a particle is zero the sum of the forces acting upon it is zero. The particle is then said to be in *equilibrium* under the action of the applied forces and $F = 0$. If the particle also has zero velocity it is in *static equilibrium.*

2.2   Units

If we choose a set of units such that one unit of mass multiplied by one unit of acceleration gives one unit of force, k will have a value of one. Under these conditions we may express the Second Law as

$$F = ma \tag{2.2}$$

A system of units in which the product of any two unit quantities is the unit of the resultant quantity is called a *coherent* system of units. The SI system (Système International d'Unites), which will be used throughout this text, is a coherent system of metric units. It is now becoming established in many countries as the only legal system of units.

We have already come across two basic units in our study of kinematics, the metre, which is the basic unit of length, and the second, the basic unit of time. All the other units in kinematics, i.e. those of velocity and acceleration, are expressed in terms of these two basic units. In dynamics we need to introduce the basic unit of mass, the kilogram. These three units, the kilogram (kg), the metre (m), and the second (s) are the only *basic* units required in the study of dynamics. All other units can be expressed in terms of the basic units and are therefore *derived* units. For example, the unit of velocity is the $m\,s^{-1}$, and may be considered as that uniform velocity which a point would have if it were to travel a distance of 1 metre in a time of 1 second. Similarly, the unit of force may be defined as that force which when acting on a particle of mass 1 kg causes it to accelerate at $1\,m\,s^{-2}$. From Newton's Second Law,

$$1 \text{ unit of force} = 1\,kg \times 1\,m\,s^{-2} = 1\,kg\,m\,s^{-2}.$$

The unit of force is therefore the $kg\,m\,s^{-2}$. This unit of force is called the *Newton*. The Newton (N) is therefore a *derived* unit and can always be expressed in terms of the three basic units since $1\,N = 1\,kg\,m\,s^{-2}$. As will be seen later, other groups of basic units used in dynamics have been given names, although some, such as the units of velocity and acceleration, are always referred to in terms of the basic units.

3

Statics

Basic Theory

3.1 Introduction

The study of the special case of Newton's Second Law in which the acceleration of the particle is zero is referred to as Statics. In these circumstances the resultant force acting on the particle must be zero so that, from eqn (2.1),

$$\sum F = 0. \tag{3.1}$$

Usually statics is concerned with systems which are at rest, although the results do apply to systems which are moving with constant velocity. (Remember that velocity is a vector quantity and for it to remain constant neither its direction nor its magnitude must change.)

We will first consider the case of a particle, to which Newton's Laws apply directly, and then consider the statics of rigid bodies.

3.2 Equilibrium of a particle in a plane

Consider a particle P in the XY plane of Fig. 3.1 with forces $F_1 \cdots F_i \cdots F_n$ applied to it as shown. For the particle to be in a state of equilibrium, i.e. to

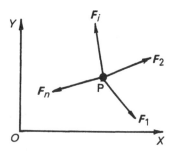

FIG. 3.1

have zero acceleration, eqn (3.1) gives

$$\sum F_i = 0. \tag{3.2}$$

Each of the forces F_i may be expressed in components along OX and OY as

$$F_i = F_{xi} + F_{yi}. \tag{3.3}$$

Hence eqn (3.2) may be rewritten in component form as

$$\sum F_{xi} = 0, \quad \text{and} \quad \sum F_{yi} = 0. \tag{3.4}$$

It has been shown in Chapter 1 that a particle moving in a plane has two degrees of freedom. Equations (3.4) show that the motion of the particle along the two co-ordinate directions corresponding to these two degrees of freedom is suppressed. Its equilibrium is therefore defined by two equations, one corresponding to each of the degrees of freedom.

If a particle P is in equilibrium under the action of the four forces shown in Fig. 3.2(a), the resultant of these forces must be zero, i.e.

$$F_1 + F_2 + F_3 + F_4 = 0. \tag{3.5}$$

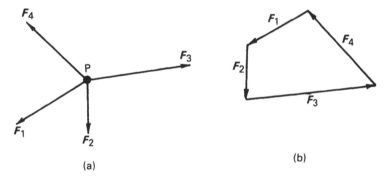

(a)

(b)

Fig. 3.2

These forces can be added together using the rules of vector addition given in Chapter 1 (p. 2). The polygon of Fig. 3.2(b) represents a graphical solution of eqn (3.5). If the particle is in equilibrium, the polygon will close as shown. If polygon does not close, there will be a resultant force acting on the particle and the particle will accelerate in the direction of the resultant force.

The forces $F_1 \cdots F_4$ may alternatively be given in component form along axes OX and OY, and eqn (3.5) rewritten as

$$\begin{aligned} F_{x1} + F_{x2} + F_{x3} + F_{x4} &= 0, \\ F_{y1} + F_{y2} + F_{y3} + F_{y4} &= 0, \end{aligned} \tag{3.6}$$

where F_{xi}, F_{yi} are the components of F_i along OX and OY respectively.

Thus the sums of the components of F_1, F_2, F_3 and F_4 along OX and OY must also be zero.

Example 3.1   Consider the arrangement shown in Fig. 3.3 in which a particle P, of mass m, is supported by the two wires AP and BP. Let us obtain expressions for the tensions in the two wires caused by the weight of the particle.

The particle P is in equilibrium under the action of the forces applied to it and these forces must add vectorially to zero. Let us identify these forces by drawing the *free body diagram* of the particle. To do this we imagine the particle to be separated from the two wires. We first draw the particle as a point, and then draw in the forces which act upon it. In this case only three forces act:

(i)   the weight, W of the particle acting vertically downwards,
(ii)   the force in the wire PA,
(iii) the force in the wire PB.

Figure 3.4 shows the free body diagram of the particle in which T_1 and T_2 are the forces which are applied to the particle by the wires AP and BP.
Since the particle is in equilibrium the sum of the forces acting on it must be zero, i.e.

$$T_1 + T_2 + W = 0, \tag{i}$$

where $W = mg$.

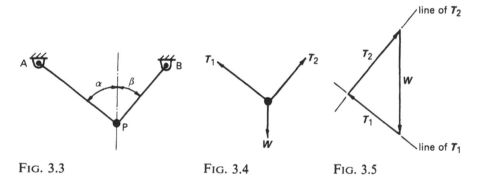

FIG. 3.3   FIG. 3.4   FIG. 3.5

In eqn (i) W will be known in both magnitude and direction. The lines of action of T_1 and T_2 are known, i.e. along PA and PB, but their magnitudes are not. Equation (i) can therefore be solved graphically as shown in Fig. 3.5. The weight vector W is first drawn to some convenient scale with an arrow showing its direction. From its tip a line parallel to AP, which is the line of action of T_1, is drawn. The third vector representing T_2 must close the vector diagram if eqn (i) is to be satisfied. It must therefore pass through the tail end of the vector W.

A line parallel to PB, which is the line of action of T_2, can now be drawn passing through the tail end of W as shown. These two lines of action intersect and, with the vector W, form a closed polygon, in this case a triangle. Arrows can now be added to the vector diagram as shown so that it satisfies eqn (i). The magnitudes of T_1 and T_2 can now be obtained by direct measurement of the vectors.

Alternatively we may consider the horizontal and vertical equilibrium of the particle using eqn (3.4). Again using the free body diagram of Fig. 3.4 and resolving the three forces horizontally and vertically, we have

$$\rightarrow T_2 \sin \beta - T_1 \sin \alpha = 0,$$

and $\quad \uparrow T_1 \cos \alpha + T_2 \cos \beta - W = 0.$ \hfill (ii)

The arrows at the beginning of the equations are used to define the positive directions used. The magnitudes of T_1 and T_2 can then be obtained from eqns (ii) as

$$T_1 = \frac{W \sin \beta}{\sin(\alpha + \beta)} \quad \text{and} \quad T_2 = \frac{W \sin \alpha}{\sin(\alpha + \beta)}. \tag{iii}$$

Note that if α and β both approach 90°, $\alpha + \beta \rightarrow \pi$, so that T_1 and $T_2 \rightarrow \infty$. It is therefore impossible to support any vertical load on a *horizontal* wire.

Overconstrained systems

We have demonstrated that a particle moving in a plane has two degrees of freedom. To maintain the particle in a fixed position we must remove these two degrees of freedom. In Fig. 3.3 the constraints provided by the two wires AP and BP achieve this result by applying the two forces T_1 and T_2 to the particle. Let us now consider the effect of adding an additional constraint, e.g. the wire CP, as shown in Fig. 3.6(a).

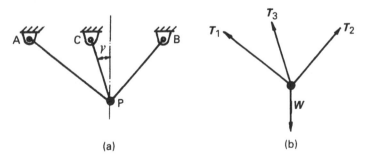

(a) (b)

FIG. 3.6

The corresponding free body diagram is shown in Fig. 3.6(b) where the force T_3 is the tension in the additional wire CP. Using eqn (3.4) to describe the equilibrium of the particle, we can obtain only *two* equations of equilibrium, i.e.

$$\rightarrow T_s \sin \beta - T_1 \sin \alpha - T_3 \sin \gamma = 0$$

and $\uparrow T_1 \cos \alpha + T_2 \cos \beta + T_3 \cos \gamma - W = 0.$ (3.7)

These two equations contain *three* unknown forces. Thus, although the particle is in equilibrium the magnitudes of the forces T_1, T_2 and T_3 cannot be found from equilibrium considerations alone. The wire CP could be installed so that it is just taut and therefore carries very little load, or it could be installed so that it carries a large proportion of the load.

This problem can only be solved by considering the extensions of the wires produced by the displacement of P due to the load W. Equations (3.7) still apply, but an additional equation relating the extensions of the wires to the displacement of P is also necessary. This third equation is called the compatibility equation. Systems of this type, in which the number of unknown forces exceeds the number of equilibrium equations, are said to be *overconstrained* or *statically indeterminate*.

In this book we will only consider systems which are *statically determinate*, i.e. those in which the number of equations of equilibrium is equal to the number of unknown quantities in the equations.

3.3 Equilibrium of rigid bodies

A rigid body moving in a plane has three degrees of freedom and its displacement in the plane is specified by three independent co-ordinates. It is convenient to choose our three co-ordinates as the x and y co-ordinates of a point in the body and the angular position, relative to the axis OX, of a line in the body. For the body to be in equilibrium these three degrees of freedom must be eliminated.

The form in which Newton's Laws of motion are presented in Chapter 2 applies only to particles. A rigid body is made up of a large number of individual particles held together by internal forces. By applying Newton's Laws to each of the particles in the body we can obtain equations of motion for the complete body, or for the static equilibrium of the body.

The assumption that the body is rigid means that the relative positions of the particles making up the body do not change when loads are applied. We are therefore neglecting any strains which occur in the material of the body.

Let us consider a rigid body with external forces F_1, F_2, $F_3 \cdots F_i \cdots F_n$ applied to it as shown in Fig. 3.7(a). The resultant external force F on the body will be the vector sum of the individual applied forces, i.e.

$$F = \sum F_i.$$ (3.8)

The body will be assumed to be composed of a large number of particles $P_1, P_2, \cdots P_i, \cdots P_n$. Internal stresses in the body hold these particles together and cause internal forces to act on the particles. Consider a typical particle

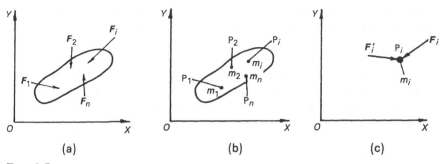

FIG. 3.7

P_i as shown in Fig. 3.8 in contact with a number of other particles. The particle P_j will exert a force F'_{ji} on the particle P_i, and the particle P_i, from Newton's Third Law, will exert an *equal* and *opposite* force F'_{ij} on particle P_j, i.e. $F'_{ij} = -F'_{ji}$. The other particles in contact will also exert forces on particle P_i. Let the sum of all of these internal forces on particle P_i be F'_i. If we now add together the individual internal forces over the whole body we see that, since $F'_{ij} = -F'_{ji}$, they will cancel out in pairs, so that the sum of the internal forces over the whole of the body

$$\sum F'_i = 0. \tag{3.9}$$

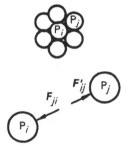

FIG. 3.8

For the typical particle P_i with net *external* force F_i, and net internal force F'_i as shown in Fig. 3.7(c), the total force acting on the particle will be $F_i + F'_i$. For the particle P_i to be in equilibrium, eqn (3.2) gives

$$F_i + F'_i = 0.$$

This must apply to all particles in the body, so that by summing over all of the particles we obtain

$$\sum F_i + \sum F'_i = 0. \tag{3.10}$$

Now we have seen from eqn (3.9) that the internal forces cancel out when summed over the whole body, so that eqn (3.10) can be rewritten as

$$\sum F_i = 0, \tag{3.11}$$

i.e. the vector sum of all of the external forces acting on the body must be zero.

This equation is identical with that for a single particle. However, there is a major physical difference. In the case of a single particle, all of the forces must act at the same point since the particle has no dimensions, but the forces on a rigid body can act at different points on the body and their lines of action need not intersect at a common point.

Let us consider the case of two equal and parallel forces acting in opposite directions on a body, as shown in Fig. 3.9. Since the forces are equal and opposite, eqn (3.11) is satisfied. However, from our everyday experience we know that the body would not be in equilibrium, but would rotate in an anticlockwise direction. The condition given by eqn (3.11) is therefore not sufficient on its own to define the equilibrium of the rigid body. A further condition, associated with the rotational degree of freedom of the body, must be introduced.

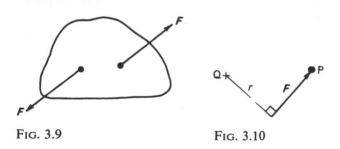

FIG. 3.9   FIG. 3.10

Moment of a force

Consider the force F acting at a point P as shown in Fig. 3.10. We define the *moment* M_Q, of the force F about any point Q, as

$$M_Q = rF, \tag{3.12}$$

where r is the *perpendicular* distance from Q to the line of action of F. Moments are vector quantities and have both magnitude and direction. For the two dimensional case shown in Fig. 3.10, it is sufficient to define the direction of the moment by considering the direction in which the line PQ would rotate if P moved in the direction of F. In Fig. 3.10, PQ would rotate in an anticlockwise direction and the direction of the moment would therefore be anticlockwise. We will use a sign convention in which anticlockwise moments are positive and clockwise moments are negative.

When we consider the total moment about Q produced by a number of forces, such as shown in Fig. 3.11, then the total moment is determined by

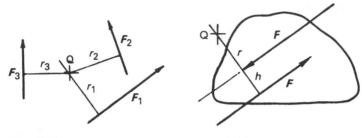

FIG. 3.11   FIG. 3.12

the addition of the individual moments about Q; i.e.

$$M_Q = r_1 F_1 + r_2 F_2 - r_3 F_3.$$

Note that by the above convention $r_1 F_1$ and $r_2 F_2$ are positive and $r_3 F_3$ is negative.

If two parallel forces of equal magnitude but opposite directions are applied to a body as shown in Fig. 3.12, such that the perpendicular distance between the force vectors is h, then their total moment about any point Q is given by

$$M_Q = (r + h)F - rF$$

$$= hF. \tag{3.13}$$

For this case the resulting moment is positive and therefore anticlockwise. Note that the magnitude of the moment is *independent* of the position of Q. This type of force system with equal and opposite parallel, but non-coincident, forces is sometimes referred to as a *couple*.

Let us now return to our original rigid body and again consider the body to be made up of a large number of individual particles. The external force F_i and internal force F_i' are assumed to act on the typical particle P_i, as shown in Fig. 3.13. We can now take the moments of these forces about any point Q. If the perpendicular distances from Q to the lines of action of F_i and F_i'

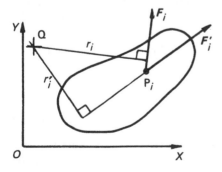

FIG. 3.13

are r_i and r_i' as shown, then their moment M_{Qi} about Q is given by

$$M_{Qi} = r_i F_i + r_i' F_i'. \tag{3.14}$$

The total moment M_Q of all the external and internal forces acting on the body is therefore

$$M_Q = \sum M_{Qi} = \sum r_i F_i + \sum r_i' F_i'. \tag{3.15}$$

Since at each contact point between adjacent particles the internal forces are equal and opposite, their moments about Q will cancel. Hence the summation of the moments of the internal forces over the whole body must be zero, i.e.

$$\sum r_i' F_i' = 0. \tag{3.16}$$

Thus from eqn (3.15) the total moment about Q is determined only by the moments of the external forces about Q, i.e.

$$M_Q = \sum r_i F_i.$$

For the body to be in equilibrium the total moment about Q of these external forces must be zero, i.e.

$$\sum r_i F_i = 0, \tag{3.17}$$

or

$$M_Q = 0. \tag{3.18}$$

The condition given by eqn (3.18) eliminates the third (rotational) degree of freedom of the body.

Equation (3.18) has not here been obtained in a rigorous manner. For a full derivation we need to consider the angular motion of the body under the action of external forces and then to set its *angular acceleration* to zero. A rigorous treatment of rigid body motion is given in Chapter 4. Equation (4.39) describes the rotational motion of the body and the static case is obtained by setting the accelerations $\ddot{x}_Q$, $\ddot{y}_Q$ and $\ddot{\theta}$ to zero. For this special case the summation of the moments of the external forces about any point Q must be zero.

Using eqn (3.11), the conditions for static equilibrium of a rigid body may be written as

$$\sum F_i = 0,$$

and $\sum M_Q = 0,$ \hfill (3.19)

i.e. the vector sum of all the external forces acting on the body must be zero, and the sum of their moments about any arbitrary point Q must also be zero.

In component form these equations may be expressed as

$$\sum F_{xi} = 0,$$

$$\sum F_{yi} = 0,$$

and $\quad \sum M_Q = 0,$ (3.20)

where $\sum F_{xi}$ and $\sum F_{yi}$ are the sums of the components of the external forces in two perpendicular directions and $\sum M_Q$ is the sum of the moments of the external forces about *any* point Q.

3.4 Alternative formulations of equilibrium conditions

The three equations (3.20), which describe the equilibrium of a rigid body in a plane, may be formulated in different ways. In certain applications, these alternatives can provide a more elegant solution to a problem.

Let us consider the rigid body shown in Fig. 3.14(a) with a number of external forces $F_1 \cdots F_i \cdots F_n$ applied to it.

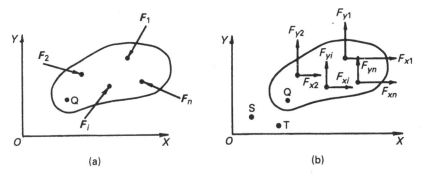

(a)   (b)

FIG. 3.14

If the body is in equilibrium then the sum of the moments of these forces about any point Q must be zero. It is convenient to consider the forces F_i in components F_{xi}, F_{yi} along OX and OY as shown in Fig. 3.14(b). Taking moments about point Q, which has coordinates $x_Q, y_Q,$

$$M_Q = \sum F_{yi}(x_i - x_Q) - \sum F_{xi}(y_i - y_Q)$$ (3.21)

where x_i, y_i are the coordinates of the point of application of F_i.

If we now take the moments of the external forces about two other points S and T with coordinates (x_S, y_S) and (x_T, y_T) we obtain

$$M_S = \sum F_{yi}(x_i - x_S) - \sum F_{xi}(y_i - y_S),$$

and $\quad M_T = \sum F_{yi}(x_i - x_T) - \sum F_{xi}(y_i - y_T).$ (3.22)

Subtracting eqn (3.21) from eqn (3.22) gives

$$M_S = M_Q + (x_Q - x_S) \sum F_{yi} - (y_Q - y_S) \sum F_{xi},$$

and $$M_T = M_Q + (x_Q - x_T) \sum F_{yi} - (y_Q - y_T) \sum F_{xi}. \qquad (3.23)$$

If we consider the proposition that M_Q, M_S and M_T are all zero, then from eqn (3.23)

$$(x_Q - x_S) \sum F_{yi} = (y_Q - y_S) \sum F_{xi},$$

and $$(x_Q - x_T) \sum F_{yi} = (y_Q - y_T) \sum F_{xi}. \qquad (3.24)$$

By rearranging eqn (3.24) we obtain

$$\sum F_{yi} = \left(\frac{y_Q - y_S}{x_Q - x_S} \right) \sum F_{xi} = \left(\frac{y_Q - y_T}{x_Q - x_T} \right) \sum F_{xi}. \qquad (3.25)$$

For eqn (3.25) to be satisfied either

$$\left(\frac{y_Q - y_S}{x_Q - x_S} \right) = \left(\frac{y_Q - y_T}{x_Q - x_T} \right), \qquad (3.26)$$

or $$\sum F_{xi} = \sum F_{yi} = 0. \qquad (3.27)$$

The left-hand side of eqn (3.26) is the gradient of the line QS and its righthand side is the gradient of line *QT*. Thus eqn (3.26) is true only if the three points Q, S and T lie on the same straight line. If the points, Q, S and T are *not* on the same straight line then eqn (3.27) must apply, i.e. the sum of the external forces on the body must be zero.

This gives us an *alternative* formulation of the equilibrium condition which can now be expressed as

$$M_Q = 0,$$

$$M_S = 0,$$

and $$M_T = 0, \qquad (3.28)$$

provided that the points Q, S and T *do not lie on the same straight line.*

To obtain a second alternative formulation, let us again consider the body in Fig. 3.14(b) and take the moments of the external forces about the points Q and S.

Thus, as before,

$$M_S = M_Q + (x_Q - x_S) \sum F_{yi} - (y_Q - y_S) \sum F_{xi}. \qquad (3.29)$$

Now let us resolve each external force along the direction of a line which makes an angle α with *OX*, as shown in Fig. 3.15. The total force F_α in this direction will be given by

$$F_\alpha = \sum F_{\alpha i} = \cos \alpha \sum F_{xi} + \sin \alpha \sum F_{yi}. \qquad (3.30)$$

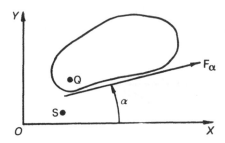

FIG. 3.15

If we now consider the proposition that M_Q, M_S and F_α are all zero, then from eqns (3.29) and (3.30),

$$\sum F_{yi} = \left(\frac{y_Q - y_S}{x_Q - x_S}\right) \sum F_{xi} = -\frac{\cos \alpha}{\sin \alpha} \sum F_{xi}. \tag{3.31}$$

For eqn (3.31) to be satisfied either

$$\frac{y_Q - y_S}{x_Q - x_S} = -\frac{1}{\tan \alpha}, \tag{3.32}$$

or $\quad \sum F_{yi} = \sum F_{xi} = 0.$ (3.33)

If $\tan \alpha = -(x_Q - x_S / y_Q - y_S)$, our chosen direction would be perpendicular to the line joining points Q and S. For all other values of α the condition of eqn (3.33) must apply.

This gives the second *alternative* formulation of the conditions of equilibrium of a body in a plane as

$$M_Q = 0,$$

$$M_S = 0,$$

and $\quad F_\alpha = 0,$ (3.34)

where the direction chosen for F_α must *not* be perpendicular to the line joining points Q and S.

It must be remembered that there are only *three independent equations* which describe the equilibrium of a body in a plane. We can use eqns (3.20) *or* (3.28), *or* eqns (3.34). If we decide to solve a problem using one of the formulations discussed above, any attempt to derive additional equilibrium equations for the body, based on the other formulations, will only yield equations which could have been derived from the original three. The solution of these additional equations will produce the result that $0 = 0$ which, although correct, is not very helpful.

Example 3.2   A weightless beam ABC is suspended horizontally from three wires as shown in Fig. 3.16. Find the tension in each of the wires when the beam carries a load *W* at its mid-point.

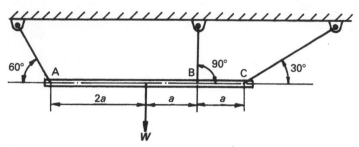

FIG. 3.16

The first step in any problem is to draw a *free body diagram* of the body, in this case of the beam. Only when this diagram has been drawn can we decide which formulation of the equilibrium conditions is the most convenient.

The free body diagram shows only the beam itself, completely detached from the supporting wires. All external forces which act *on* the beam are drawn in, together with the forces *applied to* the beam by the wires.

There are four forces acting on our beam, the tensions T_1, T_2 and T_3 in the wires, and the externally applied force *W*, as shown in Fig. 3.17.

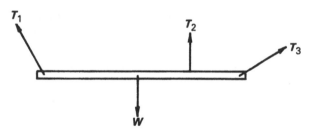

FIG. 3.17

If we apply eqn (3.20) to the beam we obtain, by resolving forces horizontally and vertically,

$$\rightarrow -T_1 \cos 60 + T_3 \cos 30 = 0 \tag{i}$$

and $$\uparrow \ T \sin 60 + T_2 + T_3 \sin 30 - W = 0. \tag{ii}$$

Taking moments about point A gives

$$\circlearrowleft 3aT_2 + 4aT_3 \sin 30 - 2aW = 0. \tag{iii}$$

These are the three equations which must be satisfied for the beam to be in equilibrium. They contain three unknown quantities T_1, T_2 and T_3 and can be solved to give

$$T_1 = \frac{\sqrt{3}}{4} W, \qquad T_2 = \frac{W}{2} \quad \text{and} \quad T_3 = \frac{W}{4}. \tag{iv}$$

Let us now reconsider the problem using eqn (3.34), by taking moments about two points Q and R and summing the components of the forces in a direction *not* perpendicular to the line joining Q and R.

Since Q and R can be chosen arbitrarily it is convenient to locate them at the intersections of the lines of action of forces, thereby reducing the number of terms in the moment equations. For example, in this case, if we choose Q at the intersection of the lines of action of T_1 and T_3, and R at the intersection of the lines of action of W and T_3 as shown in Fig. 3.18, then the moment equations become

$$\text{ⓠ} \quad -aW + 2aT_2 = 0.$$

$$\text{and} \quad \text{ⓡ} \quad aT_2 - \frac{a}{\cos 30} T_1 = 0. \tag{v}$$

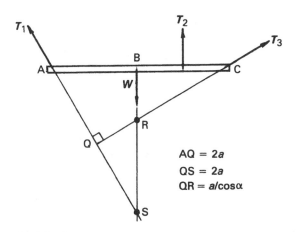

FIG. 3.18

We must now choose the direction in which to sum the components of the external forces, noting that it must *not* be perpendicular to QR. In this case the horizontal direction is a convenient choice as W and T_2 have no components in this direction. The third of eqn (3.22) now gives

$$\rightarrow \quad T_3 \cos 30 - T_1 \cos 60 = 0. \tag{vi}$$

Equations (v) and (vi) will yield the same results as given in eqn (iv).

The third method, using eqn (3.28), requires us to take moments about three points which are not co-linear. We have already selected two convenient points, Q and R above, and the point of intersection S of lines of action of T_1 and

W represents a third convenient point. For this case eqn (3.28) may be written as

$$\circlearrowleft \quad aW + 2aT_2 = 0,$$

$$\circlearrowleft \quad aT_2 - \frac{aT_1}{\cos 30} = 0,$$

and $\circlearrowleft$ $\quad aT_2 - 2aT_3 = 0.$ (vii)

This again gives the results of eqn (iv).

Comparison of the three methods shows that by performing some simple geometrical constructions the three simultaneous equations (i), (ii) and (iii) have been reduced to three equations, (v) and (vi), or (vii), which can be solved consecutively. In this case the second and third methods offer advantages compared with the first method. This is not always the case. Each problem has to be approached individually and an appropriate choice of method made. However, the reader should not spend too much time worrying about which approach to use. If there is no obvious advantage it is usual to employ eqn (3.20).

3.5   Special cases

Two cases of particular interest are the equilibrium of a rigid body under the action of two forces or of three forces.

Two forces

Consider the rigid body shown in Fig. 3.19(a) with forces F_1 and F_2 applied to it at points A and B as shown. For equilibrium, we require that

$$F_1 + F_2 = 0,$$ (3.35)

and, by taking moments about O, that

$$\circlearrowleft \quad -r_1 F_1 - r_2 F_2 = 0.$$ (3.36)

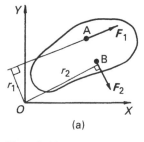
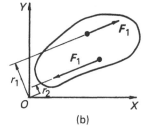
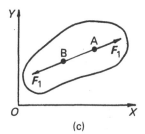

(a)   (b)   (c)

FIG. 3.19

For the two vectors F_1, F_2 to satisfy eqn (3.35) they must have the same magnitude. Also their lines of action must be parallel to one another, and the forces must act in opposite directions as shown in Fig. 3.19(b). The moment equilibrium now gives

$$(r_1 - r_2)F_1 = 0.$$

Since $F_1 \neq 0$ it therefore follows that r_1 must equal r_2, i.e. the forces must have the same line of action.

Thus when a body is in equilibrium under the action of only *two* forces the forces must be equal in magnitude, opposite in direction and have the same line of action as shown in Fig. 3.19(c).

Three non-parallel forces

Consider the body shown in Fig. 3.20 with forces F_1, F_2, F_3 applied at points A, B, C as shown. If the body is in equilibrium the moments of these three forces about any point must be zero. Let us choose the intersection I of the lines of action of F_1 and F_2 as the point about which we take moments. Since, by definition, F_1 and F_2 produce no moment about I, equilibrium can only be achieved if F_3 also exerts zero moment about I. This is possible only if F_3 also has a line of action which passes through I. Hence for equilibrium the lines of action of the three forces must *all* intersect at a single point.

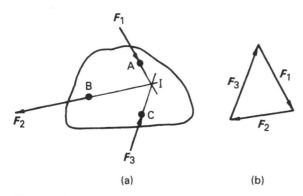

(a)   (b)

FIG. 3.20

If the points of application A, B, C are specified and if the lines of action of two of the three forces are known, then the line of action of the third force can be found immediately by locating I. This satisfies eqn (3.19), i.e. $\sum M_Q = 0$. If the magnitude of one of the forces is known, the magnitudes of the others may then be obtained by solving the force equilibrium equation

$$F_1 + F_2 + F_3 = 0,$$

graphically as shown in Fig. 3.20(b).

A systematic application of eqns (3.19) or (3.20) would, of course, yield a solution in the above cases but the use of these special conditions for the equilibrium of bodies subjected to two or three forces can often provide a quicker solution.

Example 3.3   The light rod AB of length $4a$ carries a load W as shown in Fig. 3.21(a) and has to be supported by two wires AP and BQ such that it lies in a horizontal position a distance a below a horizontal surface. If the position of the end B of the rod has to be located relative to the support point Q as shown, how far from Q must the point of attachment P be positioned?

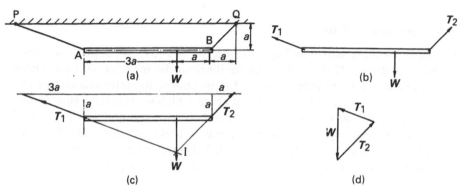

FIG. 3.21

To solve the problem we must first draw the free body diagram of the rod. The only forces acting on the rod are the tensions T_1 and T_2 in the wires and the load W as shown in Fig. 3.21(b). We see therefore that there are only three forces acting on the body AB. In this case we know the directions of two of these forces W and T_2, but not that of T_1. However, the lines of action of all three must pass through the same point I. Using the free body diagram we can find I immediately from the intersection of the lines of action of W and T_2. The direction of the line of action of T_1 can thus be found by construction, as shown in Fig. 3.21(c). The geometry of Fig. 3.21(c) is very simple and it can be seen that P must be a distance $8a$ from Q. Use of eqns (3.20) would yield the same result but their solution would involve much more algebra.

We have not needed to find the magnitudes of T_1 and T_2, but they can be found if required, from our basic vector equation, eqn (3.19), i.e.

$$W + T_1 + T_2 = 0. \qquad (i)$$

The solution of eqn (i) is shown in Fig. 3.21(d) from which T_1 and T_2 may be found by drawing the figure to scale, or by calculation.

3.6 Distributed forces

So far we have considered only forces which act at a particular point on a body. Some forces are, however, distributed across the surface of a body, e.g. wind loads, hydrostatic pressure due to liquids, gas forces, and the weight of the body itself.

Let us consider a force distributed from O to a along the axis OX and acting in the OY direction, as shown in Fig. 3.22(a). Over some elemental length δx of the axis OX the force in the direction OY can be expressed as

$$\delta F_y = f_y\,\delta x, \tag{3.37}$$

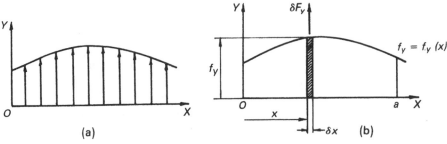

FIG. 3.22

where f_y is the force/unit length at the element. The magnitude of f_y will be a function of x and the curve $f_y = f_y(x)$ of Fig. 3.22(b) shows how f_y varies along OX.

The total force, F_y, due to the distributed force $f_y(x)$ will now be given by summing the values of δF_y, i.e.

$$F_y = \sum \delta F_y,$$

or $$F_y = \int_0^a f_y(x)\,\mathrm{d}x. \tag{3.38}$$

This integral is equal to the area enclosed by the curve $f_y(x)$.

We can now replace the distributed force $f_y(x)$ by a single equivalent force F_y, whose magnitude is given by eqn (3.38), provided that we position the force F_y such that its moment about any point Q is the same as the moment of the distributed force about Q.

Let the equivalent force F_y be positioned a distance x_F from the origin as shown in Fig. 3.23. Its moment about any point Q is given by

$$M_Q = (x_F - x_Q)F_y. \tag{3.39}$$

The moment about Q of the distributed force $f_y(x)$ has to be found by first considering the moment of the elemental force δF_y, shown in Fig. 3.22, about

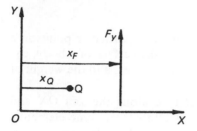

Fig. 3.23

Q, i.e.

$$\delta M_Q = (x - x_Q)\,\delta F_y,$$

or $\qquad \delta M_Q = (x - x_Q)f_y(x)\,\delta x.$

The total moment about Q of the distributed force is therefore

$$M_Q = \sum (x - x_Q)f_y(x)\,\delta x.$$

Taking the limit as $\delta x \to 0$ and integrating from $x = 0$ to $x = a$ gives

$$M_Q = \int_0^a (x - x_Q)f_y(x)\,\mathrm{d}x$$

$$= \int_0^a xf_y(x)\,\mathrm{d}x - x_Q \int_0^a f_y(x)\,\mathrm{d}x. \qquad (3.40)$$

The position at which F_y must be placed in order to produce a moment about Q equal to that of the distributed force is obtained by equating the expressions for M_Q given by eqns (3.39) and (3.40), i.e.

$$(x_F - x_Q)F_y = \int_0^a xf_y(x)\,\mathrm{d}x - x_Q \int_0^a f_y(x)\,\mathrm{d}x.$$

Now from eqn (3.38)

$$F_y = \int_0^a f_y(x)\,\mathrm{d}x,$$

so that $\quad x_F = \dfrac{\int_0^a xf_y(x)\,\mathrm{d}x}{\int_0^a f_y(x)\,\mathrm{d}x}. \qquad (3.41)$

This expression is the same as that for the position of the centroid of the area enclosed by the curve $f_y(x)$ and OX. Thus the distributed force represented by $f_y(x)$ per unit length can be replaced by a single equivalent force F_y, given by eqn (3.39), positioned at x_F given by eqn (3.41). We can therefore say that the magnitude of the equivalent force is equal to the *area* enclosed by the curve $f_y(x)$, and that the line of action of the equivalent force must pass through the *centroid* of that area.

Example 3.4 Let us consider a uniformly distributed load as drawn in Fig. 3.24(a). This distributed load is represented by

$$f_y(x) = -w,$$

and can be replaced by an equivalent force F_y given, from eqn (3.38), by

$$F_y = \int_0^a -w \, dx.$$

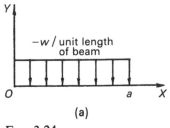 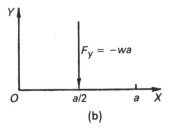

(a)   (b)

FIG. 3.24

Since w is a constant, eqn (i) may be rewritten as

$$F_y = -w \int_0^a dx = -wa.$$

The position at which F_y must act is given by eqn (3.41) as

$$x_F = \frac{-\int_0^a xw \, dx}{-\int_0^a w \, dx} = \frac{w\left[\dfrac{x^2}{2}\right]_0^a}{w[x]_0^a} = \frac{a}{2}. \tag{iii}$$

The equivalent force F_y must therefore be placed at the mid-point of the uniformly distributed load, and will act in the direction shown in Fig. 3.24(b).

Example 3.5 Now let us consider the case in which the magnitude of the load/unit length increases linearly from zero at $x = 0$ to w at $x = a$, as shown in Fig. 3.25(a).

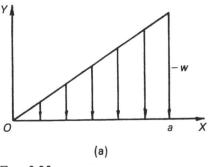 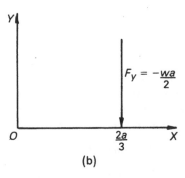

(a)   (b)

FIG. 3.25

Using eqn (3.38) the equivalent force

$$F_y = \int_0^a f_y(x)\, dx.$$

In this case the linear increase in load can be written as

$$f_y(x) = -w\frac{x}{a}. \tag{i}$$

Thus $F_y = -\int_0^a \frac{wx}{a}\, dx = -\frac{w}{a}\left[\frac{x^2}{2}\right]_0^a$

$$= -w\frac{a}{2}. \tag{ii}$$

From eqn (3.41)

$$x_F = \frac{-\int_0^a x \cdot \frac{wx}{a}\, dx}{-\int_0^a \frac{wx}{a}\, dx}$$

$$= \frac{2a}{3}.$$

The equivalent force F_y must therefore be placed a distance $2a/3$ from O as shown in Fig. 3.25(b).

Note that in both of the cases considered the magnitude of the equivalent force is equal to the area enclosed by the curve $f_y(x)$ and that its line of action passes through the centroid of that area.

3.7   Centres of mass

Each particle in a rigid body is subject to a vertically downward gravitational force. A general particle P_i of mass m_i will therefore experience a downward force of $m_i g$, referred to as the weight of m_i. These forces will be distributed throughout the body and, as in Section 3.6, the distributed load may be replaced by a single equivalent force acting through some particular point. The point through which this equivalent force acts is called the centre of mass (sometimes centre of gravity) and is usually denoted by G.

Let us consider the rigid body of Fig. 3.26. The general particle P_i of mass m_i, will experience a downward gravitational force $m_i g$. Let the position of P_i in the body be defined by coordinates x_i, y_i relative to an arbitrary set of axes OXY in which OY makes an arbitrary angle α with the vertical as shown.

The component of the force $m_i g$ along OX is given by

$$\delta F_x = m_i g \sin \alpha.$$

This can be summed for all the particles in the body to give the total force in

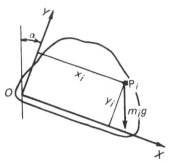

FIG. 3.26

the direction OX as

$$F_x = \sum m_i g \sin \alpha,$$

or $$F_x = mg \sin \alpha,$$ (3.42)

where m is the total mass of the body.

Similarly along the direction OY the total force is given by

$$F_y = -mg \cos \alpha.$$ (3.43)

If we add F_x and F_y vectorially, the total resultant force mg is equal to the total weight of the body and acts vertically downwards as shown in Fig. 3.27. Thus we can replace the distributed mass of the body by an equivalent single particle of mass m. The position of the equivalent particle must be chosen, such that the moment, about any point, of the gravitational force on the equivalent particle is the same as the sum of the moments about the same point, of the gravitational forces acting on the individual particles in the body.

Let the equivalent particle be located at point G which is assumed to have coordinates x_G, y_G, as shown in Fig. 3.28.

The moment M_Q of the gravitational force mg on the equivalent particle about some point Q is given by

$$M_Q = mg \cos \alpha (x_Q - x_G) + mg \sin \alpha (y_Q - y_G).$$ (3.44)

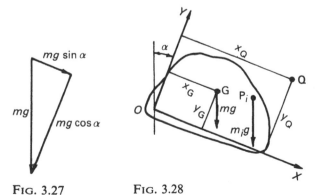

FIG. 3.27   FIG. 3.28

Particle P_i will experience a gravitational force m_ig which will exert about Q a moment

⊚ $M_Q = m_ig \cos \alpha (x_Q - x_i) + m_ig \sin \alpha (y_Q - y_i).$ (3.45)

Summing eqn (3.45) for all the particles in the body

⊚ $M_Q = g \cos \alpha \sum m_i(x_Q - x_i) + g \sin \alpha \sum m_i(y_Q - y_i).$ (3.46)

For equations (3.44) and (3.46) to be identical the coefficients of the terms in $\sin \alpha$ and $\cos \alpha$, for arbitrary values of α, must be the same in both equations. Thus

$$x_Q \sum m_i - \sum m_ix_i = m(x_Q - x_G),$$

and $$y_Q \sum m_i - \sum m_iy_i = m(y_Q - y_G).$$

Since $\sum m_i = m$ these equations give

$$x_G = \frac{\sum m_ix_i}{m} \quad \text{and} \quad y_G = \frac{\sum m_iy_i}{m}.$$ (3.47)

If Q were chosen to be coincident with G, substitution of eqn (3.47) into eqn (3.46) would give $M_G = 0$, i.e. the sum of the moments about G caused by the gravitational forces on the individual particles in the body is zero.

The point G, at which the equivalent particle of mass m must be located, is called the *centre of mass* of the body. Its position is given by eqn (3.47) and depends upon how the mass is distributed within the body.

[Note that the result of eqn (3.47) is independent of g. It is therefore preferable to refer to centre of mass rather than centre of gravity, as its position is a property solely of the body and is independent of the forces acting on the body or of its position in space.]

The position of the centre of mass is an important property of the body when considering the statics of systems in which gravitational effects are not negligible.

We will also see, in Chapters 4, 5 and 6, that it is always necessary to know the position of a body's centre of mass when considering its motion.

In statics problems it is usual to represent the gravitational forces acting on a rigid body by a single concentrated force $W = mg$ acting vertically downwards at the centre of mass of the body.

Example 3.5. Uniform rectangular plate   Let us find the position of the centre of mass of the uniform rectangular plate, ABCD, of thickness h, shown in Fig. 3.29. It is convenient to choose the origin O of our reference frame OXY at A.

All of the particles contained in the elemental strip of width δx are distant x from OY. We may therefore, instead of considering individual particles in

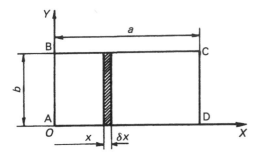

FIG. 3.29

the body, take an elemental strip of mass m_i where

$$m_i = \rho h b \, \delta x, \qquad \text{(i)}$$

in which ρ is the density of the material of the plate.

From eqn (3.47), the x coordinate of the position of the centre of mass is given by

$$x_G = \frac{\sum \rho h b x \, \delta x}{m} = \frac{\rho h b \sum x \, \delta x}{\rho h b a},$$

since the parameters ρ, h and b are constants and can be taken outside the summation. Expressing the summation as a definite integral from $x = 0$ to $x = a$ gives

$$x_G = \frac{\rho h b \int_0^a x \, dx}{\rho h b a} = \frac{a}{2}. \qquad \text{(ii)}$$

Similarly, by taking an elemental strip of width δy and integrating from $y = 0$ to $y = b$ eqn (3.47) gives

$$y_G = b/2. \qquad \text{(iii)}$$

Thus the centre of mass of a uniform rectangular plate coincides with its geometric centre.

Example 3.6. Uniform triangular plate   Now let us consider a triangular plate ABC of uniform thickness h, as shown in Fig. 3.30, in which $\hat{CAB}$ is 90°. It is convenient to choose our reference frame OXY such that its origin O is at A. We again consider an elemental strip of width δx distant x from OY. The mass of this strip, which has length y, is

$$m_i = \rho h y \, \delta x. \qquad \text{(i)}$$

From eqn (3.47) the x coordinate of the centre of mass of the plate is given by

$$x_G = \frac{\sum \rho h y x \, \delta x}{m}. \qquad \text{(ii)}$$

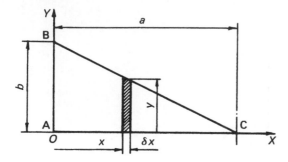

FIG. 3.30

The total mass m of the plate is given by

$$m = \tfrac{1}{2}\rho hab.$$

Substituting for m in eqn (ii) and expressing the summation as a definite integral from $x = 0$ to $x = a$ gives

$$x_G = \frac{\int_0^a yx\,dx}{\tfrac{1}{2}ab}. \tag{iii}$$

Before we can evaluate the integral in eqn (iii) y must be expressed as a function of x. From simple geometry

$$\frac{y}{a-x} = \frac{b}{a},$$

so that $y = \dfrac{b}{a}(a - x).$ $\tag{iv}$

Substituting eqn (iv) into eqn (iii) gives

$$x_G = \frac{\dfrac{b}{a}\displaystyle\int_0^a (ax - x^2)\,dx}{\tfrac{1}{2}ab}$$

$$= \frac{2}{a^2}\left[\frac{ax^2}{2} - \frac{x^3}{3}\right]_0^a$$

$$= \frac{a}{3}. \tag{v}$$

By repeating the procedure along OY,

$$y_G = \frac{b}{3}. \tag{vi}$$

Again we find that the centre of mass of the uniform plate is at its centroid. Although we have only considered two simple examples, this result is true for all shapes of *uniform* plate.

This is confirmed by substituting into eqn (3.47) the condition that the body is of uniform thickness and density. In this case

$$m_i = A_i \rho h \quad \text{and} \quad m = A \rho h,$$

where A_i is an elemental area corresponding to particle P_i, and A is the total area of the plate.

Thus

$$x_G = \frac{\rho h \sum A_i x_i}{\rho h A} = \frac{\sum A_i x_i}{A}.$$

Similarly

$$y_G = \frac{\sum A_i y_i}{A}.$$

These equations define the position of the centroid of an area.

Example 3.7. Semicircular plate   As a further example let us take the semi-circular plate shown in Fig. 3.31. Again let the plate have uniform thickness h and be of material with density ρ.

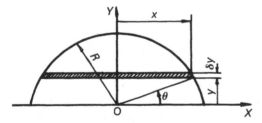

FIG. 3.31

With axes OXY as shown we can consider an elemental strip of length $2x$ and width δy. The strip is parallel to, and distant y from, OX. Its mass

$$m_i = \rho h 2x\, \delta y. \tag{i}$$

Using eqn (3.47) the y coordinate of the centre of mass of the plate is given by

$$y_G = \frac{\sum \rho h 2xy\, \delta y}{m}. \tag{ii}$$

The total mass of the plate is

$$m = \frac{\rho h \pi R^2}{2},$$

and expressing the summation in eqn (ii) as a definite integral, we obtain

$$y_G = \frac{\int_0^R xy \, dy}{\pi R^2}.$$ (iii)

To integrate eqn (iii) we must express x in terms of y using the geometry of the plate. In this case, evaluation of the integral is more convenient if we express x and y in terms of the polar coordinates R and θ.

Thus from the geometry of Fig. 3.31,

$$x = R \cos \theta,$$ (iv)

and $y = R \sin \theta.$ (v)

By differentiation,

$$dy = R \cos \theta \, d\theta.$$

The integration limits for θ, corresponding to $y = 0$ and $y = R$, are, from eqn. (v), $\theta = 0$ and $\theta = \pi/2$.

Substitution into eqn (iii) gives

$$y_G = 4R^3 \frac{\int_0^{\pi/2} \cos^2 \theta \sin \theta \, d\theta}{\pi R^2}$$

$$= -\frac{4R}{\pi} \int_0^{\pi/2} \cos^2 \theta \, d(\cos \theta)$$

$$= -\frac{4R}{\pi} \left[\frac{\cos^3 \theta}{3} \right]_0^{\pi/2} = \frac{4R}{3\pi}.$$

Thus $y_G = 4R/3\pi$, and since the plate is symmetric about OY, $x_G = 0$. The reader should verify this latter result.

Example 3.8. Composite shapes   The positions of the centres of mass of plates with more complex shapes can be considered by breaking them up into simple shapes. First we determine the positions of the centres of mass of each simple shape from a convenient origin, and then combine them to give the position of the centre of mass of the whole body.

Let us, as an example, consider the trapezoidal plate ABCD shown in Fig. 3.32 and choose axes OXY with origin O at A. The plate may be broken down into two simple shapes, a rectangle and a triangle, as shown by the broken line. We have already obtained, in example 3.5, the position of the centre of mass of the rectangular portion as

$$x_{GR} = a/2, \qquad y_{GR} = b/2.$$ (i)

Using the result of example 3.6 the coordinates of the centre of mass of the triangular portion will be

$$x_{GT} = a/3 \quad \text{and} \quad y_{GT} = c/3 + b.$$ (ii)

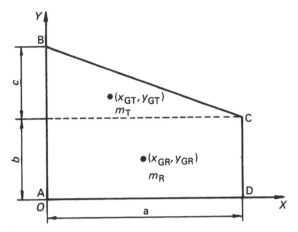

FIG. 3.32

We can treat the rectangular portion of the plate as a particle of mass m_R at its centre of mass, and the triangular portion as a second particle of mass m_T at its centre of mass. The complete plate can now be considered as a single rigid body consisting of these two particles alone, and eqn (3.47) will give the position of the centre of mass of the whole plate as

$$x_G = \frac{\sum m_i x_i}{m} = \frac{m_R \frac{a}{2} + m_T \frac{a}{3}}{m_R + m_T}. \tag{iii}$$

Now $m_R = \rho hab$, and $m_T = \frac{1}{2}\rho hac$ so that

$$x_G = \frac{a}{3}\left(\frac{3b + c}{2b + c}\right). \tag{iv}$$

Similarly

$$y_G = \frac{m_R \frac{b}{2} + m_T \left(\frac{c}{3} + b\right)}{m_R + m_T},$$

which on substitution for m_R and m_T gives

$$y_G = \frac{b^2 + c\left(\frac{c}{3} + b\right)}{2b + c}. \tag{v}$$

Other complex shapes can be treated in a similar manner. For example, to find the position of the centre of mass of a plate with a hole we could:

(i) find the mass and position of the centre of mass of the plate without the hole;

(ii)   find the mass and position of the centre of mass of a plate the same size as the hole;

and then

(iii)   use eqn (3.47) to find the position of the centre of mass of the plate with the hole.

In this case the term which refers to the particle representing the hole would be *negative*.

3.8 Internal forces

The fundamental equations, (3.19), which describe the static equilibrium of a rigid body may also be applied to a part of a rigid body. Consider the rigid body shown in Fig. 3.33 which is in equilibrium under the action of forces $F_1, F_2 \cdots F_i \cdots F_n$. From eqn (3.19)

$$\sum F_i = 0,$$

and $$\sum M = 0,$$

where $\sum M$ is the sum of the moments of the forces F_i about any point. Now

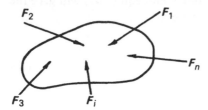

FIG. 3.33

imagine the body cut into two parts as shown in Fig. 3.34(a). Each part of the body will be in equilibrium under the action of the external forces F applied to that part plus the action of the internal forces distributed over the boundary at the cut. In general these internal forces have tangential and normal components which will vary from point to point across the cut.

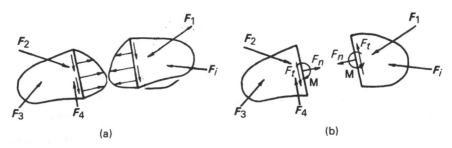

(a)   (b)

FIG. 3.34

In accordance with Newton's Third Law, the internal forces on the right-hand section will be equal and opposite to those on the left-hand section as shown in the free body diagrams of Fig. 3.34(a). Using the results given in Section 3.6 these distributed forces can be replaced by their resultant equivalent forces F_n, F_t, and a moment M acting at a *defined point* on the boundary. The free body diagrams can therefore be redrawn as shown in Fig. 3.34(b).

If we know the magnitudes, directions and the points of application of the external loads eqn (3.19), for static equilibrium, applied to either portion of the divided body, will yield values for F_n, F_t and M. In some special cases the resultant of the normal forces may be zero, and yet, because of their distribution across the surface, there could still be a resultant moment.

For example, if the part of the body in Fig. 3.35(a) were of uniform thickness and the distributed normal force f_n varied linearly across the section as shown, the resultant F_n would be zero. The distributed force would, however, exert an anticlockwise moment on that part of the body. Equations (3.19) would not be able to show how f_n is distributed across the section, but they would enable the total moment M exerted by the distributed forces to be obtained.

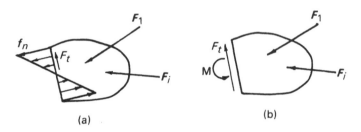

(a)

(b)

FIG. 3.35

Example 3.9 Consider a light rectangular bar resting horizontally on knife-edge supports at A and B, as shown in Fig. 3.36(a), carrying a vertical load W at each end. Let us find out how the internal forces vary along the length of the bar.

First we must draw a free body diagram of the bar and determine the magnitudes and directions of the external forces acting on the bar. In addition

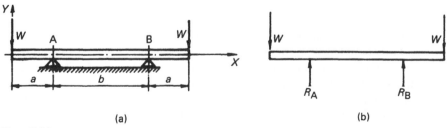

(a)

(b)

FIG. 3.36

to the two loads W there will be forces applied to the bar by its supports. Let these be R_A and R_B and we will assume that they act as shown in the free body diagram of Fig. 3.36(b). For the bar to be in equilibrium under the action of these forces, eqns (3.20), i.e.

$$\sum F_x = 0, \qquad \sum F_y = 0, \quad \text{and} \quad \sum M = 0,$$

must be satisfied.

No forces act on the bar along OX. Along OY we have the condition that

$$\uparrow R_A + R_B - 2W = 0. \tag{i}$$

It is convenient to take moments about one of the supports, so eliminating the unknown force applied at that point from the moment equation. Taking moments about A

$$\text{Ⓐ} \; aW + bR_B - (a + b)W = 0, \tag{ii}$$

so that

$$R_B = W. \tag{iii}$$

Since R_B is positive it is in the direction shown in Fig. 3.36(b). Substituting for R_B in eqn (i) gives

$$R_A = W, \tag{iv}$$

which is also in the direction shown in Fig. 3.36(b).

We now know the magnitudes and directions of all of the external forces acting on the bar. To find the internal forces at any section through the bar we must draw the free body diagrams of the bar when cut at that section. Let us consider a section through the bar between the left-hand end and support A, distant x from the end, as shown in Fig. 3.37.

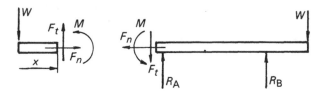

FIG. 3.37

At the mid point of the section at x we shall assume that there will be tangential internal force F_t in the vertical direction, a horizontal normal internal force F_n and an internal moment M, as shown in Fig. 3.37. The assumed directions of these forces and the moment do not matter, as long as they are equal and opposite on both parts of the bar. We can now apply the equations of static equilibrium eqn (3.20) to each section of the bar to find the internal forces.

Taking the left-hand part of the bar we obtain, for the horizontal direction

$$\rightarrow F_n = 0, \tag{v}$$

so that the resultant normal force at the section is zero. This does *not* mean that there is no normal force distributed across the section, only that its resultant is zero.

In the vertical direction the equation of equilibrium for the left-hand part of the bar gives

$$\uparrow -W + F_t = 0,$$

so that $F_t = W$. (vi)

Thus, since F_t is positive, the internal force on the left-hand part of the bar at the chosen section is in the direction shown. We can also see from eqn vi that F_t is independent of x, and so will be constant for any value of x *for which our free body diagram of Fig. 3.37 applies.*

Note that if $x > a$ the free body diagram of Fig. 3.37 does *not* apply.

We can take moments, about any point, of the forces on the left-hand part of the bar to give the third equation of equilibrium. Choosing the mid point in the plane of the section at x we obtain

$$\circlearrowright xW + M = 0,$$

so that

$$M = -xW. \tag{vii}$$

The negative sign in equation (vii) shows that the moment will physically act in the opposite sense from that shown in Fig. 3.37 and that its magnitude will increase linearly with x, for values of x for which the free body diagram is appropriate. Since $F_n = 0$, the distribution of internal forces across the section will be as in Fig. 3.35(a).

Values of F_t and M could also have been obtained from the free body diagram of the right-hand part of the bar. The reader should check that the results obtained are the same as given by eqns (vi) and (vii).

Let us now consider the internal forces at sections of the bar between A and B. We have seen above that we only require to consider the equilibrium of one part of the beam and so we will draw the free body diagram for the left-hand part only.

The free body diagram is shown in Fig. 3.38. Since no horizontal external forces act on the bar we have seen in eqn (v) that there will be no resultant horizontal force at the section so only the moment and the tangential force are shown at the section in Fig. 3.38. We now have two external forces, W and R_A, acting on the part of the bar under consideration.

For vertical equilibrium

$$\uparrow -W + R_A + F_t = 0,$$

so that $F_t = W - R_A$.

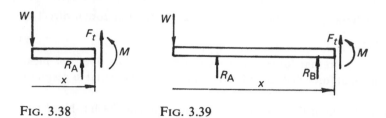

FIG. 3.38   FIG. 3.39

Substituting for R_A from eqn (iv) gives

$$F_t = 0. \tag{viii}$$

The internal tangential force at the section is therefore zero for all values of x for which the free body diagram of Fig. 3.48 applies, i.e. for all sections between the supports A and B. Taking moments about a point in the section,

$$\circlearrowright \quad xW - (x - a)R_A + M = 0.$$

Substituting from eqn. (iv) for R_A gives

$$M = - Wa. \tag{ix}$$

The moment M is again in the opposite sense from that shown in the free body diagram, but is now independent of x for $a < x < a + b$.

If we now consider the internal forces between B and the right-hand end of the bar we obtain the free body diagram of Fig. 3.39.

For vertical equilibrium,

$$\uparrow -W + R_A + R_B + F_t = 0,$$

so that $F_t = W - R_A - R_B$.

Substituting for R_A, R_B from eqns (iii) and (iv) gives

$$F_t = - W, \tag{x}$$

which shows that F_t acts in the opposite direction from that shown in Fig. 3.39, and is again independent of x.

Taking moments about a point in the section gives

$$\circlearrowright \quad xW - R_A(x - a) - R_B(x - (a + b)) + M = 0.$$

Substituting for R_A and R_B yields

$$M = W(x - (2a + b)). \tag{xi}$$

Since $(2a + b) > x$, M will be negative and will also vary with x, becoming zero at the end of the bar when $x = 2a + b$.

Identical results could have been obtained by considering the equilibrium of the right-hand part of the bar.

We now have developed expressions which show how the internal tangential force and moment acting at a section vary along the length of the bar. Graphs showing the variation of F_t and M along the bar are plotted in Fig. 3.40.

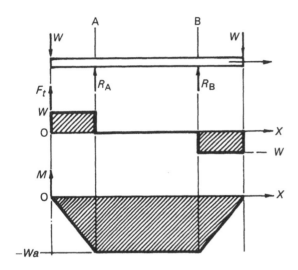

FIG. 3.40

Thus by drawing a free body diagram of part of the bar and applying the conditions of static equilibrium (eqns 3.10 or 3.20) we can obtain a full picture of how the resultant internal forces and moments vary along its length.

3.9 Connected systems

So far we have considered only the static equilibrium of single bodies. Engineers are, however, normally concerned with structures or machines which consist of a number of bodies connected together in some manner. The individual bodies may be shafts, links, beams, gear wheels, pistons, or any other engineering component. When designing a machine or structure the engineer needs to know how the external forces are transmitted through the system, and the forces acting on the individual elements. The force transmission through the system depends upon the types of connection used. We will now consider mathematical models of the connections used in planar systems.

Frictionless contact at a point

The body A shown in Fig. 3.41 is connected to a fixed body B. The connection is provided by the contact between the two frictionless surfaces, S_A and S_B, of the bodies. Contact will occur at some point P (point contact) or along a line through P perpendicular to the plane of Fig. 3.41 (line contact) depending

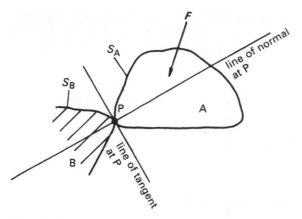

Fig. 3.41

upon the shape of the surfaces at P. The contact between gear teeth is an example of line contact between two surfaces.

Let the body A be in equilibrium under the action of some external force *F*. If body *A* is to remain in contact with surface S_B, the point at P on the body must have zero displacement in a direction normal to the common tangent at the point of contact. Any displacement of the point along the normal direction would either cause separation of the bodies, or penetration of the surfaces S. A force will therefore be generated at the contact point. This force must be compressive as a surface contact cannot sustain a tensile load. If we consider the force in terms of its components along the common normal and along the common tangent at P we find that the tangential component of the force must be zero since we have assumed frictionless conditions. The force *R* at the contact must therefore be along the normal direction as shown in the free body diagram of Fig. 3.42.

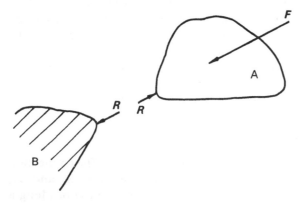

Fig. 3.42

For the body A to be in equilibrium the vector sum of the forces acting on it and the sum of the moments of these forces about any point Q must be zero. Thus if a single external force F is applied to the body it must also act along the normal at P as shown in Fig. 3.42 so that

$$F + R = 0,$$

and $\sum M_Q = 0.$

If more than one external force acts on the body, the resultant representing the sum of these external forces must also act along the normal.

Surface contact with friction

If the contact surface is not frictionless the contact force will have both tangential and normal components. As the body attempts to slide over the surface S along the line of the common tangent, the friction force on the body will always oppose the incipient motion.

In the free body diagram of Fig. 3.43 the normal force is denoted by R_N and the friction force tangential to the surfaces at the point of contact by R_F. The forces on surface S_B are equal in magnitude but opposite in direction from those on the body A. Here we have assumed that the point of contact on body A is about to move in the opposite direction from that of R_F on A.

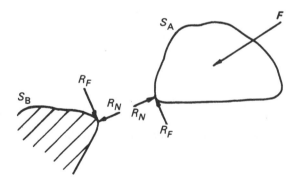

Fig. 3.43

Usually R_F is limited in magnitude by the nature of the contacting surfaces and the value of the normal force R_N. If the equilibrium calculation yields a value for R_F greater than the limiting value the assumption that the system is in equilibrium would then be invalid. The limiting value of R_F will be greater for rough surfaces than for surfaces which have good surface finish.

The *limiting* value of R_F is related to the magnitude of R_N by the coefficient of friction, μ, of the contacting surfaces, i.e.

$$R_F \leq \mu R_N. \tag{3.48}$$

It must be remembered that R_F reaches a value equal to μR_N only as slip is about to occur. When slip does not occur R_F could have any value between zero and the limiting value, and would be independent of the nature of the surfaces and the value of R_N.

If we combine R_F and R_N vectorially into a single contact force $\boldsymbol{R}$ such that

$$\boldsymbol{R} = \boldsymbol{R}_F + \boldsymbol{R}_N,$$

it can be seen in Fig. 3.44 that $\boldsymbol{R}$ will make an angle λ with the normal direction. In the *limiting* case when $R_F = \mu R_N$, the angle λ reaches a maximum value and

$$\tan \lambda = \frac{R_F}{R_N} = \mu. \tag{3.49}$$

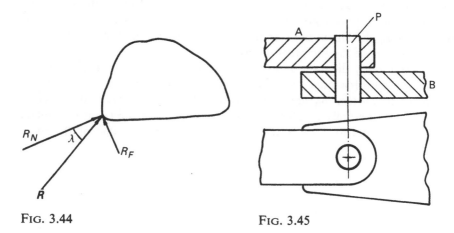

FIG. 3.44　　　　　　　　　　FIG. 3.45

This limiting value of λ is called the friction angle, and its tangent is equal to the coefficient of friction.

For equilibrium of the body the sum of the external forces and the contact force must be zero, i.e.

$$\sum \boldsymbol{F} + \boldsymbol{R} = 0.$$

If only one external force is applied to the body $\boldsymbol{R}$ and $\boldsymbol{F}$ must again be equal in magnitude, opposite in direction and colinear. In some situations friction can be neglected and calculations made without much loss of accuracy whilst in others it would not be sensible to proceed with an analysis which did not include friction.

Pin joint or bearing

The pin joint is probably the most commonly used method of connecting two bodies. It is sometimes referred to as a revolute joint and allows only relative rotation of the two connected bodies (see Chapter 1 p. 30). A typical example is shown in Fig. 3.45. The bodies A and B are connected together by a cylindrical pin P. The pin may, for example, be fixed rigidly to body A. An infinitesimally small amount of clearance between the pin and its mating hole in body B allows the bodies to rotate relative to one another about an axis defined by the centre line of the pin, but prevents any relative translation at the point of connection.

If we separate the bodies A and B at the connection, we can represent the forces on each of the bodies as shown in Fig. 3.46, the force on body A being equal in magnitude and opposite in direction from that on body B. In the absence of friction the resultant force R is normal to the contacting surfaces and must therefore pass through the axis of the pin. It is often convenient to consider the force at a pin a connection in terms of two orthogonal components R_1 and R_2. A ball or roller bearing has very low friction and can usually be regarded as equivalent to a frictionless pin joint when considering how forces are transmitted at the connection.

If friction is present at the connection the normal force will have an associated friction force at the point of contact. Figure 3.47 shows a pin joint

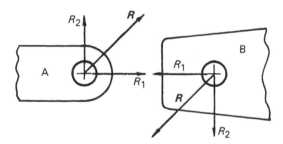

Fig. 3.46

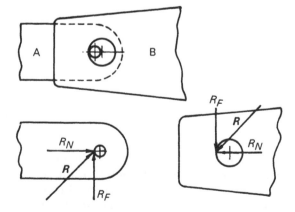

Fig. 3.47

with exaggerated clearance in which the pin is part of body A. (Normally the clearance would be approximately one thousandth of the bearing's diameter.) The forces at the point of contact in the joint will be as shown, where R_N is the normal component and R_F the tangential frictional component. If the value of R_F becomes equal to the limiting value of μR_N slip will occur, allowing relative rotation of the bodies. Again R_F is chosen to be in a direction which would oppose incipient sliding of the pin surface relative to the surface of the hole. In Fig. 3.47 we have assumed that body A is tending to rotate in an anticlockwise direction relative to body B. Hence the friction forces R_F on A and B have directions which would oppose this motion. It can be seen from Fig. 3.47 that the effect of the frictional component is to alter the direction of the resultant contact force so that it no longer passes through the centre of the pin.

For this direction of R_F the resultant force $\boldsymbol{R}$ will exert a moment about the pin centre line equal to $-dR_F/2$ on body A and $dR_F/2$ on body B. It is possible to replace the force $\boldsymbol{R}$ shown in Fig. 3.47 by an equivalent system consisting of a force $\boldsymbol{R}$ passing through the centre of the connection together with a moment $M_F = dR_F/2$ as shown in Fig. 3.48. This moment will reach a limiting value of $(d/2)\,\mu R_N$ just as slip is about to occur. If the direction of incipient sliding were reversed the direction of $\boldsymbol{R}$ would change and the direction of M_F would be reversed.

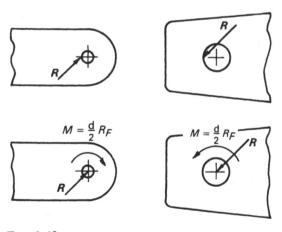

FIG. 3.48

Special case

A commonly encountered situation in many structures and machines is that of a body whose only connections are two frictionless pin joints. If the weight of the body itself is negligible, and no external forces are applied to it, we have the first special case of Section 3.5, i.e. that of a body subject to two forces. The forces on the body therefore lie along the line joining the centres of the two pins as shown in Fig. 3.49. The reader should review Section 3.5 to confirm this result.

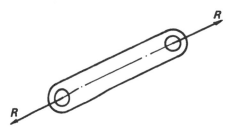

FIG. 3.49

'Built-in' connection

A connection which prevents both relative translation and rotation of the connected bodies is termed 'built-in'. The most common forms of this type of connection are welded and bolted joints and the resulting connected pair of bodies is effectively a new single rigid body. The forces and moments at the connection can be found as outlined in Section 3.8.

Engineering Applications

When applying the results of this Chapter to engineering problems we first have to set up a mathematical model of the physical system. The basic assumptions we will make are that bodies are rigid and that ropes or chains are flexible but inextensible. In addition we will sometimes neglect the mass of certain bodies where their weights are much smaller than the forces which are applied to the system. Friction may be taken into account or neglected depending upon the circumstances.

In *all* cases we first draw a free body diagram of the whole, or of part of the system and then apply our basic equations of equilibrium, eqn (3.20), or *one* of the alternative formulations eqns (3.28) *or* (3.34).

3.1 Frameworks

The A frame shown in Fig. 3.50 is made up of three rigid bodies AC, CE and BD. It supports a horizontal force, $F = 1000\,\text{N}$, at the point P. The connections at A, B, C and D are pin joints, whilst the ground connection at

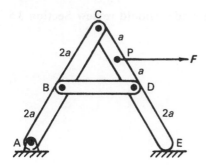

FIG. 3.50

E takes the form of a point contact. It is required to determine the forces at the connections A, B, C, D and E. We will assume that friction at the pinned connections, and at E, is negligible. The distance AE is 4a.

We first consider the equilibrium of the structure as a whole. Fig. 3.51(a) shows the free body diagram for the structure with the ground connections A and E removed. Since E is a frictionless contact the contact force R_E is normal to the contacting surfaces. The ground at E is assumed to be horizontal so that R_E will therefore be in the vertical direction.

The horizontal force F at P is known in both magnitude and direction and a force R_A, whose magnitude and direction are unknown, acts on the structure at A. Thus we have a rigid structure with three forces acting upon it. The complete structure may be considered as a rigid body and the result of Section 3.5 applies. The line of action of R_A must therefore pass through the intersection I of the lines of action of F and R_E. By drawing the framework to scale and locating I the line of action of R_E can be found. We now know the directions of all of the forces and the magnitude of one of them.

The moment equation for static equilibrium has already been satisfied from the condition that the lines of action of three forces all pass through one point. Thus, for the equilibrium eqn (3.19) to be satisfied, we require that

$$\sum F_i = 0,$$

giving, in this case

$$F + R_E + R_A = 0. \tag{i}$$

A graphical solution of eqn (i) can be carried out by first drawing F to scale as in Fig. 3.51(b). A vector representing R_E must then be added to F. This vector will be in the vertical direction, but its magnitude is unknown. The line representing the line of action of R_E can therefore be drawn through the tip of the vector F. The third vector R_A, whose direction is known, must close the vector diagram to give $\sum F_i = 0$. We can therefore draw in the line of action of R_A through the tail of vector F as shown. The intersection of the lines of action of R_A and R_E now defines their magnitudes and arrows can be added

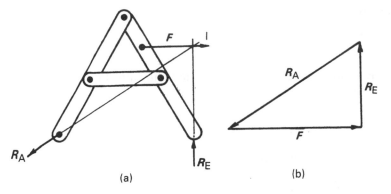

FIG. 3.51

as shown to ensure that eqn (i) is satisfied. The magnitudes of R_A and R_E can now be obtained from the vector diagram as $R_A \approx 1.19F$ and $R_E \approx 0.65F$.

Alternatively, if an analytical approach is used, it is generally convenient to represent the reaction at A in terms of its horizontal and vertical components, R_{HA} and R_{VA}, and to use the equilibrium conditions of eqn (3.20). It is not possible to know beforehand the directions in which R_{HA} and R_{VA} act. We will therefore assume directions as shown in the free body diagram, Fig. 3.52. If, after all the calculations have been performed, R_{HA} and R_{VA} turn out to be positive then the assumed directions are correct. A negative value would mean that the force acts in the opposite direction from that assumed.

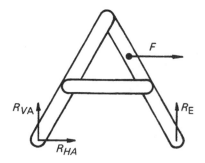

FIG. 3.52

From eqn (3.20)

$$\rightarrow R_{HA} + F = 0,$$

$$\uparrow R_{VA} + R_E = 0, \tag{ii}$$

and taking moments about point A to exclude R_{HA} and R_{VA} from our moment equation,

$$\circlearrowleft \!\!\!\!A \quad 4aR_E - (3a\cos 30)F = 0. \tag{iii}$$

Thus from eqn (iii)

$$R_E = \frac{3\sqrt{3}}{8} F = 0.65 \, F.$$

R_E is positive and is therefore in the direction shown in Fig. 3.52. From eqns (ii)

$$R_{HA} = -F \text{ and } R_{VA} = -\frac{3\sqrt{3}}{8} F. \tag{iv}$$

These are both negative and will act in the *opposite* directions from those shown in Fig. 3.52.

The magnitude of the force R_A is given by

$$R_A = \sqrt{R_{HA}^2 + R_{Va}^2}$$
$$= F\sqrt{1 + \tfrac{27}{64}} = 1.19 \, F. \tag{v}$$

To determine the internal forces at the pins B, C and D, we must break the structure down into its basic elements and consider the equilibrium of the links BD and ABC. Since link BD is pinned at B and D, and does not carry any external load, (see Section 3.5) the forces on BD at B and D must be equal and opposite and have a line of action defined by the line BD. At this stage we do not know whether the link is in tension or compression. To determine the direction of the force in BD and the internal force at C we need to consider the partially complete free body diagram, Fig. 3.53(a), of link ABC.

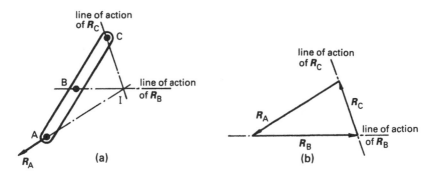

FIG. 3.53

The force at pin A has already been found and can be drawn on the free body diagram. Since the force R_B on link ABC at B must act along BD, a line representing its line of action can be drawn horizontally through point B. This line intersects the known line of action of R_A at I. The link ABC is in equilibrium under the action of three non parallel forces (see Section 3.5) so that the line of action of the force R_C at C must also pass through I. A graphical solution of the equilibrium equation for link ABC, i.e.

$$R_A + R_B + R_C = 0, \qquad \text{(vi)}$$

can now be obtained by scale drawing. The magnitude and direction of R_A has already been determined and the lines of action of R_B and R_C are obtained by drawing link ABC to scale and locating I. The magnitudes and directions of the forces R_B and R_C may be determined by first drawing R_A to scale and then drawing, through the tip and the tail of R_A, lines parallel to the lines of action of R_B and R_C.

The lines of action of R_B and R_C intersect and define the vector triangle shown in Fig. 3.53(b). Arrows are added to this diagram so that eqn (vi) is satisfied. From Fig. 3.53(b) the magnitudes of R_B and R_C are found to be 1.25 F and 0.7 F.

Alternatively we can obtain R_B and R_C using eqn (3.20). First we draw the free body diagram of link ABC as shown in Fig. 3.54. The forces acting on the link are the unknown force at pin C, the force at B which must act along the line BD, and the components of the force at pin A which have already been calculated in eqn (iv).

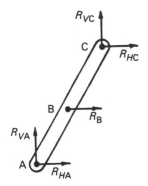

FIG. 3.54

The forces at A and C are again represented by their horizontal and vertical components.

The equations of equilibrium for the link can now be written as

$$\rightarrow R_{HC} + R_B + R_{HA} = 0, \qquad \text{(vii)}$$

$$\uparrow R_{VC} + R_{VA} = 0, \qquad \text{(viii)}$$

and taking moments about point C,

$$\circlearrowleft \quad (2a \cos 30)R_B + (4a \cos 30)R_{HA} - (4a \sin 30)R_{VA} = 0. \qquad \text{(ix)}$$

From eqn (ix)

$$R_B = \frac{2R_{VA}}{\sqrt{3}} - 2R_{HA}.$$

Substituting from eqn (iv) for R_{VA} and R_{HA} gives

$$R_B = \tfrac{5}{4} F. \tag{x}$$

R_B is positive and is therefore in the direction shown in Fig. 3.54.
From eqn (viii),

$$R_{VC} = -R_{VA} = \frac{3\sqrt{3}}{8} F,$$

and from eqn (vii)

$$R_{HC} = -R_B - R_{HA}$$
$$= -\tfrac{5}{4}F + F = -F/4.$$

R_{HC} is negative and therefore acts in a direction opposite from that assumed in Fig. 3.54.

The magnitude of R_C can now be obtained as

$$R_C = \sqrt{R_{HC}^2 + R_{VC}^2} = \sqrt{\frac{F^2}{16} + \frac{27\,F^2}{64}}$$

$$= 0.7\,F.$$

The forces acting on links ABC, BD and CDE are shown in the free body diagrams of Fig. 3.55 and we can now see that the link BD is in tension.

Although we have not considered link CDE in the solution of the problem the link is in equilibrium under the action of the forces shown. The reader is recommended to verify this result using both graphical and analytical methods.

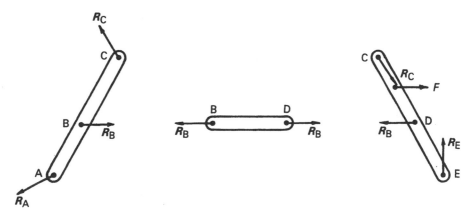

FIG. 3.55

3.2 Toggle linkage

Figure 3.56 shows the schematic layout of a toggle linkage. It consists of a special form of slider crank mechanism (see application 1.6). The slider is

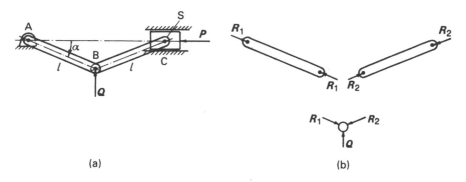

Fig. 3.56

constrained by a guide and is connected back to the frame by two links AB, BC with pin joints at A, B and C. A force *Q*, whose direction is perpendicular to AC, is applied to the pin at B and a horizontal force *P* is applied to the slider. The problem is to determine *P* for given values of *Q* and α when the system is in equilibrium. We will neglect friction and the weights of the links.

Let us first consider the equilibrium of pin B. The forces R_1, R_2 in the links AB and BC have lines of action along AB and BC respectively since, as we are neglecting the weights of the links, no external load is applied to either link (see Section 3.5). Therefore, the forces acting on the pin at B can be drawn as shown in the free body diagram of Fig. 3.56(b). Forces R_1 and R_2 on the pin are assumed positive in the directions shown. For horizontal and vertical equilibrium of the pin we have

$$\rightarrow R_1 \cos \alpha - R_2 \cos \alpha = 0, \tag{i}$$

and $\uparrow \ Q - (R_1 + R_2) \sin \alpha = 0.$ (ii)

Thus from eqns (i) and (ii)

$$R_1 = R_2 = R, \tag{iii}$$

and $Q = 2R \sin \alpha.$ (iv)

Now let us consider the equilibrium of the slider. The forces acting on the slider are the external force *P*, that applied at the bearing *C* from link BC and a contact force between the slider and its guide. Since friction is being neglected the contact force will be normal to the contact surface, i.e. perpendicular to the guide. Now *R* acts along BC and *P* along CA so that the resultant of the distributed force on the slider from the guide will have a line of action passing through point C (Section 3.5, three forces acting on a rigid body).

The free body diagram of the slider is therefore as shown in Fig. 3.57. It has been *assumed* that the slider is in contact with the top surface of the guide. The force on the slider due to the guide reaction *N* is therefore assumed positive in the direction shown.

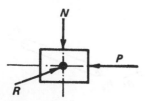

FIG. 3.57

Using the free body diagram of Fig. 3.57 the conditions for horizontal and vertical equilibrium of the slider may be written as

$$\rightarrow R \cos \alpha - P = 0, \qquad \text{(v)}$$

and $\quad\uparrow\ R \sin \alpha - N = 0.$ (vi)

Hence, using eqn (iv), we find that

$$Q = 2P \tan \alpha \quad \text{and} \quad N = Q/2.$$

The relationship between the magnitudes of P and Q shows that, if the angle α is small, a large force P at the slider can be produced by the application of a small force, Q, at B. This type of linkage is often used in presses, crushers and injection moulding machines, where large forces have to be generated.

We should also note that R_1, R_2 and N are positive. This shows that the assumed directions are correct. Links AB and BC are therefore in compression and the slider is in contact with the upper surface of the guide.

3.3 Lifting tongs

A set of symmetrical lifting tongs is shown in Fig. 3.58 lifting a load W. Let us determine

(i)   the forces applied to the arm DEF at the points D, E, and F, and
(ii) the conditions which determine when slip will occur between the block and the contact points F and G.

Friction will be neglected at all the connections in the tongs assembly except at the points F and G where the tongs support the load.

The system we are analysing is geometrically symmetrical about the vertical centre line, and the external loading, W, which acts along the centre line, is also symmetrical about the vertical centre line. When viewed as in Fig. 2.58 points C and F lie in the left-hand half plane and the corresponding points D and G lie in the right-hand half plane. Now let us take an imaginary walk around the tongs and view them from the back. The tongs will look the same as before and we can introduce letters to define points in the system as before with C and F in the new left-hand half plane and D and G in the new right-hand half plane.

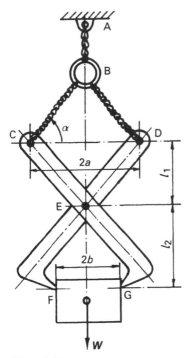

FIG. 3.58

The physical points that were originally labelled C and F have now become
D and G, but the act of viewing from the other side will not have changed
the forces acting at these points. The force acting at F when viewed from the
front becomes the force acting at G when viewed from the back. This applies
at all points in the system so that the forces in the system must also be
symmetrical about the vertical centre line.

Note that if **W** were offset from the centre line symmetry would be lost.

To determine the force in the main supporting chain AB we remove the
connection at A and consider the equilibrium of the whole system. The free
body diagram of the whole system is shown in Fig. 3.59 in which V is the
tension in the chain. If it is assumed that the weights of the links are small
compared with W, then, for equilibrium in the vertical direction,

$$\uparrow V - W = 0. \tag{i}$$

The tension in the chain AB is therefore equal to the load W.

The loads in side chains BC and BD may be found by considering the forces
applied to the ring. The free body diagram of the ring is shown in Fig. 3.60.
W represents the load applied by the main chain and **T** represents the tension
in the side chains. (From symmetry equal loads are applied by BC and BD—the
reader can check that this is so by using T_1 and T_2 in the free body diagram
and from horizontal equilibrium showing that $T_1 = T_2$.)

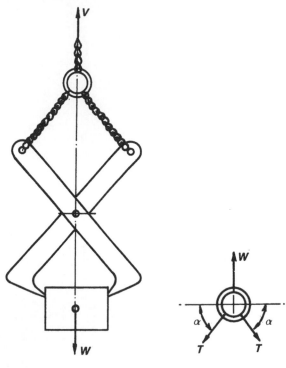

Fig. 3.59   Fig. 3.60

For vertical equilibrium of the ring we require that

$$\uparrow W - 2T \sin \alpha = 0,$$

so that

$$T = \frac{W}{2 \sin \alpha}. \tag{ii}$$

Before we can proceed further we need to consider the forces transmitted to the pin at E. The arms CEG and DEF will apply forces **P** to the pin. These forces must again act symmetrically about the vertical centre line as shown in the free body diagram of Fig. 3.61(a). Horizontal equilibrium of the pin is assured by the symmetry, and for vertical equilibrium

$$\uparrow 2P \sin \theta = 0. \tag{iii}$$

Thus θ must be zero so that the line of action of the forces **P** must be horizontal as shown in Fig. 3.61(b), but the directions of the forces are not known. We are now in a position to determine the forces acting on the arms CEG and DEF. Fig. 3.62 shows arm CEG drawn to scale. The force **T** at point C is due to the side chain BC. Its magnitude is known from eqn (ii) and it acts along CB at an inclination α to the horizontal, as shown. The line of action of the

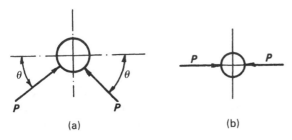

Fig. 3.61

force P at the pin connection E is horizontal. This enables us to determine the point of intersection, I, between the lines of action of the force at C and the force at E.

Link GEC is under the action of three non-parallel forces (see Section 3.5), so that the line of action of the force R at G must also pass through I.

For equilibrium

$$T + P + R = 0. \tag{iv}$$

Since we know the lines of action of all three vectors in eqn (iv), and the magnitude of one, a graphical solution can be obtained by scale drawing. P and R can then be determined by measurement as shown in Fig. 3.63. The direction of the arrows in Fig. 3.63 must be such that eqn (iv) is satisfied. The free body diagrams for the links CEG and DEF, the pin E and the block can now be constructed using Fig. 3.63, (see Fig. 3.64).

For equilibrium of the block

$$W + R_{(F)} + R_{(G)} = 0. \tag{v}$$

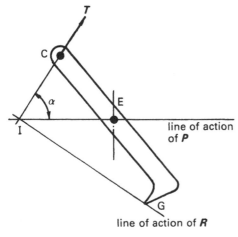

Fig. 3.62

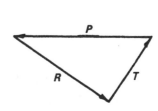

Fig. 3.63

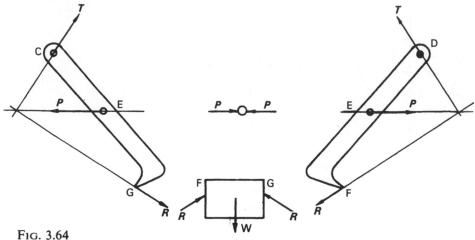

FIG. 3.64

We should therefore check that the forces on the block at F and G add vectorially with the weight **W** to satisfy eqn (v) as shown in Fig. 3.65.

If we decide upon an analytical solution the free body diagram of the link CEG may be drawn as shown in Fig. 3.66. The forces at C, E and G are now expressed in component form. Since the directions of the components at E and G are, as yet, unknown we shall assume the directions shown.

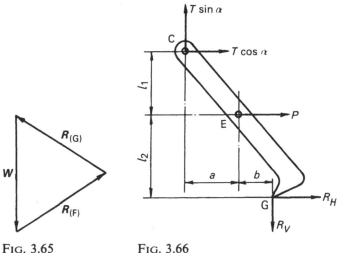

FIG. 3.65   FIG. 3.66

For horizontal and vertical equilibrium we have

$$\rightarrow T \cos \alpha + P + R_{\mathrm{H}} = 0, \tag{vi}$$

$$\uparrow T \sin \alpha - R_{\mathrm{V}} = 0, \tag{vii}$$

and by taking moments about the point G,

$$\circlearrowleft \quad -(a+b)\,T\sin\alpha - (l_1+l_2)\,T\cos\alpha - l_2 P = 0. \qquad\text{(viii)}$$

By substituting for T from eqn (ii), these equations give

$$P = -\frac{W}{2}\left[\frac{(a+b)}{l_2} + \left(1+\frac{l_1}{l_2}\right)\cos\alpha\right], \qquad\text{(ix)}$$

$$R_H = \frac{W}{2}\left[\left(\frac{a+b}{l_2}\right) + \left(\frac{l_1}{l_2}\right)\cot\alpha\right], \qquad\text{(x)}$$

and $\quad R_V = \dfrac{W}{2}.$ $\qquad\qquad\qquad\qquad\qquad\text{(xi)}$

Since P is negative the force on the link at E from the pin acts in the opposite sense from that shown on the free body diagram, i.e. it acts to the left. This, as would be expected, agrees with the result obtained from Fig. 3.63. R_H and R_V are positive and therefore act in the directions shown in Fig. 3.66.

Once R_H and R_V have been determined the ability of the tongs to lift the load W may be assessed by considering the nature of the force transmission at F and G. The free body diagram of the block is shown in Fig. 3.67.

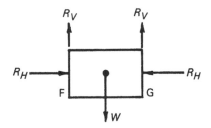

Fig. 3.67

Since the force transmission at F and G is provided by contacting surfaces with friction, the magnitude of R_V will be limited by the coefficient of friction μ for the surfaces, i.e.

$$R_V \leqslant \mu R_H. \qquad\text{(xii)}$$

If we now substitute for R_H and R_V using eqns (x) and (xi) inequality (xii) may be written,

$$1 \leqslant \mu\left[\left(\frac{a+b}{l_2}\right) + \left(\frac{l_1}{l_2}\right)\cot\alpha\right]. \qquad\text{(xiii)}$$

It is important to note that this condition is *independent of W*. Therefore, provided that the geometry and frictional properties of the tongs and the block, as specified by α, l_1, l_2, a, b and μ, satisfy inequality (xiii) the tongs will lift the load W irrespective of its magnitude. Such a system is said to be self-locking. If inequality (xiii) is *not* satisfied the tongs *will not* lift the load.

However, if the form of the connection between the load and the tongs were changed to provide a more positive force transmission by, for example, changing to pin joints as shown in Fig. 3.68, it would not be necessary to satisfy inequality (xiii). We should note that the loads, as determined by eqns (ii), (ix), (x) and (xi) would remain unchanged.

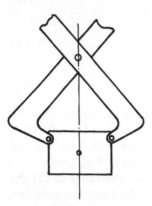

FIG. 3.68

3.4   Wall crane

Figure 3.69 shows the details of a wall crane. It consists of a jib PQ along the horizontal upper surface of which runs a trolley. The jib is connected to the

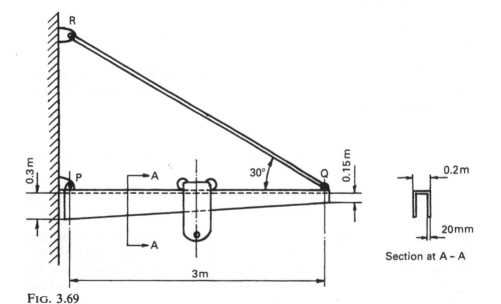

FIG. 3.69

wall by a pin joint at P and a tie rod RQ with pin joints at R and Q as shown. The jib is fabricated from 20 mm thick steel plate of density 8×10^3 kgm^{-3}. A trolley of mass $m_T = 50$ kg is free to run on wheels along the top section of the jib. If the centre of mass of the trolley lies on its vertical centre line determine the forces in the pinned support P, and in the tie rod RQ, due to the weight of the crane when the trolley is 1.5 m from P.

We first need to find the position of the centre of mass of the jib so that we can replace its distributed mass by a single equivalent particle. The jib can conveniently be divided into two rectangular elements, A and B, and one triangular element C, having centres of mass at G_A, G_B and G_C respectively, as shown diagrammatically in Fig. 3.70. The mass of each element is given by

$$m_A = 0.02 \times 0.2 \times 3 \times 8 \times 10^3 = 96 \text{ kg},$$

$$m_B = 2(0.15 \times 3 \times 0.2 \times 8 \times 10^3) = 144 \text{ kg},$$

$$m_C = 2(0.5 \times 0.15 \times 3 \times 0.02 \times 8 \times 10^3) = 72 \text{ kg},$$

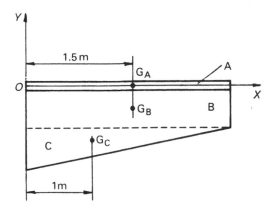

FIG. 3.70

so that the total mass of the jib

$$m = 312 \text{ kg}.$$

G_A, G_B and G_C lie at the geometric centroids of their respective areas as shown in Fig. 3.70. The position of the centre of mass of the jib may be obtained from eqn (3.47) (see also example 3.8).

Using the coordinate system OXY with origin at P as shown

$$x_G = \frac{m_A x_{G_A} + m_B x_{G_B} + m_C x_{G_C}}{m_A + m_B + m_C}. \tag{i}$$

Thus

$$x_G = \frac{96 \times 1.5 + 144 \times 1.5 + 72 \times 1}{312} = 1.38 \text{ m.} \qquad \text{(ii)}$$

The y coordinate of G is not required in this case.

We can now consider the loading due to the distributed mass of the jib as equivalent to that due to a single particle of mass 312 kg at a distance 1.38 m from P. The forces at the supports can now be determined using the free body diagram of the jib and trolley in Fig. 3.71. Since the trolley is symmetrical about its vertical centre line, its centre of mass will also be on this line. The force at P is shown by the two components R_X and R_Y, and, if we neglect the weight of the tie rod, the force T at Q must be along the line PQ (see Section 3.5). The only other forces acting on the jib are its weight, which will act through G, and that of the trolley.

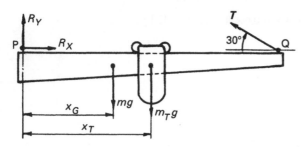

FIG. 3.71

From the free body diagram, the three equations of equilibrium can be written:

$$\rightarrow R_X - T \cos 30 = 0,$$

$$\uparrow \ R_Y + T \sin 30 - mg - m_T g = 0,$$

$$\circlearrowright 3T \sin 30 - mg x_G - m_T g x_T = 0$$

from which $R_X = 2.87$ kN, $R_Y = 1.89$ kN and $T = 3.32$ kN.

The forces acting on the wall and on the tie rod are shown in Fig. 3.72.

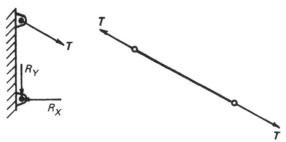

FIG. 3.72

3.5 Power transmission systems

(a) Friction clutch

Many power transmission systems use a friction clutch to connect the prime mover, e.g. an electric motor or internal combustion engine to the load. The clutch allows the prime mover to be disengaged from the load so that the load may come to rest without the prime mover coming to rest. A very common example is in motor car transmission systems where the engine can be disconnected from the road wheels when the car is at rest, or when changing gear. The clutch is necessary in this case because the internal combustion engine cannot generate torque at zero speed.

Another reason for using a friction clutch is to protect a prime mover from overloads which may arise on the driven machine. The torque which can be transmitted by the clutch is limited by the frictional conditions at its driving surfaces and slip will occur if the load torque exceeds a set value.

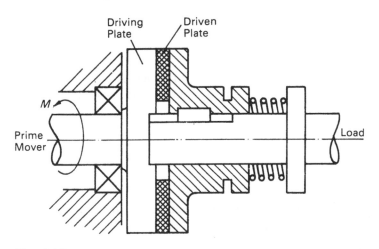

Fig. 3.73

The basic layout of a simple single plate clutch is shown in Fig. 3.73. A circular plate is fixed to the output shaft of the prime mover and is maintained in contact with a second concentric plate mounted on the input shaft to the load. The second plate can slide axially on its shaft so that it can be moved out of contact with the first plate so disengaging the drive. Normally springs are used to provide the contact force when the clutch is engaged, i.e. transmitting power, and the sliding plate is moved axially against the spring force by a lever to disengage the clutch. Special materials are used at the contact surfaces to provide high friction and good wear characteristics. In Fig. 3.73 this material is shown in the form of an annulus attached to the driven plate.

The maximum torque that can be transmitted by the clutch is limited and occurs at the onset of slip between the plates. At slip, the tangential friction forces generated at the contact will depend upon the coefficient of friction μ and the normal force at the contacting surfaces. We will assume that μ is constant across the contact area, but the normal force will vary. Before we can calculate the limiting torque for the clutch it is necessary to determine how the normal force varies, i.e. how the pressure distribution $p(r)$ on the friction pads varies with r as a result of the force F_0 applied to the driven plate by the spring in Fig. 3.73.

Let us assume that all the wear occurs on the friction material attached to the driven plate. In this case the axial displacement of the driven plate due to wear will be the same at all radii. Hence the wear will be uniform across the contact surface, and the rate of removal of material will be the same at all points.

It has been found experimentally that the rate, dm/dt, at which material is removed from a pair of sliding surfaces, is related to the contact pressure p between the surfaces and their relative sliding velocity v_r by a relationship of the form

$$\frac{dm}{dt} = Cpv_r, \tag{i}$$

where C is a constant.

If the input shaft has angular velocity ω_1 and the output shaft has angular velocity ω_2, the relative velocity v_r of the two surfaces at some radius r is

$$v_r = (\omega_1 - \omega_2)r. \tag{ii}$$

Hence from eqns (i) and (ii)

$$p = \left[\frac{dm/dt}{C(\omega_1 - \omega_2)}\right]\frac{1}{r}. \tag{iii}$$

Since dm/dt is the same at all values of r we can express the radial pressure distribution, for any given value of $(\omega_1 - \omega_2)$, as

$$p(r) = A/r, \tag{iv}$$

where A is independent of r.

We can now relate the axial force F_O to this pressure distribution by considering the axial equilibrium of the driven plate.

The free body diagram of this arrangement is shown in Fig. 3.74.

The axial force produced by the pressure $p(r)$ over a small elemental ring of radius r and width δr is given by

$$\delta F = 2\pi r\, \delta r\, p(r).$$

Integration over the surface of the plate gives the total force due to the contact pressure, and for axial equilibrium of the driven plate

$$\rightarrow \quad \int_{r_1}^{r_2} 2\pi r p(r)\, dr - F_O = 0. \tag{v}$$

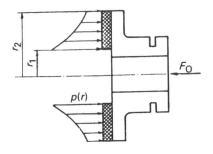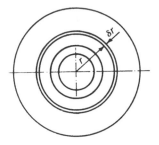

Fig. 3.74

Substituting for $p(r)$ using eqn (iv) and evaluating the integral, we find that

$$F_O = 2\pi(r_2 - r_1)A. \qquad \text{(vi)}$$

Substituting for A from eqn (vi) into eqn (iv) we see that the contact pressure during slip is related to the clamping force by the relationship

$$p(r) = \frac{F_O}{2\pi(r_2 - r_1)r}. \qquad \text{(vii)}$$

The maximum torque M_{max} which can be transmitted by the clutch can now be determined by considering the frictional forces acting between the plates when slip is just about to occur.

Consider an elemental area of surface at radius r. This area has dimensions δr and $r\,\delta\theta$ as shown in the free body diagram of Fig. 3.75. The frictional

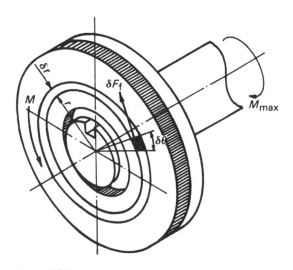

Fig. 3.75

force acting on this elemental area is

$$\delta F_f = \mu p(r) r \, \delta\theta \, \delta r.$$

The torque δM_f exerted by δF_f about the axis of rotation is therefore

$$\delta M_f = \mu p(r) r^2 \, \delta\theta \, \delta r.$$

The friction forces acting on an elemental ring of radius r will therefore transmit a torque

$$\delta M_r = \mu p(r) r^2 \, \delta r \int_0^{2\pi} d\theta,$$

which on integrating with respect to θ gives

$$\delta M_r = \mu p(r) 2\pi r^2 \, \delta r.$$

The total torque M transmitted by the friction forces acting on whole of the contact surface can now be obtained by integrating δM_r from r_1 to r_2 so that

$$M = \int_{r_1}^{r_2} \mu p(r) 2\pi r^2 \, dr.$$

Substituting for $p(r)$ using eqn (vii) and integrating gives

$$M = \frac{\mu F_0 (r_1 + r_2)}{2},$$

or $M = \mu F_0 r_m,$ (viii)

where r_m is the mean radius of the contact area. This is the maximum frictional torque that can be applied to the driven plate by the driving plate. If we assume that the driven shaft runs at constant speed, the equations for static equilibrium can be applied. For equilibrium of the driven shaft in Fig. 3.75

$$\circlearrowleft M - M_{max} = 0,$$

i.e. $M_{max} = \mu F_0 r_m$ (ix)

Thus if the load torque exceeds this value, which is the torque which would be exerted by the total friction force μF_0 acting at the mean radius r_m of the contact area, slip will occur.

Belt and pulley

Let us consider a thin inextensible belt passing over a pulley of radius R and subjected to tensions T_1 and T_2 as shown in Fig. 3.76(a). To maintain equilibrium a moment M must be applied to the pulley as shown. For a given value of T_1 let us find the values of T_2 and M at which slip will occur between the belt and the pulley.

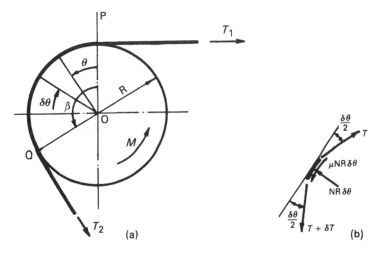

FIG. 3.76

Let the angle of contact between the belt and pulley be β and the coefficient of friction between belt and pulley be μ.

First we will consider the equilibrium of a general elemental length of the belt situated at some position θ from the datum line OP and subtending an angle $\delta\theta$ at the centre of the pulley.

The free body diagram of this element is shown in Fig. 3.76(b). The tension T in the belt will vary around the circumference of the pulley and it is assumed that over the elemental length, $R\,\delta\theta$, the tension increases by an amount δT. There is also a normal contact pressure between the belt and the pulley. If this is represented as a force N per unit a length of belt, then the radial force acting on the element is $NR\,\delta\theta$. N will vary around the pulley and is therefore a function of θ. Friction between the belt and pulley causes a tangential friction force on the pulley which opposes its tendency to slip. Since we are considering the limiting condition when the pulley is just about to slip relative to the belt this tangential force will have magnitude $\mu NR\,\delta\theta$ in a direction which opposes the incipient motion of the pulley.

For radial and tangential equilibrium of the element,

$$\nwarrow\ NR\,\delta\theta - T\sin\frac{\delta\theta}{2} - (T+\delta T)\sin\frac{\delta\theta}{2} = 0. \tag{i}$$

and

$$\swarrow (T+\delta T)\cos\frac{\delta\theta}{2} - T\cos\frac{\delta\theta}{2} + \mu NR\,\delta\theta = 0. \tag{ii}$$

Using the small angle approximations $\sin\delta\theta/2 \simeq \delta\theta/2$, $\cos\,\delta\theta/2 \approx 1$, and taking the limit as $\delta\theta \to 0$, eqns (i) and (ii) reduce to

$$NR - T = 0, \tag{iii}$$

and $\dfrac{\mathrm{d}T}{\mathrm{d}\theta} + \mu NR = 0.$ (iv)

Substituting eqn (iii) into eqn (iv) gives

$$\frac{\mathrm{d}T}{\mathrm{d}\theta} + \mu T = 0.$$ (v)

This is a first order differential equation in T and the variables can be separated. Integrating gives

$$\int \frac{\mathrm{d}T}{T} = -\int \mu \, \mathrm{d}\theta + C,$$

where C is a constant of integration. Hence

$$\ln T = -\mu\theta + C,$$

or $T = e^{-\mu\theta + C} = De^{-\mu\theta},$ (vi)

where D is a constant.

D can be found from the condition that at $\theta = 0$, $T = T_1$, so that $D = T_1$ and

$$T = T_1 e^{-\mu\theta}.$$

T therefore *decreases* as θ increases.

The tension T_2 at $\theta = \beta$ i.e. at the point Q where contact between belt and pulley ceases, is therefore

$$T_2 = T_1 e^{-\mu\beta}.$$ (vii)

Equation (vii) relates the tensions T_1 and T_2 in the 'tight' and 'slack' sides of the belt at the instant when slip is just about to occur between the belt and the pulley.

If we now consider the free body diagram of the belt and pulley together, shown in Fig. 3.76(a), we can find the torque M required to maintain the equilibrium of the pulley.

Thus

$$M + (T_2 - T_1)R = 0,$$

i.e. $M = (T_1 - T_2)R.$

The maximum value of M occurs just before the onset of slip. Substitution for T_2 from eqn (vii) gives

$$M_{max} = T_1 R(1 - e^{-\mu\beta}).$$ (viii)

Since $e^{-\mu\beta} < 1$, M_{max} will be positive and therefore will be in the direction shown in Fig. 3.76(a).

Capstan

A capstan is a device used to produce large tensions in ropes. The most common application is in mooring ships to docks. The capstan consists of a rotating drum, usually with a vertical axis, which is driven at a slow speed. A rope is looped around the drum as shown in Fig. 3.77. When $T_2 = 0$, or is very small, slip occurs between the rope and drum. If a small tension is now applied by hand, so increasing T_2, the resulting value of T_1 will depend upon the coefficient of friction between the rope and the drum and the angle of wrap of the rope on the drum.

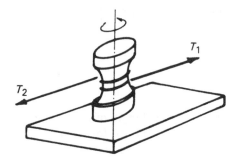

FIG. 3.77

Let us consider the effect of the number of turns of rope on the drum. From eqn (vii)

$$T_2 = T_1 e^{-\mu 2\pi n},$$

where n is the number of turns of rope on the drum. If we assume that $\mu = 0.2$ then

$$T_1/T_2 = e^{0.4n\pi}$$

at the instant when slip is about to occur.

Table 3.1 shows the values of T_1/T_2 for different number of turns of rope on the drum. It can be seen that very large tensions can be generated with only a small pull T_2 on the rope.

An interesting illustration is that of the cowboy hitching his horse to the rail outside a saloon bar as shown in Fig. 3.78. If he wraps the rope five times

Table 3.1

n	T_1/T_2
1	3.5
2	12.4
3	43.8
4	154.0
5	544.0
6	1920.0

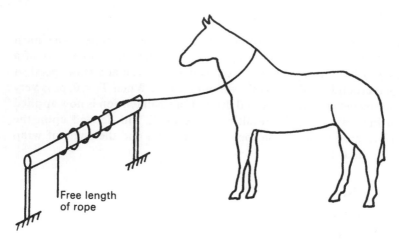

Free length
of rope

FIG. 3.78

round the rail as shown in the figure and leaves a free length of rope hanging, then T_2 is equal to the weight of the free length of rope. If, for example, the length of rope weighs 5 N and $\mu = 0.2$, the horse will have to exert a force of 2720 N before the rope will slip.

Toothed gearing

We have already seen in Chapter 1, application 3, that a pair of toothed gears behaves kinematically like a pair of cylinders in rolling contact. The torque which can be transmitted by a pair of cylinders is limited by the coefficient of friction between the two surfaces and the allowable normal force which can be applied to force the cylinders together.

The addition of suitably shaped teeth to the cylinders allows a much greater torque to be transmitted between the cylinders. Before we can analyse the forces transmitted between a pair of meshing gears, and to their supporting bearings, we need to have some basic knowledge about the shape of the gear teeth. There is a vast literature on gears, gear teeth and gear design. Here we will consider only those properties which affect the force transmission between two geometrically correct gears.

The simplest and most common form of gear is the spur gear which has teeth cut parallel to the axis of rotation. An involute curve is used for the working part of the tooth profile where the teeth of the two gears in mesh make contact. A meshing pair of teeth is shown in Fig. 3.79. The gears G_1, G_2 have centres O_1, O_2 and pitch circle radii R_1, R_2. P is the point of contact of the two pitch circles.

The involute profiles of the teeth are generated relative to two base circles of radii r_1 and r_2 and the pair of teeth we are considering make contact at the contact point C shown. Note that the actual tooth may undercut the base circle

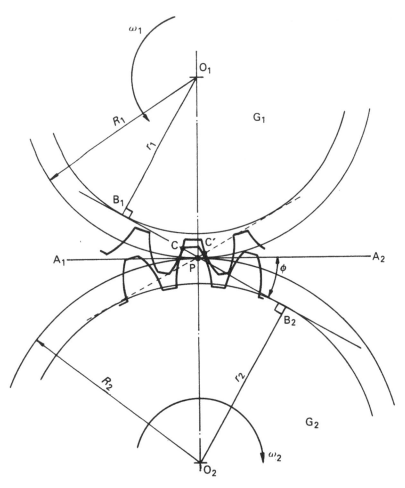

FIG. 3.79

as shown in Fig. 3.79; both the pitch and base circles are geometric rather than physical properties. The properties of the involute profile are such that if we draw a common tangent, B_1B_2, to the base circles of the gears it will pass through the pitch point P. Also the point of contact, C, between the teeth will always lie on the common tangent, and as the gears rotate C will move along the common tangent in the direction B_1B_2. A further property of the involute form is that the common tangent at the point of contact C is normal to the tooth surfaces for all points C along B_1B_2.

Thus, even when more than one pair of teeth are in mesh, the points of contact will all lie on the common tangent. The angle ϕ between the common tangent B_1B_2 and the common tangent to the pitch circles A_1A_2 is called the pressure angle. For most applications $\phi = 20°$. During the contact period, if G_1 is driving G_2 as shown, the point C will move over the tooth flank from the tip of the tooth towards the root on gear G_2 and from root to tip on gear G_1. Relative sliding therefore occurs between the teeth during the contact period. In our elementary analysis the effects of friction will be neglected since gears are usually well lubricated. The contact conditions therefore represent frictionless contact at a point and the direction of the forces on the bodies at the contact will be along the common normal to the teeth flanks (see Section 3.9).

We can now calculate the forces acting on a simple pair of spur gears mounted on shafts at O_1 and O_2 and having pitch circle radii R_1 and R_2 as shown in Fig. 3.80. The gears are mounted in a gear box which is bolted to the ground at A and B.

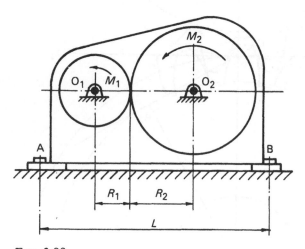

Fig. 3.80

Let the gears be in equilibrium under the action of external torques M_1, M_2 applied to the input and output shafts as shown. Remember that the equilibrium condition applies when a system is not accelerating, i.e. it may be at rest or moving with constant linear or angular velocity.

We can now separate the two gears and draw free body diagrams for each as shown in Fig. 3.81.

The forces acting on each gear, in addition to the externally applied torques M_1, M_2, are the forces at the support bearings and the tooth contact forces. At the bearings O_1 and O_2 the forces are shown in component form as V_1, H_1 and V_2, H_2.

All that is known about the tooth contact forces is that they have lines of action along the common normal at the contact points, i.e. along the line of the common tangent to the base circles which is at an angle ϕ to the common tangent at the pitch point P. There are two possible common tangents to the base circles. In Fig. 3.79 the force on gear 1 from gear 2 will act along CB_1

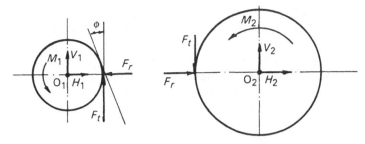

FIG. 3.81

and the force on gear 2 from gear 1 will act along CB_2, since only compressive forces can be transmitted at the contact. If the contact were at C' on the other flank of the tooth on gear 2, the contact forces would again be compressive but would act along the other common tangent to the base circles. For this case G_2 would drive G_1.

Whichever contact occurs, the force can be resolved into a radial component and a tangential component along the tangent to the pitch circles.

Figure 3.82 shows the two possible cases with the contact force F resolved into tangential and radial components F_t and F_r, where

$$F_r = |F_t| \tan \phi. \tag{i}$$

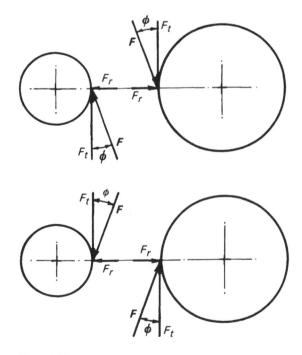

FIG. 3.82

No matter which direction is chosen for F_t, the component F_r must always act towards the centre of the gear. The radial component therefore tries to push the gears apart. If F_r acted in the opposite sense it would try to pull the gears *towards* each other—obviously this is not physically possible since contacting surfaces cannot attract one another. The forces F, on each gear and hence their components F_t and F_r must, by Newtons Third Law, be equal in magnitude and opposite in direction. Let us assume that they act as shown in Fig. 3.81.

We can now write our three equations of static equilibrium for each body. For gear G_1,

$$\rightarrow \quad H_1 - F_r = 0,$$

$$\uparrow \quad V_1 + F_t = 0,$$

$$\circlearrowleft_{O_1} \quad M_1 + R_1 F_t = 0, \tag{ii}$$

and for gear G_2,

$$\rightarrow \quad F_r + H_2 = 0,$$

$$\uparrow \quad V_2 - F_t = 0,$$

$$\circlearrowleft_{O_2} \quad M_2 + R_2 F_t = 0. \tag{iii}$$

If the input torque M_1 is known, then from eqn (ii)

$$F_t = -\frac{M_1}{R_1},$$

and is therefore in the opposite direction from that shown in Fig. 3.81. Also

$$V_1 = \frac{M_1}{R_1},$$

and $H_1 = F_r.$

Equation (i) now gives the magnitude of F_r as

$$F_r = F_t \tan \phi,$$

i.e. $$F_r = \frac{M_1 \tan \phi}{R_1}. \tag{iv}$$

The direction of F_r is as shown in Fig. 3.82, since from the geometry of the contact F_r must *always* force the gears apart for both positive and negative values of F_t.

Thus

$$H_1 = \frac{M_1}{R_1} \tan \phi, \tag{v}$$

and is in the direction shown in the free body diagram of Fig. 3.81.

From eqn (iii),

$$H_2 = -F_r = -\frac{M_1}{R_1} \tan \theta$$

and is therefore in the opposite sense from that shown in Fig. 3.81. Also

$$V_2 = F_t = -\frac{M_1}{R_1}, \tag{vii}$$

again in the opposite direction from that shown,

and

$$M_2 = -R_2 F_t = \frac{M_1 R_2}{R_1}. \tag{viii}$$

Thus, for equilibrium, the torque M_2 applied at gear G_2 is equal to the torque M_1 applied at gear G_1 multiplied by the gear ratio R_2/R_1, and is in the *same* sense as M_1.

Note that the forces on the gears at their centres O_1 and O_2 are equal in magnitude, but opposite in direction. The forces on the frame, or gearbox, at O_1 and O_2 are shown in Fig. 3.83, and are equal and opposite to those on the gears. It can be seen by inspecting Fig. 3.83 that the forces at O_1 and O_2 are equivalent to a couple applied to the gearbox of magnitude $M_1(1 + R_2/R_1)$.

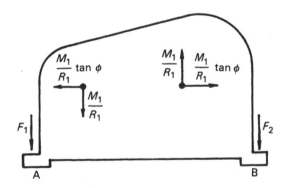

FIG. 3.83

By substituting for R_2/R_1 using eqn (viii), the couple becomes $(M_1 + M_2)$, i.e. the sum of the input and output torques. For the gearbox to be in equilibrium the bolts holding down the gearbox at points A and B must produce forces F_1 and F_2 such that

$$-F_1 - F_2 - \frac{M_1}{R_1} + \frac{M_1}{R_1} = 0.$$

Thus

$$F_1 = -F_2. \tag{ix}$$

By taking moments about A

$$\text{\textcircled{A}} \; M_1 + M_2 - F_2 L = 0. \tag{x}$$

Thus

$$F_2 = \frac{M_1 + M_2}{L},$$

and is in the direction shown in Fig. 3.84. F_1, from eqn (ix) will be in the opposite direction from that shown.

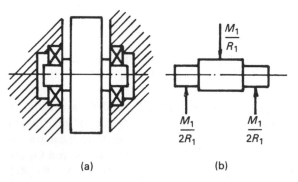

(a)   (b)

FIG. 3.84

The intermediate steps in obtaining the moment equation (x) above have been omitted. If the reader does not see how this equation was derived at he should work it out from first principles using the free body diagram of Fig. 3.83.

If the gears are mounted symmetrically on shafts, as shown in Fig. 3.84(a), the forces at each of the support bearings will be half of that on the gear. This is shown in the free body diagram of the shaft in Fig. 3.84(b).

3.6   Shear forces and bending moments in beams

The stresses which arise in simple machine structures, for example shafts and pin jointed frameworks, are determined from the values of the internal forces and moments within the structure. These internal forces and moments are calculated from basic equilibrium requirements. An important example is the calculation of stresses in beams. In this text we will only calculate the internal forces and moments. The use of these in the determination of stresses is dealt with in texts on stress analysis and strength of materials.

Let us consider the beam shown in Fig. 3.85. Its length is $4a$ and its weight is assumed to be negligibly small compared with the loads applied. The beam is loaded over half its length with a uniformly distributed load w/metre and also carries a concentrated load W positioned as shown. Simple knife edges, located at A and B, are used to support the beam. We shall determine how the internal forces and moment vary along the length of the beam.

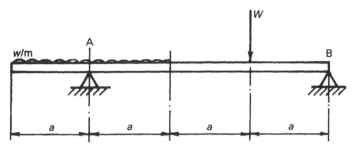

FIG. 3.85

The first step is to draw the free body diagram of the whole beam, Fig. 3.86, so that the support forces R_A and R_B at A and B may be determined. The surface contact between the beam and the knife edge supports will produce a contact force along the common normal at the contact point. If the deflection of the beam is small the slope of the beam at the supports will also be small, so that the contact forces can be assumed to be vertical. Friction at the supports is also neglected. From section 3.6 the distributed load can be replaced by a single equivalent load $W_e = 2wa$. This equivalent force will act at the mid-point of the distributed load. There are no forces in the horizontal direction, so for

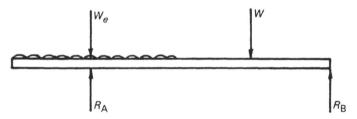

FIG. 3.86

equilibrium we obtain,

$$\uparrow R_A + R_B - W_e - W = 0,$$

and   Ⓐ $3aR_B - 2aW = 0.$ (i)

Thus $R_B = \dfrac{2}{3}W,$

and $R_A = \dfrac{W}{3} + 2wa.$ (ii)

Now that all the external forces acting on the beam are known we can determine the internal forces and moments using the results of Section 3.8. To find the internal forces and moment at a particular section of the beam we

have to imagine the beam cut at the section. We then draw a free body diagram
of the right or left-hand part of the divided beam. Because the load distribution
is discontinuous, due to the concentrated loads W, R_A and R_B, it is convenient
to start from the left-hand end and consider sections which are located between
the points of application of these concentrated loads, so covering the full
length of the beam.

Let us first consider a section between the left-hand end and support A.
The free body diagram of this section is shown in Fig. 3.87(a), where F and
M are the internal force and moment at the section distant x from the end of
the beam. There will be no net horizontal force at the section, but an internal
moment is necessary for equilibrium. Fig. 3.87(a) is valid for $0 < x < a$. The
distributed load on this section can now be replaced by an equivalent load
$W_e = wx$ a distance $x/2$ from the end of the beam. For equilibrium

$$F - wx = 0,$$

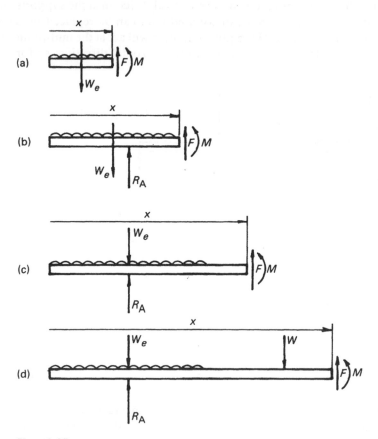

FIG. 3.87

and taking moments about the right-hand end of our section of beam,

$$\circlearrowright \quad M + \frac{wx^2}{2} = 0.$$

Thus for $0 < x < a$,

$$F = wx \quad \text{and} \quad M = \frac{-wx^2}{2}. \tag{iii}$$

When $a < x < 2a$ the free body diagram of the section of beam is as shown in Fig. 3.87(b). For equilibrium

$$\uparrow \quad F + R_A - wx = 0,$$

and $\quad \circlearrowright M + \dfrac{wx^2}{2} - R_A(x - a) = 0.$

Thus for $a < x < 2a$,

$$F = w(x - 2a) - W/3,$$

and

$$M = \frac{W}{3}(x - a) - w\left(\frac{x^2}{2} - 2ax + 2a^2\right). \tag{iv}$$

Fig. 3.87(c) shows the free body diagram of the section of beam when $2a < x < 3a$.

In the case, for equilibrium,

$$\uparrow \quad R_A + F - 2wa = 0,$$

and $\quad \circlearrowright M + 2wa(x - a) - R_A(x - a) = 0.$

Thus $\quad F = -\dfrac{W}{3}, \text{ and } M = \dfrac{W}{3}(x - a).$ $\qquad$ (v)

Over the final section of the beam, $3a < x < 4a$, the free body diagram of Fig. 3.87(d) applies

Now we have

$$\uparrow \quad R_A + F - W - 2wa = 0,$$

and $\quad \circlearrowright M + 2wa(x - a) + W(x - 3a) - R_A(x - a) = 0.$

Upon substitution for R_A we obtain

$$F = \tfrac{2}{3}W, \text{ and } M = \tfrac{2}{3}W(4a - x). \tag{vi}$$

Equations (iii), (iv), (v) and (vi) give a complete analysis of the internal shear forces and bending moments in the beam.

The variation of F and M along the beam can be plotted as shown in Fig. 3.88. These graphs are referred to as shear force and bending moment diagrams

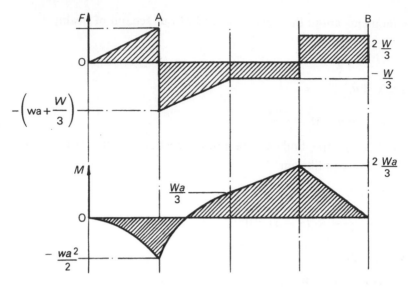

Fig. 3.88

and enable the positions at which the maximum values of F and M occur to be located. With practice shear force and bending moment diagrams can be drawn fairly quickly without writing down all of the equations given here, but when the reader is in doubt about the shear force or bending moment at a particular section he should *draw the free body diagram* of the appropriate section of the beam. The equations of equilibrium will then give both the shear force and bending moment at the section.

Exercises

NOTE   Free body diagrams are *essential* when solving these problems

1   A disc of radius r and weight W is supported by a light rod AB of length l with pin joints at A and B which may be assumed frictionless. The disc rests against a vertical wall as shown in Fig. 3.89. If the disc's position is such that its centre C lies vertically below A, find the force at pin A.

   What would be the equilibrium position of the disc if friction at the wall were negligible?

2   Determine the force on the ground at pin joint A and on the wall at B for the structure of Fig. 3.90. Neglect friction and the mass of the structure.

3   A rod ABD is connected to a vertical wall by a pin joint at A and a wire CB as shown in Fig. 3.91. The weight of the rod may be neglected in comparison to the vertical load W applied at D. Find the tension in the wire and the force at pin A.

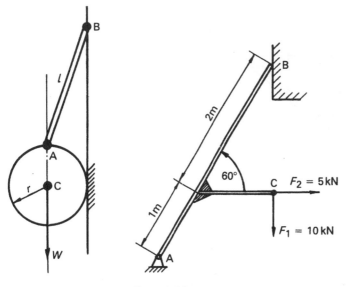

FIG. 3.89   FIG. 3.90

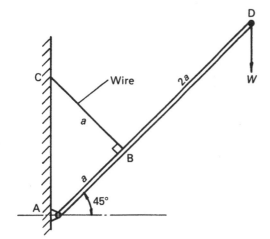

FIG. 3.91

4 A pin jointed structure is shown in Fig. 3.92 in which AC = BF = BC = CD = 1 m. It is pinned to a vertical wall at A and rests against the wall at D, and supports a load W at F as shown. Neglecting the masses of the links and friction find the force in link AC, and the force on the wall at D.

5 A pulley system used in lifting ships out of the water is shown in Fig. 3.93. The upper fixed pulley block has three pulleys and the lower moving

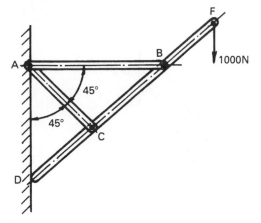

FIG. 3.92

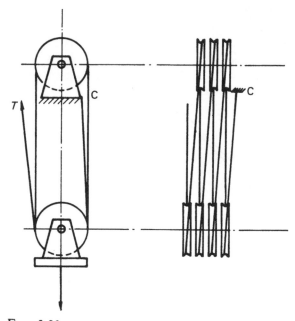

FIG. 3.93

block has four pulleys. A tension T is applied to the free end of the rope as shown. The rope passes first around the end lower pulley, then passes over the other pulleys as shown. The other end of the rope is clamped at C. Neglecting friction and the small angles made with the vertical by the ropes, what is the load W which can be lifted by a tension T in the rope?

6   A light bar AB rests upon a cylinder as shown in Fig. 3.94 and supports a vertical load P at A. What is the direction of the force on the rod at C when the system is in equilibrium?

Can the system be in equilibrium if there is no friction at C and D?

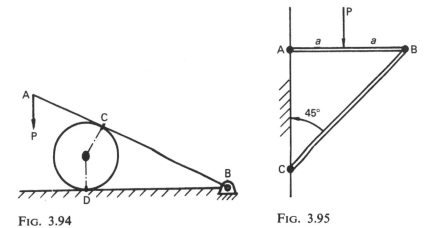

FIG. 3.94   FIG. 3.95

7   Find the forces on the wall at A and C for the pin jointed structure shown in Fig. 3.95. Neglect the weight of the structure.

8   A uniform circular roller of weight W is held in equilibrium in contact at P with a rough inclined plane by a vertical cable as shown. Show that the frictional component of the force acting on the roller at P must act *up* the plane, and derive an expression for the smallest coefficient of friction that makes equilibrium possible.

9   A simple screw clamp is shown in Fig. 3.97. If the required clamping force is 200 N what are the forces in the screws A and B.

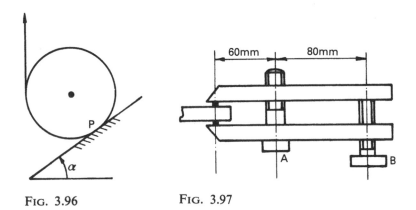

FIG. 3.96   FIG. 3.97

10   Fig. 3.98 shows a device used to limit the tension in a cable. It consists of a pair of jaws pivoted at B which are maintained in position by pin A. If the tension T in the cable rises above a value determined by the shear strength of pin A, the pin shears allowing the jaws to rotate, so releasing the left-hand cable. Find the maximum allowable tension in the

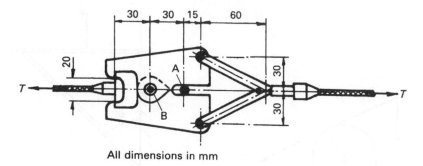

All dimensions in mm

FIG. 3.98

cable if the pin A will resist up to 12 kN shear force. Assume that the cable attachment at the right-hand end does not affect the rotation of the jaws. Neglect friction.

11 Fig. 3.99 shows a heavy roller being lifted up a step of height h. Determine the angle θ at which the force F must be applied for its magnitude to be a minimum. Assume that the roller does not slip on the step.

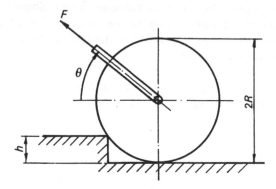

FIG. 3.99

12 A door closing mechanism is shown in plan view in Fig. 3.100. The door is maintained in contact with the wall at E by a torsion bar at pin joint B. If the contact force between the closed door and the wall at E is required to be 50 N what torque must be provided by the torsion bar? The angle BCD is 15°.

13 A simple press mechanism is shown in Fig. 3.101. Determine the forces at pins A and B and the force on the surface S when a force of 400 N is applied in a direction perpendicular to the press handle as shown.

14 Determine the forces in the bars of the truss loaded as shown in Fig. 3.102 where $a = 2$ m, $F_1 = 2500$ N, and $F_2 = 5000$ N.

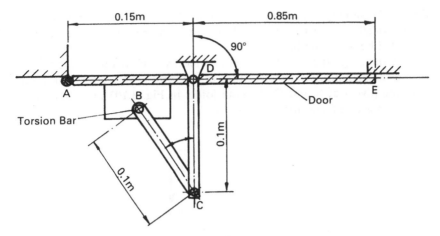

Fɪɢ. 3.100

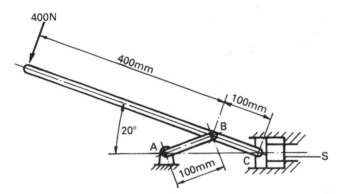

Fɪɢ. 3.101

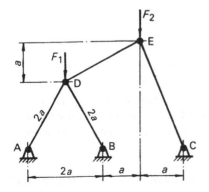

Fɪɢ. 3.102

15   Fig. 3.103 shows a pair of bolt cutters. Forces F_1 of 200 N applied symmetrically to the handles as shown are required to cut a small bolt held at E between the blades.

   Find the magnitude of the forces exerted by each blade on the bolt at E, and the magnitudes and directions of the forces exerted by the top jaw 1 on the pivots at B and C. Assume joints in Fig. 3.103 are frictionless. Try using a graphical method initially.

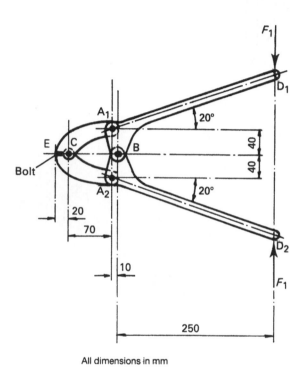

All dimensions in mm

FIG. 3.103

16   Fig. 3.104 shows (to scale) the mechanism of a hand-operated rivetting press. Arm ABC is pivoted freely at A to the machine frame, and is operated by the toggle action of link DE and cranked lever BEG as shown, to exert a vertical force F_2 on the rivet head at C. For the position shown, find the force F_2 generated by a hand pull of $F_1 = 200$ N applied as shown. What is the force in link DE?

17   A simple lever brake mechanism is shown in Fig. 3.105. Find the force P on the lever which is necessary to just prevent rotation of the wheel when

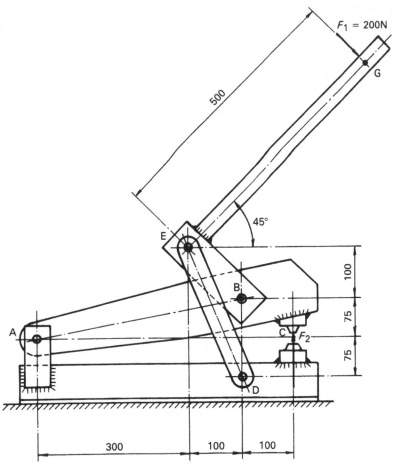

All dimensions in mm

Fig. 3.104

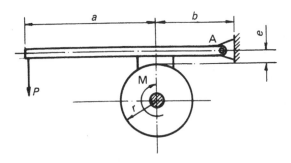

Fig. 3.105

a torque M is applied in the direction shown. What is then the force at the pin joint A. The coefficient of friction between the block and the wheel is μ.

18　Fig. 3.106 shows a simple planetary gear box. Determine the ratio M_2/M_1 between the output and input torques in terms of r_1 and r_2.

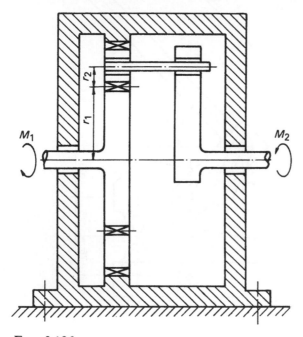

FIG. 3.106

19　A concial thrust bearing is shown in Fig. 3.107. It supports a vertical load P and the coefficient of friction between the rubbing surfaces is μ. Obtain

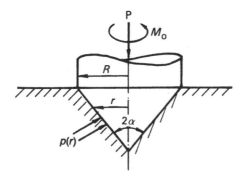

FIG. 3.107

an expression for the moment M required to rotate the shaft against the friction torque generated on the bearing. [Assume that the product of the normal pressure p at any point and the radial distance r of the point from the axis of rotation is constant, i.e. $rp(r) = $ constant. This is equivalent to assuming constant wear rate in the bearing.]

20 A differential band brake is shown in Fig. 3.108. The flexible brake band passes over the rotating wheel and is connected to a pivoted lever as shown. If the coefficient of friction between the band and the wheel is μ determine the force P on the lever which is required to just prevent rotation of the wheel when a clockwise torque M_0 is applied to it. Discuss what happens if $a_1/a_2 = e^{-\mu\beta}$.

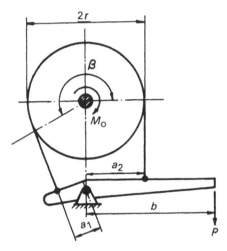

FIG. 3.108

21 Find the coordinates of the centre of mass of the flat uniform sheet shown in Fig. 3.109.

22 A uniform wire is bent into a circular arc of radius r which subtends an angle of $135°$ at the centre O of the circle. Find the distance of its centre of mass from O.

23 A uniform solid body has the shape of a thin wedge with a semicircular front elevation of radius R as shown in Fig. 3.110. Its thickness t varies such that $t = x/10$, where x is the half-width at height y above the base. Find the coordinates of its centre of mass.

24 A truck bed with centre of mass G and weighing $10\,\text{kN}$ is raised slowly by a hydraulic actuator, as shown in Fig. 3.111. Determine the actuator force for the position shown and the force at the rear pivot.

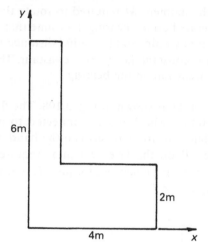

Fig. 3,109

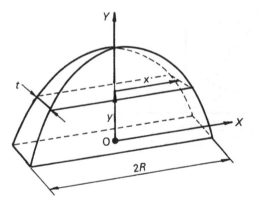

Fig. 3.110

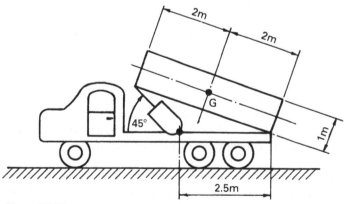

Fig. 3.111

25 Fig. 3.112 shows the design of a wind power turbine. The turbine is mounted at the top of a pivoted column OB which can be rotated about O by a hydraulic ram AC to allow maintenance of the turbine at ground level. At maximum design wind speed v_{max}, measurements show that the wind load on the turbine assembly is $F_0 = 2$ kN as shown, and that the wind load on the column varies linearly from zero at D to w_{max} at B. The weight of the turbine is 1.5 kN and its centre of mass is at B. Neglect the mass of the structure.

Find the force exerted by the hydraulic ram and the magnitude and direction of the force on the structure at O.

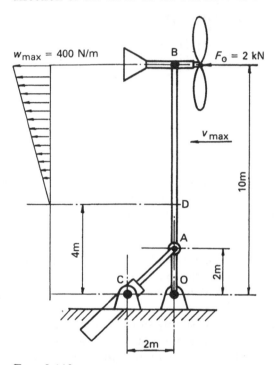

FIG. 3.112

26 Fig. 3.113 shows a crawler tractor at rest on horizontal ground. The length of ground contact on each side is 2 m as shown, and each track is 0.4 m wide. The ground contact pressure varies approximately linearly as shown from P_{min} at the rear to P_{max} at the front.

(a) If the tractor weighs 48 kN with centre of mass G as shown, find P_{min} and P_{max}.

(b) Maximum tractive effort is obtained when the pressure under the tracks is uniformly distributed, in which case the maximum pull that the tractor can achieve is 0.8 W. What must be the height h of the drawbar at A to achieve this if the drawbar pull is horizontal?

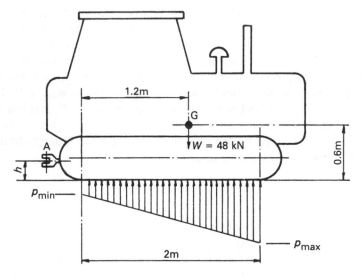

Fɪɢ. 3.113

27 A beam carries loads of 200 N at its ends and four loads of 60 N at distances 2, 2.5, 3.5 and 5 m from one end. The beam is 8 m long and rests on two supports that are 1 m from the ends of the beam. Sketch the variation in the internal shear force and bending moment along the beam and find their maximum values.

28 A cantilever beam of length 2*a* carries two loads *W*, one at its mid-point and the other at its free end. Find the maximum internal shear force in the beam and the bending moment at its mid-point.

29 A horizontal beam AB, 9 m long is simply supported at A and C where C is 6 m from A. A load of 35 kN is positioned at B and one of 50 kN mid-way between A and C. There is also a uniformly distributed load of 5000 N/m between A and C. Sketch diagrams showing the variation in internal shear force and bending moment along the beam. What is the maximum value of the shear force and at what point is the bending moment zero?

30 A beam of length *L* carries a uniformly distributed load over its total length and rests on two symmetrically disposed simple supports. How far from the ends must the supports be placed if the greatest bending moment is to be as small as possible?

31 A simply supported beam of span *L* carries a distributed load which increases uniformly from zero at the left hand support to a maximum of

ω/unit length at the right hand support. Find the maximum bending moment in the beam and the distance from the left hand end of the beam of the section at which it occurs.

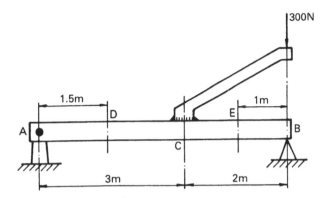

FIG. 3.114

32 The angled bracket shown in Fig. 3.114 is welded to the beam AB at C and supports a load of 300 N. What is the value of the shear force in the beam at sections D and E, and what is the internal bending moment at E?

4
Dynamics

Basic Theory

4.1   Dynamics of particles and rigid bodies

Newton's Laws refer only to particles. We will therefore consider first the motion of a particle and later develop methods for dealing with the motion of rigid bodies. The approach is similar to that used in our study of kinematics where we started with the motion of a point, and then proceeded to consider rigid body motion.

4.2   Dynamics of a particle

The motion of a particle under the action of applied forces obeys Newton's Second Law. Hence, if we know the magnitudes and directions of the forces acting on a particle, its acceleration can be found. Conversely, if we know the acceleration of a particle from kinematic considerations, it is possible to find the total force acting on the particle.

For a particle moving in a plane the general equation,

$$F = ma$$

may be expressed in components which are at right angles, e.g.

$$F_x = ma_x = m\ddot{x},$$

and $F_y = ma_y = m\ddot{y},$ (4.1)

where F_x, F_y and $\ddot{x}$, $\ddot{y}$ are the components of F and a along the directions OX, OY of a Cartesian frame of reference.

Example 4.1   Consider a particle P of mass m moving on a smooth horizontal table. Let the particle be restrained by a string which has one end fixed, so that the particle moves in a circular path of radius r, as shown in Fig. 4.1. If v, the magnitude of the velocity of the particle, is maintained constant, let us obtain an expression for the force in the string.

In this case we can obtain the acceleration of the particle from kinematic considerations. A point moving in a circular path has an acceleration (see p. 16)

$$a = a_r + a_t$$

where $a_r = -\dot{\theta}^2 r$ and $a_t = \ddot{\theta} r.$ (i)

In this case the magnitude of the velocity of the particle is constant so that $\ddot{\theta} = 0$. The acceleration of the particle is therefore towards the centre of the path and its magnitude

$$a = \dot{\theta}^2 r = v^2/r.$$

We now draw the *free body diagram* of the particle showing *only* the forces *acting on* the particle. Let us assume that there is a tension T in the string in the direction shown in Fig. 4.2. In this simple example we know that the string will be in tension. In most cases we do not know the direction of the forces. This need not worry us, because if the actual force is in the opposite sense from that chosen in the free body diagram, its value at the end of the calculation will be negative.

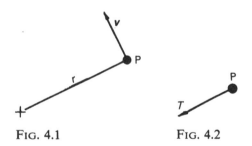

FIG. 4.1   FIG. 4.2

There are no other forces acting on the particle in the plane of the motion. (The particle will be in static equilibrium in the vertical direction with the force on the particle from the table being equal and opposite to the weight of the particle.)

Applying Newton's Second Law to the particle along the line of the string, we obtain from eqn (2.2)

$$\swarrow T = m\dot{\theta}^2 r = mv^2/r.$$ (ii)

Note that the radially inward direction has been taken as positive so that both the force and the acceleration are positive.

It is important to realise that *the only force acting on the particle in the plane of motion is the inward force T*, since only the string can apply a force to the particle in this plane.

The point on the table to which the end of the string is fixed will experience an *outward* force as shown by the three free body diagrams of Fig. 4.3. As we have applied Newton's law to the *particle*, only the forces acting *on the particle* are taken into account in eqn (ii).

FIG. 4.3

D'Alembert's principle

The use of D'Alembert's principle allows a dynamic problem to be treated as a problem in statics. In this book we will not normally use D'Alembert's approach to dynamic problems as it does not usually offer any great advantages over the direct application of Newton's Laws. Most readers will have had some limited experience of the use of D'Alembert's principle, but unfortunately seldom have a clear understanding of its application. The resulting confusion is one of the major sources of error in solving simple problems. Even though the method will not be used in this book the reader is recommended to study the following section carefully so that he can distinguish clearly between the direct application of Newton's Laws and the use of D'Alembert's principle.

Newton's Second Law, eqn (2.2), may be rewritten as

$$\boldsymbol{F} - \boldsymbol{ma} = 0. \tag{4.2}$$

If we now consider the term $(-\boldsymbol{ma})$ as an *imaginary* force (or *inertia* force) of magnitude ma, which acts in a sense *opposite* from that of the acceleration of the particle, then the particle may be considered to be in *static equilibrium* under the action of the applied force $\boldsymbol{F}$ and the imaginary *inertia* force, $(-\boldsymbol{ma})$. The sum of these forces will then be zero. This representation of Newton's Second Law is referred to as D'Alembert's principle and its effect is to convert the real problem in dynamics into an equivalent problem in statics.

Let us now reconsider example 4.1 using D'Alembert's principle. We again draw the free body diagram of the particle, but this time we have to include the imaginary inertia force $(-\boldsymbol{ma})$ which, since the accelerationof the particle is radially inwards, will be radially outwards as shown in Fig. 4.4. We now apply the equation of *static equilibrium*, i.e. $\sum F = 0$, to the particle in our free body diagram.

Hence $T - mv^2/r = 0$,

or $T = mv^2/r$ as before.

It is very important to decide at the *outset* of the solution which method is to be used. A common mistake is to include the inertia forces as real applied forces, and then to apply Newton's Laws, i.e. to use a mixture of the two methods. This will obviously give an incorrect result.

Example 4.2   A smooth cone, of half angle 30°, rotates about its vertical axis with a constant angular velocity ω. A small mass, m, which may be considered as a particle, lies on the surface of the cone and is held by a string fixed to the apex as shown in Fig. 4.5. The length of the string is 2 m, and the mass of the particle is 2 kg. We will assume that the cone and particle rotate together. Let us find the tension in the string and the force between the particle and the surface of the cone.

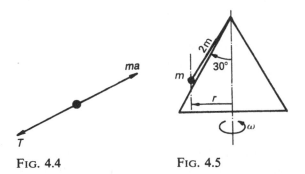

FIG. 4.4   FIG. 4.5

The particle is moving in a circular path of radius r and the angular velocity of the line from the particle to the axis of rotation of the cone is constant. Hence, from kinematic considerations, the acceleration of the particle is $\omega^2 r$ towards the centre of rotation. We will, in this case, obtain the solution by both direct application of Newton's Law and by using D'Alembert's principle. In both cases we first draw the free body diagram.

Let T be the tension in the string and R be the contact force acting on the particle. The forces acting on the particle will be its weight mg, the tension in the string, and the contact force R which, since the cone is smooth, will be perpendicular to the surface of the cone.

Newton's Law

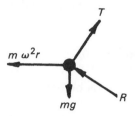

D'Alembert's Principle

FIG. 4.6

The free body diagram Fig. 4.6(a) shows applied forces only.	The free body diagram Fig. 4.6(b) shows the applied forces and the imaginary inertia force $(-ma)$.
Now apply Newton's Second Law to the particle in the horizontal and vertical directions.	Now consider the *static equilibrium* of the particle in the horizontal and vertical directions.
Thus $F = ma$,	Thus $\sum F = 0$,
or, in component form,	or, in component form,
$\rightarrow T \sin 30 - R \cos 30 = m\omega^2 r,$ <div align="right">(i)a</div>	$\rightarrow T \sin 30 - R \cos 30 - m\omega^2 r = 0,$ <div align="right">(i)b</div>
$\uparrow T \cos 30 + R \cos 60 - mg = 0.$ <div align="right">(ii)a</div>	$\uparrow T \cos 30 + R \cos 60 - mg = 0.$ <div align="right">(ii)b</div>

Equations (i) and (iia) are, of course, the same as eqns (i) and (iib). Solving the two simultaneous equations and substituting the numerical values gives

$$R = 4.93 - \frac{\sqrt{3}}{4}\omega^2,$$

and $T = 4.93\sqrt{3} + \tfrac{1}{4}\omega^2,$ (iii)

where the units of R and T are Newtons. Let us now consider how R and T vary with angular velocity. Both T and R have positive values when $\omega = 0$.

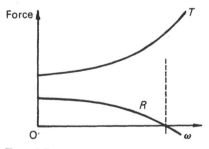

FIG. 4.7

As ω increases, T increases continuously as shown in Fig. 4.7. R decreases with angular velocity until, when $4.93 = \sqrt{3}\omega^2/4$, R becomes zero. Beyond this speed R goes negative. However, if the particle is lying on the surface and there is no adhesive force between the particle and the surface, R cannot become negative. At speeds where R is predicted to be negative the particle would, in a real case, lift off the surface so that R would always be zero.

This means that, for angular velocities, above the limiting value where R becomes zero, our mathematical model of the system is incorrect. Hence our eqns (iii) are only applicable when contact is maintained between the cone and the particle. Care must always be exercised when setting up mathematical models of systems to ensure that the models are valid for the range of variables under consideration. It may be necessary to have two, or more different models of a system to cover the whole range of interest.

As an exercise, the reader should set up the equations to determine T after contact has been lost. Note that in the above solution, r can be obtained from the geometry of the cone and R and T are the variables obtained from the equations of motion. After contact has been lost there is no normal force and r cannot be obtained from the geometry. The variables in the two equations will now be T and r, and are given by

$$T = ml\omega^2,$$

and $$r = l\left(1 - \frac{g^2}{\omega^4 l^2}\right)^{1/2}.$$

4.3 Application of Newton's Laws to rigid bodies

Newton's Laws as stated in Chapter 2 refer only to particles. In engineering we are concerned with the motion of bodies such as shafts, pistons, rotors and other machine components which effectively consist of particles held together by internal cohesive forces. It is therefore necessary to extend the application of Newton's Laws to such bodies.

All bodies are elastic to some extent and change their shape under load. In an engineering analysis we can normally consider the gross motion of a body separately from the effects of elastic deformation. Elastic motion usually appears as vibration superimposed upon the gross motion. Vibration is a very important topic in engineering but, apart from some simple examples, is beyond the scope of this text.

Let us consider a rigid body with external forces $F_1, F_2 \cdots F_i \cdots F_n$ applied to it as shown in Fig. 4.8(a). The body itself will consist of a number of elemental particles $P_1, P_2 \cdots P_i \cdots P_n$ of mass $m_1, m_2 \cdots m_i \cdots m_n$, as shown in Fig. 4.8(b), and the total mass m of the body will be equal to the sum of the masses of the individual particles so that

$$m = \sum m_i. \tag{4.3}$$

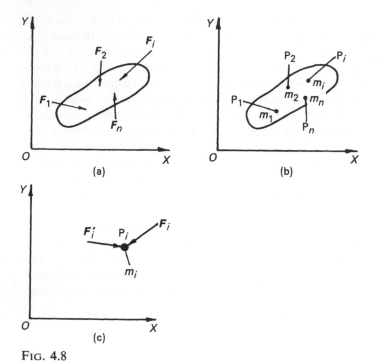

FIG. 4.8

The resultant external force F on the body will be the vector sum of the individual applied forces, i.e.

$$F = \sum F_i. \tag{4.4}$$

Internal stresses in the body will cause internal forces to act on the particles. Consider a typical particle P_i as shown in Fig. 4.9 in contact with a number of other particles. Thus particle P_j will exert a force F'_{ji} on the particle P_i, and from Newton's Third Law the particle P_i will exert an *equal* and *opposite* force F'_{ij} on particle P_j. Thus $F'_{ij} = -F'_{ji}$.

The other particles in contact with P_i will also exert forces on it. If we add together all the individual internal forces acting within the whole body we see

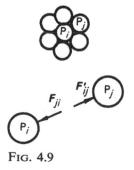

FIG. 4.9

that, since $F'_{ij} = -F'_{ji}$, they will cancel out in pairs, so that the sum of the internal forces over the whole of the body is zero.

The total internal force acting on an individual particle need not, however, be zero. Let the total internal force on our particle P_i be F'_i. Now if we add together the internal forces F'_i for all particles in the body we again have the total internal force acting on the whole body so that

$$\sum F'_i = 0. \tag{4.5}$$

For the typical particle P_i of mass m_i, with total external force F_i and total internal force F'_i, as shown in Fig. 4.8(c), the total force acting on the particle will be $F_i + F'_i$.

Applying Newton's Second Law to the particle gives,

$$F_i + F'_i = m_i a_i. \tag{4.6}$$

The directions of a_i and $(F_i + F'_i)$ will be the same.

To obtain the equations of motion for the whole body we sum eqn (4.6) for all the particles in the body, i.e.

$$\sum F_i + \sum F'_i = \sum m_i a_i.$$

Since each particle is part of a rigid body the accelerations a_i are not independent. The requirement that the distance between any two particles is constant imposes a kinematic constraint on each a_i which must be taken into account when evaluating the sum $\sum m_i a_i$.

We will follow a similar pattern to that used in the study of the kinematics of rigid bodies by starting with the case of pure translation and then going on to consider pure rotation and general plane motion. For each case we have to consider the appropriate kinematic constraint to which the particles are subjected.

Pure translation

The kinematic constraint in this case is that all the particles in the body will have the same acceleration, i.e.

$$a_i = a \tag{4.7}$$

Taking eqn (4.6), and summing the forces over the whole of the body, we obtain

$$\sum F_i + \sum F'_i = \sum m_i a_i. \tag{4.8}$$

Using eqn (4.7), a_i can be extracted from the summation, and the substitution of eqns (4.3), (4.4) and (4.5), allows eqn (4.8) to be written in terms of the total mass m of the body, and the total external force F as

$$F = ma, \tag{4.9}$$

i.e. the total external force is equal to the total mass of the body multiplied by its acceleration. In components along OX, OY, eqn (4.9) may be expressed as

$$F_x = m\ddot{x}$$

and $$F_y = m\ddot{y}.$$ (4.10)

When the body moves in pure translation no rotation is permitted and the resultant F of the external forces must act through a particular point in the body.

Let us consider the forces $F_i + F_i'$ acting on the particle P_i in the body, and take the moments about some point Q of these forces. If the coordinates of P_i and Q_i are x_i, y_i and x_Q, y_Q respectively, as shown in Fig. 4.10, then, taking anticlockwise moments to be positive,

$$M_{Qi} = (F_{yi} + F_{yi}')(x_i - x_Q) - (F_{xi} + F_{xi}')(y_i - y_Q).$$ (4.11)

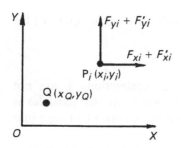

FIG. 4.10

By summing the moments of the forces on every particle in the body we obtain

$$M_Q = \sum M_{Qi} = \sum (F_{yi} + F_{yi}')(x_i - x_Q) - \sum(F_{xi} + F_{xi}')(y_i - y_Q).$$ (4.12)

Since all particles in the body have the same acceleration eqn (4.6) can be expressed in components as

$$F_{xi} + F_{xi}' = m_i\ddot{x}$$

and $$F_{yi} + F_{yi}' = m_i\ddot{y}.$$ (4.13)

Equation (4.12) can now be rewritten as

$$M_Q = \ddot{y} \sum m_i(x_i - x_Q) - \ddot{x} \sum m_i(y_i - y_Q)$$
$$= \ddot{y}(\sum m_i x_i - m x_Q) - \ddot{x}(\sum m_i y_i - m y_Q).$$ (4.14)

If x_G and y_G define the position of the centre of mass of the body,

$$\sum m_i x_i = m x_G \quad \text{and} \quad \sum m_i y_i = m y_G.$$

Substituting these expressions into eqn (4.14) gives

$$M_Q = m\ddot{y}(x_G - x_Q) - m\ddot{x}(y_G - y_Q).$$ (4.15)

Using eqn (4.10), eqn (4.15) can be rewritten in terms of the total external force as

$$M_Q = F_y(x_G - x_Q) - F_x(y_G - y_Q). \tag{4.16}$$

The right-hand side of eqn (4.16) is the moment about Q of the resultant *external* force applied to the body.

Let us now return to the right-hand side of eqn (4.12). The terms which represent the sum of the moments about Q of the internal forces will be zero, i.e.

$$\sum F'_{yi}(x_i - x_Q) = \sum F'_{xi}(y_i - y_Q) = 0,$$

since the moments of the internal forces cancel in pairs when summed.

Equation (4.12) therefore reduces to

$$M_Q = \sum F_{yi}(x_i - x_Q) - \sum F_{yi}(y_i - y_Q). \tag{4.17}$$

The right-hand side of eqn (4.17) is the sum of the moments about Q of the *external* forces acting on the body. If the resultant F of these external forces has a line of action which passes through some point in the body which has coordinates x_P and y_P, the moment M_Q may be expressed as

$$M_Q = F_y(x_P - x_Q) - F_x(y_P - y_Q), \tag{4.18}$$

where F_x, F_y are the components of F along OX and OY.

We now have two expressions, eqns (4.16) and (4.18), for M_Q. Both can be satisfied only if $x_P = x_G$ and $y_P = y_G$, i.e. only if the line of action of the resultant of the external forces passes through G.

Thus the *resultant force* on a body moving with *pure translation must* act through the *centre of mass* of the body, and hence the *moment* of the applied forces *about the centre of mass* must be *zero*.

The equations of motion for a rigid body in pure translation are therefore

$$F = ma$$

and $M_G = 0.$ \hfill (4.19)

Example 4.3   Let us consider the case of a body being moved along a horizontal straight track by the application of a horizontal force. Fig. 4.11 shows a

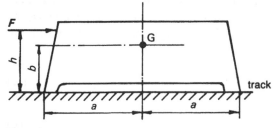

FIG. 4.11

mathematical model of such a body of mass m, in which the position of the centre of mass and the height of the point of application of the applied force F are defined. If we assume that the edges of the body always remain in contact with the track, the body will move in pure translation in the plane of the track.

The first step is to draw the free body diagram of the block, Fig. 4.12, in which it has been assumed that friction between the body and the track is negligible and that the forces R_1 and R_2 on the feet of the body are point forces at the extreme edges of the body. Since the surface is horizontal these contact forces will act in the vertical direction. R_1 and R_2 to represent the total forces at these edges. If, for example, the body is rectangular in plan and has four feet, R_1 will be the sum of two forces, as will R_2.

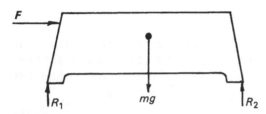

FIG. 4.12

Applying eqns (4.10) for a rigid body in pure translation, and taking the positive directions as indicated by the arrows, we obtain

$$\uparrow\ R_1 + R_2 - mg = 0, \tag{i}$$

and $$\rightarrow F = m\ddot{x}. \tag{ii}$$

Also, since the moment of the applied forces about the centre of mass must be zero,

$$\circlearrowright\quad aR_2 - aR_1 - F(h - b) = 0. \tag{iii}$$

From eqn (ii),

$$\ddot{x} = F/m,$$

so that the body will continue to accelerate as long as F is applied. If F were removed so that $F = 0$, the body would have zero acceleration and would continue to move with constant velocity. By solving eqns (i) and (iii) we obtain the contact forces,

$$R_1 = \frac{mg}{2} - \frac{F(h - b)}{2a} \tag{iv}$$

and $$R_2 = \frac{mg}{2} + \frac{F(h - b)}{2a}. \tag{v}$$

From eqns (iv) and (v) it can be seen that if F acts through the centre of mass of the body, i.e. if $h = b$, then $R_1 = R_2 = mg/2$. These forces are then independent of the magnitude of F. Also, from eqn (v), we can see that R_1 becomes negative if

$$\frac{F(h - b)}{a} > mg.$$

Thus if $(h - b) > mga/F > 0$, the rear edge of the body will lift and the assumed mathematical model will be longer be valid since, physically, R_1 cannot be negative.

Similarly, from eqn (v), R_2 will become negative if

$$\frac{F(h - b)}{a} < -\frac{mg}{2},$$

i.e. when $(b - h) > mga/F$. In this case the line of action of F must be below G. The mathematical model is again invalid when $R_2 < 0$, as shown in Fig. 4.13. The effect of friction at the contact area will, of course, influence the result. The reader should investigate this effect by drawing a revised free body diagram with frictional forces μR_1 and μR_2 in directions which oppose F.

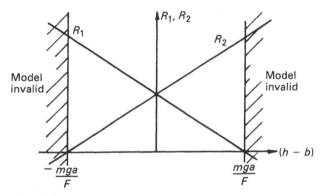

Fig. 4.13

Pure rotation (rotation about a *fixed* point)

Let us consider the body in Fig. 4.14, which has total mass m and is rotating about some fixed point O, and let us choose O to be at the origin of a set of fixed axes OXY. Again let us consider some particle P_i in the body with external and internal forces F_i and F'_i applied as before. Rewriting our equation of motion for a typical particle, eqn (4.6), in terms of components along OX, OY, we obtain

$$F_{xi} + F'_{xi} = m_i\ddot{x}_i,$$

and $F_{yi} + F'_{yi} = m_i\ddot{y}_i.$ (4.20)

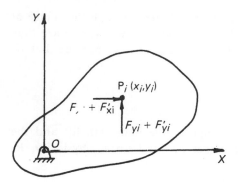

F<small>IG</small>. 4.14

When summed for all the particles in the body,

$$\sum F_{xi} = \sum m_i \ddot{x}_i,$$

and $$\sum F_{yi} = \sum m_i \ddot{y}_i,$$ (4.21)

since the internal forces cancel in pairs.

Since the position of the centre of mass of a rigid body is given by

$$mx_G = \sum m_i x_i \quad \text{and} \quad my_G = \sum m_i y_i,$$

where $m = \sum m_i$ is the total mass of the body, the right-hand sides of eqn (4.21) can be rewritten

$$\sum m_i \ddot{x}_i = m\ddot{x}_G \quad \text{and} \quad \sum m_i \ddot{y}_i = m\ddot{y}_G.$$

Substituting these expressions into eqn (4.21), and letting $\sum F_{xi} = F_x$ and $\sum F_{yi} = F_y$, we obtain

$$F_x = m\ddot{x}_G$$

and

$$F_y = m\ddot{y}_G.$$ (4.22)

Combining the two eqns (4.22) gives

$$\mathbf{F} = m\mathbf{a}_G.$$ (4.23)

Thus the resultant force on the body is equal to the mass of the body multiplied by the *acceleration of its centre of mass*, and is in the same direction as this acceleration. The force $\mathbf{F}$ *must* include the force applied to the body by the pivot at the point O.

Let us now consider the moments, about the fixed point O, of the forces acting on the particle P_i. If M_{Oi} is the moment about O of the forces on P_i, then by taking the anti-clockwise direction as positive, we obtain

$$M_{Oi} = (F_{yi} + F'_{yi})x_i - (F_{xi} + F'_{xi})y_i.$$ (4.24) .

Summing the values of M_{Oi} for each particle in the body, and noting that the moments of the internal forces about O will cancel in pairs, we obtain

$$M_O = \sum M_{Oi} = \sum (F_{yi}x_i - F_{xi}y_i),$$

so that M_O is the sum of the moments about O of the external forces acting on the body.

If we now return to eqn (4.24) and substitute for $(F_{xi} + F'_{xi})$ and $(F_{yi} + F'_{yi})$ from eqn (4.20),

$$M_O = \sum (m_i\ddot{y}_i x_i - m_i\ddot{x}_i y_i). \tag{4.25}$$

In this case the kinematic constraints of pure rotation apply. All the particles have different accelerations because the body is rotating about a fixed point. However these accelerations can all be expressed in terms of the *angular* velocity and *angular* acceleration of the body.

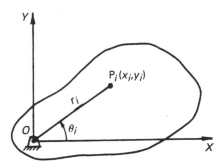

FIG. 4.15

Expressing the cartesian coordinates of P_i in terms of the polar coordinates (r_i, θ_i), we obtain, from Fig. 4.15

$$x_i = r_i \cos \theta_i,$$

and

$$y_i = r_i \sin \theta_i. \tag{4.26}$$

Differentiating eqns (4.26) twice with respect to time and noting that, for each particle, r_i is constant, we obtain

$$\ddot{x}_i = -r_i\dot{\theta}_i^2 \cos \theta_i - r_i\ddot{\theta}_i \sin \theta_i,$$

and $\quad \ddot{y}_i = -r_i\dot{\theta}_i^2 \sin \theta_i + r_i\ddot{\theta}_i \cos \theta_i. \tag{4.27}$

Substituting eqns (4.26) and (4.27) into eqn (4.25) gives

$$M_O = \sum m_i\{(-r_i\dot{\theta}_i^2 \sin \theta_i + r_i\ddot{\theta}_i \cos \theta_i)r_i \cos \theta_i$$
$$+ (r_i\dot{\theta}_i^2 \cos \theta_i + r_i\ddot{\theta}_i \sin \theta_i)r_i \sin \theta_i\}$$
$$= \sum m_i r_i^2 \ddot{\theta}_i.$$

Now $\ddot{\theta}_i$ is the angular acceleration of the line joining the particle P_i to the fixed point O. Since the body is rigid, the angular acceleration of all lines in the body must be the same, so that $\ddot{\theta}_i = \ddot{\theta}$ and can be taken outside the summation giving

$$M_O = \ddot{\theta} \sum m_i r_i^2.$$ (4.28)

The quantity $\sum m_i r_i^2$ is called the *moment of inertia* of the body about an axis which passes through O and is perpendicular to the plane of the motion. Letting $\sum m_i r_i^2 = I_O$, eqn (4.28) may be rewritten in its usual form,

$$M_O = I_O \ddot{\theta}.$$ (4.29)

Thus, for a body rotating about a *fixed point*, the moment of the applied forces about that point is equal to the moment of inertia of the body about an axis through the fixed point and perpendicular to the plane of motion, multiplied by the *angular* acceleration of the body.

The units of M_O will be Nm, from its definition I_O will have units of kg m^2, and the units of $\ddot{\theta}$ are rad s^{-2}. Substitution of these units into eqn (4.29) shows that it is dimensionally correct (remember that a Newton is a derived unit and can be expressed in terms of the three basic units).

Moments of inertia will be considered in more detail later. The equations of motion for a rigid body rotating about a fixed point are therefore

$$F = ma_G$$

and $$M_O = I_O \ddot{\theta}.$$ (4.30)

Note that the centre of mass G moves on a circular path so that a_G can be expressed in terms of $\dot{\theta}$ and $\ddot{\theta}$.

Example 4.4   Consider the uniform link of mass m and length $2l$, shown in Fig. 4.16, which can rotate about a fixed horizontal axis through O. Let us release the link from rest in the vertical position, i.e. with $\theta = 0$, and obtain expressions for the velocity and acceleration of point P at the end of the link, and the forces at the pivot, when $\theta = 90°$.

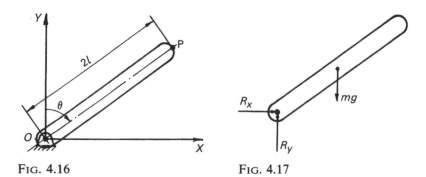

FIG. 4.16   FIG. 4.17

We first draw a free body diagram of the link in some general position, as in Fig. 4.17, showing all forces acting *on* the link. The only forces acting on the link are its weight mg and the force at the pivot which we will represent by its components R_x and R_y. We can now apply eqns (4.22) and (4.29) for a body rotating about a *fixed* point. If the link has a moment of inertia I_O about O,

$$\rightarrow R_x = m\ddot{x}_G, \tag{i}$$

$$\uparrow R_y - mg = m\ddot{y}_G, \tag{ii}$$

and $\quad \circlearrowleft mg\, l \sin \theta = I_O \ddot{\theta}. \tag{iii}$

From eqn (iii)

$$\ddot{\theta} = \frac{mgl}{I_O} \sin \theta. \tag{iv}$$

To obtain the angular velocity of the link we write in eqn (iv) $\ddot{\theta} = d\dot{\theta}/dt$, so that

$$\frac{d\dot{\theta}}{dt} = \frac{mgl}{I_O} \sin \theta. \tag{v}$$

Equation (v) cannot be integrated directly since $\sin \theta$ is not an explicit function of time. We therefore have to change the variable from t to θ by writing

$$\frac{d\dot{\theta}}{dt} = \frac{d\dot{\theta}}{d\theta}\frac{d\theta}{dt} = \dot{\theta}\frac{d\dot{\theta}}{d\theta}.$$

Substituting this change of variable into eqn (v), and integrating gives

$$\int \dot{\theta}\, d\dot{\theta} = \frac{mgl}{I_O} \int \sin \theta\, d\theta + A,$$

so that $\quad \dfrac{\dot{\theta}^2}{2} = -\dfrac{mgl}{I_O} \cos \theta + A. \tag{vi}$

Equations (iv) and (vi) are valid for all values of θ. In our particular case the constant of integration A is obtained by using the condition that at $\theta = 0$, $\dot{\theta} = 0$. (Under these conditions, if exact, the link would remain standing upright in an unstable equilibrium. We are effectively assuming an infinitesimally small positive value of θ.)

Substituting these initial values into eqn (vi) gives

$$A = \frac{mgl}{I_O},$$

so that $\quad \dot{\theta}^2 = \dfrac{2mgl}{I_O} (1 - \cos \theta). \tag{vii}$

At $\theta = 90°$,

$$\dot{\theta}^2 = \frac{2mgl}{I_O},$$

and from eqn (iv),

$$\ddot{\theta} = \frac{mgl}{I_O}.$$

We can now consider the motion of the point P. Knowing $\ddot{\theta}$ and $\dot{\theta}$ this is purely a kinematic problem (see kinematics of a rigid body rotating about a fixed point (p. 24).

From eqn (1.28)

$$\boldsymbol{a}_p = \boldsymbol{a}_r + \boldsymbol{a}_t,$$

and $a_r = 2l\dot{\theta}^2,$ $a_t = 2l\ddot{\theta},$

as shown in Fig. 4.18.

FIG. 4.18

Thus the magnitude of a_p is given by

$$a_p = 2l\sqrt{\dot{\theta}^4 + \ddot{\theta}^2} = 2\sqrt{5}ml^2g/I_O.$$

The forces R_x and R_y can now be found from eqns (i) and (ii), noting that $\ddot{x}_G$, $\ddot{y}_G$ are kinematically related to $\dot{\theta}$, $\ddot{\theta}$.

At $\theta = 90°$,

$$\ddot{x}_G = -l\dot{\theta}^2,$$

and $\ddot{y}_G = -l\ddot{\theta}.$

Hence, from eqns (i) and (ii)

$$R_x = -\frac{2m^2l^2g}{I_O}$$

and $R_y = mg\left(1 - \dfrac{ml^2}{I_O}\right).$

The negative sign of R_x shows that, at $\theta = 90°$, the direction of the force is opposite to that shown in the free body diagram. The direction of R_y depends upon the value of ml^2/I_O.

It is important to note how the solution of the dynamic equations depends upon the application of the kinematic relationships for the system. In solving dynamic problems it is first necessary to write the dynamic equations using the free body diagrams of the bodies in the system, and then to utilise any kinematic relationships obtained from the geometry of the system.

Example 4.5   Let us again consider the body in Fig. 4.16 but now we will introduce a fixed stop, as shown in Fig. 4.19, to bring the body to rest when it has reached the horizontal position with $\theta = 90°$. The effects of the position of the stop on the bearing forces at O will now be investigated.

When the link strikes the stop there will be an impact causing an unknown contact force F between the stop and the link. The magnitude of F will increase from zero at the start of the impact to a maximum value and will then decrease to zero as contact is lost. The free body diagram for the link *during the period of contact* with the stop will therefore be as shown in Fig. 4.20. We can now write the equations of motion for the link as

$$\rightarrow R_x = m\ddot{x}_G, \tag{i}$$

$$\uparrow R_y + F - mg = m\ddot{y}_G, \tag{ii}$$

and $\quad \circlearrowleft mgl - Fh = I_O\ddot{\theta}. \tag{iii}$

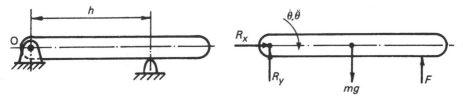

FIG. 4.19   FIG. 4.20

If we assume that the contact is very stiff so that θ remains at 90° during the impact, these equations will apply throughout the contact period. With this assumption, and the with directions of $\dot{\theta}$ and $\ddot{\theta}$ as shown, $\ddot{x}_G = -l\dot{\theta}^2$ and $\ddot{y}_G = -l\ddot{\theta}$ at $\theta = 90°$.

Using these expressions in eqns (i), (ii) and (iii) gives

$$R_x = -ml\dot{\theta}^2, \tag{iv}$$

and $\quad R_y = mg\left(1 - \dfrac{ml^2}{I_O}\right) + F\left(\dfrac{mlh}{I_O} - 1\right). \tag{v}$

The value of R_x will therefore decrease from the previous value in Example 4.4 to zero as the angular velocity reduces to zero. R_y is influenced directly by the magnitude of F which cannot be calculated from the information given since the magnitude of F will depend upon the stiffness of the contact and will vary throughout the impact. For example a rubber stop would bring the link to rest over a longer period of time than would a steel stop, and the maximum force would therefore be correspondingly lower.

However, from eqn (v) we can see that no matter what the magnitude of F may be, the value of R_y is unaffected if

$$\frac{mlh}{I_O} = 1, \quad \text{i.e. if } h = \frac{I_O}{ml}.$$

Thus if the stop is positioned such that its distance from the fixed pivot is equal to the moment of inertia of the body about O divided by the product of the mass of the body and the distance of its centre of mass from O, the force at the pivot will be unaffected by the impact. The point on the body for which this condition is satisfied is called the *centre of percussion* of the body.

If we consider a stationary pivoted link and wish to give it a sharp blow to set it in motion, the free body diagram of Fig. 4.20, and eqns (i), (ii) and (iii), will still apply. Thus, if the blow is applied at the centre of percussion, the forces at the pivot will be unaffected by the severity of the blow. It must be remembered that this is valid only if θ does not change significantly during the time over which the force is applied. One application of this result is in impact testing machines where a swinging link is used to break a specimen of a material, as shown diagramatically in Fig. 4.21. The specimen is firmly fixed and is struck and broken by the swinging pendulum. The angular displacement of the pendulum *during the impact* is very small and the force is applied perpendicular to a line passing through the pivot and the centre of mass of the pendulum.

Thus, if the point on the pendulum at which the impact occurs is at its centre of percussion, the force at its pivot will not be affected by the impact.

General plane motion

Consider the body in Fig. 4.22 to be moving with general plane motion, i.e. with combined translational and rotational motion. Using eqn (4.6) in its

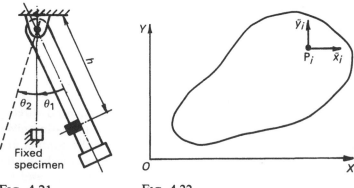

FIG. 4.21   FIG. 4.22

component form, we have, for a typical particle P_i,

$$F_{xi} + F'_{xi} = m_i\ddot{x}_i,$$

and $$F_{yi} + F'_{yi} = m_i\ddot{y}_i,$$ (4.31)

where F_{xi}, F_{yi} and F'_{xi} and F'_{yi} are the external and internal forces acting on the particle.

Summing eqn (4.31) for the whole of the body gives

$$\sum F_{xi} = \sum m_i\ddot{x}_i$$

and $$\sum F_{yi} = \sum m_i\ddot{y}_i,$$ (4.32)

since $\sum F'_{xi}$ and $\sum F'_{yi} = 0$.

Again, since the position of the centre of mass G of the body is defined by $mx_G = \sum m_i x_i$ and $my_G = \sum m_i y_i$ eqns (4.32) reduce to

$$F_x = m\ddot{x}_G$$

and $$F_y = m\ddot{y}_G,$$ (4.33)

where m is the total mass of the body and F_x, F_y are the components of the total externally applied force along OX, OY.

Combining eqns (4.33) gives the vector equation

$$\boldsymbol{F} = m\boldsymbol{a}_G$$ (4.34)

which is identical to eqn (4.23) for a rigid body rotating about a fixed point. We should note that this expression also applies to a body in pure translation (eqn 4.19) where all particles have the same acceleration, i.e. $\boldsymbol{a} \equiv \boldsymbol{a}_G$.

Let us now consider the moments, about some *moving point Q fixed in the body*, of the forces acting on the particle P_i, and let the distance QP_i be r_i, as shown in Fig. 4.23. If the coordinates of P_i and Q are x_i, y_i and x_Q, y_Q, respectively, then the moment M_{Qi} about Q of the forces on P_i, taking anticlockwise positive, is

$$M_{Qi} = (F_{yi} + F'_{yi})(x_i - x_Q) - (F_{xi} + F'_{xi})(y_i - y_Q).$$ (4.35)

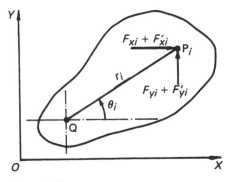

FIG. 4.23

Substituting eqns (4.31) into eqn (4.35)

$$M_{Qi} = m_i \ddot{y}_i (x_i - x_Q) - m_i \ddot{x}_i (y_i - y_Q). \tag{4.36}$$

In eqn (4.36) the coordinates x_i, y_i can be expressed in terms of x_Q, y_Q and θ_i, the angle between OX and QP_i. From Fig. 4.23

$$x_i = x_Q + r_i \cos \theta_i, \qquad y_i = y_Q + r_i \sin \theta_i.$$

Differentiating these equations with respect to time gives

$$\dot{x}_i = \dot{x}_Q - r_i \dot{\theta}_i \sin \theta_i, \qquad \dot{y}_i = \dot{y}_Q + r_i \dot{\theta}_i \cos \theta_i,$$

and

$$\ddot{x}_i = \ddot{x}_Q - r_i \ddot{\theta}_i \sin \theta_i - r_i \dot{\theta}_i^2 \cos \theta_i, \qquad \ddot{y}_i = \ddot{y}_Q + r_i \ddot{\theta}_i \cos \theta_i - r_i \dot{\theta}_i^2 \sin \theta_i.$$

Substituting these expressions into eqn (4.36) and summing M_{Qi} for all the particles in the body we obtain

$$\begin{aligned}
M_Q = \sum M_{Qi} &= \sum \{m_i (\ddot{y}_Q + r_i \ddot{\theta}_i \cos \theta_i - r_i \dot{\theta}_i^2 \sin \theta_1) r_i \cos \theta_i \\
&\quad - m_i (\ddot{x}_Q - r_i \ddot{\theta}_i \sin \theta_i - r_i \dot{\theta}_i^2 \cos \theta_i) r_i \sin \theta_i\} \\
&= \sum \{m_i \ddot{y}_Q r_i \cos \theta_i - m_i \ddot{x}_Q r_i \sin \theta_i\} \\
&\quad + \sum m_i r_i^2 \ddot{\theta}_i.
\end{aligned} \tag{4.37}$$

The angular acceleration of all lines in the body will be the same since the body is assumed rigid, so that $\ddot{\theta}_i = \ddot{\theta}$ and can be removed from the summation. Also $\sum m_i r_i^2 = I_Q$, the moment of inertia of the body about Q.

Equation (4.37) may now be expressed as

$$M_Q = I_Q \ddot{\theta} + \ddot{y}_Q \sum m_i (x_i - x_Q) - \ddot{x}_Q \sum m_i (y_i - y_Q). \tag{4.38}$$

Since $\sum m_i$ is the total mass m of the body and $m x_G = \sum m_i x_i$ and $m y_G = \sum m_i y_i$ define the position of the centre of mass G of the body eqn (4.38) can be written

$$M_Q = I_Q \ddot{\theta} + m \ddot{y}_Q (x_G - x_Q) - m \ddot{x}_Q (y_G - y_Q). \tag{4.39}$$

This equation can be greatly simplified by choosing the point Q to be coincident with the centre of mass. In this case $x_Q = x_G$ and $y_Q = y_G$ so that eqn (4.39) becomes

$$M_G = I_G \ddot{\theta}. \tag{4.40}$$

If we now reconsider eqn (4.35) we can also write

$$M_G = \sum M_{Gi} = \sum [(F_{yi} + F'_{yi})(x_i - x_G) - (F_{xi} + F'_{xi})(y_i - y_G)]. \tag{4.41}$$

Since the moments of the internal forces cancel in pairs when summed

$$\sum F'_{yi}(x_i - x_G) = \sum F'_{xi}(y_i - y_G) = 0.$$

Hence

$$M_G = \sum F_{yi}(x_i - x_G) - \sum F_{xi}(y_i - y_G) \tag{4.42}$$

and is therefore the sum of the moments of the *external* forces about G.

Equation (4.40) is applicable to any rigid body, i.e. the total moment of the external forces *about the centre of mass* of a body is equal to the moment of inertia of the body about an axis which passes through *the centre of mass* and is perpendicular to the plane of the motion, multiplied by the angular acceleration of the body.

The equations of motion of a rigid body moving with general plane motion are thus given by

$$F = ma_G$$

and $M_G = I_G\ddot{\theta}.$ (4.43)

Example 4.6   Let us consider the motion of a wheel rolling without slip along a horizontal surface under the action of a horizontal force P applied at the centre of the wheel, as shown in Fig. 4.24. The mass of the wheel is m, its moment of inertia about an axis through its centre of mass G is I_G, and we will assume that its geometric centre coincides with its centre of mass.

The forces acting on the wheel in addition to P are its weight and the contact force from the surface. The contact force will be represented by the horizontal and vertical components shown on the free body diagram of Fig. 4.25.

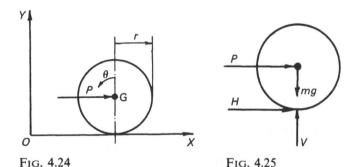

FIG. 4.24   FIG. 4.25

Using the free body diagram and eqns (4.33) and (4.40) for a body in general plane motion, we obtain

$$\rightarrow P + H = m\ddot{x}_G, \tag{i}$$

$$\uparrow V - mg = 0, \tag{ii}$$

and $\circlearrowleft Hr = I_G\ddot{\theta},$ (iii)

where θ is positive in the anticlockwise direction. (Note that if the contact were frictionless so that $H = 0$, the angular acceleration would be zero and the wheel would be in pure translation with zero angular velocity, or in general plane motion with constant angular velocity.)

Since there is no slip at the point of contact x_G and θ are not independent coordinates. Hence, from the kinematics of a rolling wheel (p. 33)

$$\dot{x}_G = -r\dot{\theta},$$

and $\ddot{x}_G = -r\ddot{\theta}.$ (iv)

From eqns (iii) and (iv),

$$H = -\frac{I_G}{r^2}\ddot{x}_G,$$ (v)

so that $\ddot{x}_G = \dfrac{P}{\left(m + \dfrac{I_G}{r^2}\right)}.$ (vi)

We see that H is negative for a positive value of $\ddot{x}_G$, and is therefore in the opposite direction from that shown in Fig. 4.25.

Equation (vi) is only valid when eqns (iv) are valid, i.e. when there is no slip at the point of contact. If the coefficient of friction between the wheel and surface is μ then the maximum possible value of H for positive $\ddot{x}_G$ is

$$H_{max} = -\mu V = -\mu mg.$$

Substituting into eqn (v), the maximum acceleration possible without slip occurring will be

$$\ddot{x}_{Gmax} = \frac{\mu mgr^2}{I_G}.$$

This corresponds to a maximum allowable value of P,

$$P_{max} = \mu mg \left(1 + \frac{mr^2}{I_G}\right).$$

At values of $P > P_{max}$ slip will occur and the kinematic eqns (iv) will be invalid. θ and x_G would then become independent coordinates and would have to be obtained from eqns (i), (ii) and (iii) by assuming that the horizontal force was some limiting frictional force related to V by the coefficient of friction.

4.4 Moment of inertia

The property of a body which we call its moment of inertia arises naturally when deriving the dynamic equations for a rigid body. It appears in eqn (4.28) and eqn (4.37) as $\sum m_i r_i^2$. When considering plane motion of rigid bodies we are concerned only with moments of inertia about axes which are perpendicular to the plane of the motion. It is convenient therefore, when considering motion in the OXY plane, to introduce a Z-axis perpendicular to the OXY plane

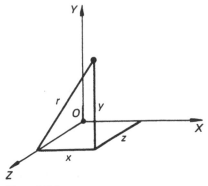

FIG. 4.26

and to consider moments of inertia I_z about this Z-axis. Thus, the moment of inertia I_z of the single particle of mass m in Fig. 4.26 about the axis OZ which is perpendicular to OXY is

$$I_z = mr^2 = m(x^2 + y^2) \tag{4.44}$$

where r is the perpendicular distance of the particle from OZ. This result is independent of the z co-ordinate of the particle.

If we now consider a system of particles, or a rigid body, the moment of inertia about OZ is given by

$$I_z = \sum m_i r_i^2 = \sum m_i(x_i^2 + y_i^2),$$

where m_i and r_i define the mass of a typical particle and its perpendicular distance from OZ.

The mass of a body tells us how much matter there is in the body and its moment of inertia about some axis tells us how the matter is *distributed* about that axis. If all the mass is situated near the axis, the r_i terms will be small and the moment of inertia will be small, but if the same mass is distributed well away from the axis, the moment of inertia will be large.

The moment of inertia of a body is sometimes expressed by considering the total mass m of the body to be concentrated as a particle at some point distant k from the axis OZ such that

$$I_z = \sum m_i r_i^2 = mk^2. \tag{4.45}$$

k is called the *radius of gyration* of the body about the axis OZ. The mass properties of a body may therefore be expressed in terms of m and I, or m and k^*.

* Moment of inertia is sometimes confused with the second moment of area of a cross-section about an axis. Second moment of area is defined as $\sum A_i r_i^2$, where A_i is the area of an infinitesimal part of the cross-section, and r_i is the distance of the infinitesimal area from the axis. This is a *geometric* property of a section and tells us how the *area* of the section is distributed about a particular axis. The second moment of area is unfortunately sometimes referred to as the moment of inertia of the section. This can cause confusion and it is important to distinguish clearly between the *mass* property of a *body*, $\sum m_i r_i^2$, and the *geometric* property of a *section*, $\sum A_i r_i^2$.

When evaluating the moments of inertia of typical bodies it is convenient to have available the results of the following two theorems.

Perpendicular axes theorem

Consider a *thin* flat plate (or lamina) in the plane OXY, as shown in Fig. 4.27, and let a typical particle P_i of mass m_i be located a distance r_i from O. By definition, the moment of inertia of the particle about OZ is

$$I_{iz} = m_i r_i^2,$$

and the moment of inertia of the whole plate about OZ is obtained by summing this equation for all the particles in the plate, i.e.

$$I_z = \sum I_{iz} = \sum m_i r_i^2. \tag{4.46}$$

Now let us consider the moment of inertia of the plate about the axis OX. If the plate is *thin* the distance of all particles P_i from OX is y_i. The moment of inertia of P_i about OX is therefore

$$I_{ix} = m_i y_i^2,$$

so that $I_x = \sum m_i y_i^2.$ \hspace{1em} (4.47)

Similarly, the moment of inertia of the plate about the axis OY is

$$I_Y = \sum m_i x_i^2. \tag{4.48}$$

Now if the plate is *flat and thin*

$$r_i^2 = x_i^2 + y_i^2. \tag{4.49}$$

If the plate is *not* assumed thin, some particles would be outside the plane OXY and eqns (4.47) and (4.48) would not be true for particles such as that at P' in Fig. 4.28.

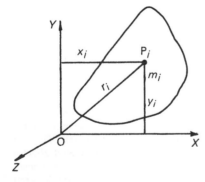

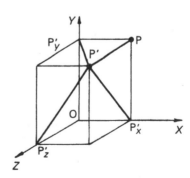

FIG. 4.27 \hspace{6em} FIG. 4.28

Multiplying eqn (4.49) by m_i and summing over the whole of the plate gives

$$\sum m_i r_i^2 = \sum m_i x_i^2 + \sum m_i y_i^2,$$

or $\qquad I_Z = I_Y + I_X.$ \hfill (4.50)

This most useful result shows that, for a *thin flat* plate, the sum of the moments of inertia about any two perpendicular axes in the plane of the plate which intersect at some point O, is equal to the moment of inertia about an axis perpendicular to the plate which also passes through O.

Parallel axes theorem

Consider a particle P_i of the rigid body shown in Fig. 4.29. Let the particle have coordinates x_i, y_i, z_i as shown. Its moment of inertia about the axis OZ is given by

$$I_{iz} = m_i(x_i^2 + y_i^2).$$

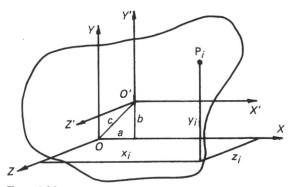

FIG. 4.29

The moment of inertia about OZ of the whole body is therefore

$$I_Z = \sum m_i(x_i^2 + y_i^2). \hfill (4.51)$$

Now let us consider a set of parallel axes $O'X'Y'Z'$ with the plane $O'X'Y'$ coincident with the plane OXY such that O' has coordinates a, b as shown in Fig. 4.29.

The moment of inertia of P_i about $O'Z'$ is given by

$$I_{iz'} = m_i(x_i'^2 + y_i'^2)$$

$$= [m_i(x_i - a)^2 + (y_i - b)^2].$$

Summing each $I_{iz'}$ to give the moment of inertia of the whole body about $O'Z'$ gives

$$I_{z'} = \sum m_i[(x_i - a)^2 + (y_i - b)^2]$$

$$= \sum m_i(x_i^2 + y_i^2) + (a^2 + b^2)\sum m_i - 2a\sum m_i x_i - 2b\sum m_i y_i. \hfill (4.52)$$

Let us consider each term in eqn (4.52) in turn.

From eqn (4.51),

$$\sum m_i(x_i^2 + y_i^2) = I_z.$$

Since $\sum m_i$ is the total mass, m, of the body, and letting the distance $OO' = c$, the term

$$(a^2 + b^2) \sum m_i = mc^2.$$

From the definition of the position of the centre of mass G of the body

$$\sum m_i x_i = m x_G \quad \text{and} \quad \sum m_i y_i = m y_G.$$

Equation (4.52) may therefore be expressed as

$$I_{z'} = I_z + mc^2 - 2m(ax_G + by_G). \tag{4.53}$$

This is a general statement of the parallel axes theorem, but if we choose OZ to pass through the centre of mass of the body, $x_G = y_G = 0$, and $I_z = I_G$. Substituting these values into eqn (4.53) gives, for this special case only,

$$I_{z'} = I_G + mc^2. \tag{4.54}$$

Equation (4.54) is the most useful form of the theorem and says that the moment of inertia of a body about any axis is equal to the moment of inertia of the body about a parallel axis *through the centre of mass* plus the total mass of the body multiplied by the square of the distance separating the axes.

Note that this refers to *any* body and is not restricted to a lamina or thin plate as in the case of the perpendicular axes theorem.

From eqn (4.54) it can be seen that $I_{z'}$ is *always* greater than I_G.

We will now use these theorems to determine the moments of inertia of typical bodies.

Example 4.7. Thin ring   A uniform thin ring (radial thickness ≪ radius) of radius r is shown in Fig. 4.30. Let us find its moment of inertia about its polar axis OZ, and about its diametral axes OX, OY.

Considering first the polar axis OZ, which is perpendicular to the plane of the ring,

$$I_Z = \sum m_i r_i^2,$$

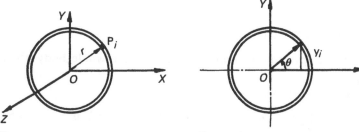

FIG. 4.30   FIG. 4.31

where m_i is the mass of a typical particle P_i in the ring. For all particles in the ring $r_i = r$ and $\sum m_i = m$, the total mass of the ring.

Hence $I_Z = mr^2$.   (i)

From considerations of symmetry, $I_X = I_Y$. Since axes OX, OY and OZ intersect at O and are perpendicular, and OX, OY are in the plane of the ring, we can apply the perpendicular axes theorem, i.e.

$$I_Z = I_X + I_Y = 2I_X.$$

Thus $I_X = I_Y = \dfrac{mr^2}{2},$ (ii)

so that the moment of inertia of a thin ring about its polar axis is equal to twice that about a diametral axis. In terms of the radius of gyration of the body we can write $I_X = mk_X^2$, so that $k_X^2 = r^2/2$.

As an alternative it is possible to determine I_X from first principles without using the perpendicular axes theorem. From Fig. 4.31

$$I_X = \sum m_i y_i^2.$$

If we consider our particle P_i as an elemental length δl of the ring and let ρ be the mass/unit length of the ring, the moment of inertia of P_i about OX is

$$\delta I_X = \rho \, \delta l (r \sin \theta)^2 = \rho r \, \delta\theta (r \sin \theta)^2$$

where the length δl subtends an angle $\delta\theta$ at the centre of the ring. Therefore, in the limit as $\delta\theta \to \mathrm{d}\theta$

$$\mathrm{d}I_X = \rho r^3 \sin^2 \theta \, \mathrm{d}\theta.$$ (iii)

If we now sum for the whole ring, i.e. integrate eqn (iii) from $\theta = 0$ to 2π, we obtain

$$I_X = \rho r^3 \int_0^{2\pi} \sin^2 \theta \, \mathrm{d}\theta = \rho \frac{r^3}{2} \int_0^{2\pi} (1 + \cos 2\theta) \, \mathrm{d}\theta$$

$$= \pi \rho r^3.$$

The total mass of the ring, $m = 2\pi r \rho$, so that

$$I_X = \frac{mr^2}{2}.$$

The use of the perpendicular axes theorem can be seen to provide a much more convenient solution than the direct approach from first principles.

Example 4.8.   Thin disc   Figure 4.32 shows a thin disc. Let us find the moments of inertia of the disc about the axes OX, OY and OZ. It is convenient to work in terms of the mass/unit area, ρ, of the disc.

Let us consider an elemental ring of radius r and radial thickness δr. From example 4.7, eqn (i), the moment of inertia δI_Z of this ring about OZ is given

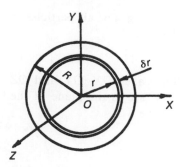

FIG. 4.32

by

$$\delta I_Z = (2\pi r \rho \, \delta r) r^2,$$

so that $\quad dI_Z = 2\pi \rho r^3 \, dr.$

Integrating this expression over the whole of the disc from $r = 0$ to R gives

$$I_Z = 2\pi\rho \int_0^R r^3 \, dr = 2\pi\rho \left[\frac{r^4}{4} \right]_0^R$$

$$= \frac{\pi\rho R^4}{2}.$$

The total mass of the disc $m = \pi R^2 \rho$, so that

$$I_Z = \frac{mR^2}{2}. \tag{i}$$

As would be expected, this value is smaller than that of a thin ring of equal mass and radius where all of the mass is distributed at the maximum radius.

From symmetry $I_X = I_Y$, and using the perpendicular axes theorem,

$$I_X + I_Y = \frac{mR^2}{2},$$

so that

$$I_X = I_Y = \frac{mR^2}{4}. \tag{ii}$$

We should note that in determining I_Z we have not had to assume that the disc is thin. The result for I_Z would therefore also apply to a long cylinder, i.e. to a series of discs with total mass m (in example 4.6 the result for I_Z would also apply to a long thin cylinder). It is only when using the perpendicular axes theorem to obtain I_X and I_Y that we are limited to thin, flat bodies.

Example 4.9. Thin rod Let us consider the uniform thin rod of length L and total mass m shown in Fig. 4.33. If we take an element of the rod of length

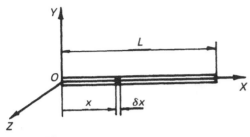

FIG. 4.33

δx which has mass δm, then because the rod is assumed thin, all particles in this element will be equidistant from OZ. Hence the moment of inertia δI_z of the element about OZ, is given by

$$\delta I_z = x^2 \, \delta m = x^2 \frac{m}{L} \, \delta x,$$

so that $dI_z = \frac{m}{L} x^2 \, dx.$

For the whole rod

$$I_z = \frac{m}{L} \int_0^L x^2 \, dx = \frac{mL^2}{3}. \tag{i}$$

In terms of the radius of gyration, $I_z = mk^2$ where $k^2 = L^2/3$.

Knowing this result we can now find the moment of inertia of the rod about an axis parallel to OZ and passing through its mid-point. Since the centre of mass of the rod lies at the mid-point, we can use the parallel axes theorem, eqn (4.54), to give

$$I_G = I_z - mc^2$$

$$= \frac{mL^2}{3} - m\left(\frac{L}{2}\right)^2 = \frac{mL^2}{12}. \tag{ii}$$

This is smaller than I_z since the mass is distributed nearer to the axis through G than to the axis through O.

These results would also apply to a thin plate of total mass m where OZ lies along one edge of the plate.

Example 4.10.   Cylinder   Consider the uniform cylinder of mass m, length L, and radius R shown in Fig. 4.34. Let us find its moment of inertia about the axes OX, OY, OZ, where O is located at the geometric centre of the cylinder. About the axis OZ,

$$I_z = \frac{mR^2}{2}$$

as shown in example 4.8.

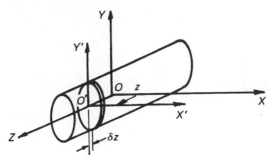

FIG. 4.34

In this case we cannot apply the perpendicular axis theorem which is for flat, thin bodies, to the whole cylinder to obtain I_X and I_Y. We can, however, apply the theorem to an elemental disc of thickness δz.

For the elemental disc shown in Fig. 4.34, the results of example 4.8 give

$$\delta I_Z = \frac{\pi R^2 \rho\, \delta z\, R^2}{2}, \qquad\qquad (\text{i})$$

where ρ is the density of the material of the cylinder. Using the perpendicular axes theorem, and noting that from symmetry $\delta I_{X'} = \delta I_{Y'}$,

$$\delta I_{X'} = \frac{\pi R^4 \rho\, \delta z}{4}. \qquad\qquad (\text{ii})$$

The axis $O'X'$ in Fig. 4.34 passes through the centre of mass of the elemental disc and so, using the parallel axes theorem, eqn (4.54), we can write

$$\delta I_X = \frac{\pi R^4 \rho\, \delta z}{4} + \pi R^2 \rho\, \delta z\, z^2. \qquad\qquad (\text{iii})$$

In the limit when $\delta z \to dz$,

$$dI_X = \pi R^2 \rho \left(\frac{R^2}{4} + z^2 \right) dz. \qquad\qquad (\text{iv})$$

Integrating eqn (iv) over the whole length of the cylinder we obtain

$$I_X = \pi R^2 \rho \int_{-L/2}^{L/2} \left(\frac{R^2}{4} + z^2 \right) dz$$

$$= \pi R^2 \rho \left[\frac{R^2 z}{4} + \frac{z^3}{3} \right]_{-L/2}^{L/2}$$

$$= \pi R^2 \rho L \left(\frac{L^2}{12} + \frac{R^2}{4} \right)$$

$$= m \left(\frac{R^2}{4} + \frac{L^2}{12} \right).$$

By expressing I_X in terms of the radius of gyration, i.e. $I_X = mk^2$, eqn (v) gives the radius of gyration of the cylinder about the axis OX as

$$k^2 = \frac{R^2}{4} + \frac{L^2}{12}.$$

We should note that if the cylinder were of very small diameter so that $R \ll L$, then

$$I_X = \frac{mL^2}{12},$$

which agrees with the result of example 4.9 for the thin rod.

Example 4.11. Complex shapes   Complex shapes can be considered as the sums or differences of simple geometric shapes. Consider for example the thin rectangular plate with a circular hole cut in it as shown in Fig. 4.35. Let us calculate its moment of inertia about the axes OX, OY, OZ as shown.

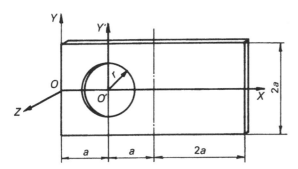

FIG. 4.35

In this case it is convenient to consider first the moment of inertia of a complete rectangle about an axis, and then to subtract from it the moment of inertia of a disc of radius r about that axis.

Consider the moment of inertia I_{RY} of the rectangle about OY. The rectangular plate is equivalent to a series of the thin rods of length $4a$ and, from example 4.9, its moment of inertia about OY will be

$$I_{RY} = \frac{m_R(4a)^2}{3},$$

where m_R is the mass of the rectangular plate.

If ρ is the mass/unit area of the plate, $m_R = \rho 8a^2$, so that

$$I_{RY} = 128\rho a^4/3. \tag{i}$$

The moment of inertia of a circular disc of mass m_D about a diameter $O'Y'$ (see example 4.7) is

$$I_{DY'} = \frac{m_D r^2}{4} = \frac{\pi \rho r^4}{4}. \tag{ii}$$

Since $O'Y'$ passes through the centre of mass of the disc, we can use the parallel axes theorem to give

$$I_{DY} = I_{DY'} + m_D a^2 = \pi \rho r^2 \left(\frac{r^2}{4} + a^2 \right). \tag{iii}$$

The moment of inertia of the plate about OY is therefore

$$I_Y = I_{RY} - I_{DY}$$

$$= \frac{128 \rho a^4}{3} - \pi \rho r^2 \left(\frac{r^2}{4} + a^2 \right). \tag{iv}$$

The moment of inertia of the plate about OX can be found in a similar manner, and shown to be

$$I_X = \frac{\rho 8 a^2 (2a)^2}{12} - \frac{\pi \rho r^4}{4}$$

$$= \frac{8}{3} \rho a^4 - \frac{\pi \rho r^4}{4}. \tag{v}$$

Since the plate is assumed *thin* and the three axes intersect at O we can now use the perpendicular axes theorem to find I_Z, i.e.

$$I_Z = I_X + I_Y.$$

From eqns (iv) and (v)

$$I_Z = \frac{136}{3} \rho a^4 - \frac{\pi \rho r^2}{2} (r^2 + 2a^2). \tag{vi}$$

Engineering Applications

When considering the dynamic behaviour of a body, or of an assembly of bodies in a machine, the first step is to reduce the actual system to an idealised mathematical model. It is often possible, without introducing large errors, to assume that certain bodies in a system are rigid and that others are non-rigid but of negligible mass. Bodies in which both mass and flexibility are important are more difficult to deal with and are beyond the scope of this book.

A typical idealisation would be to consider that ropes are massless, rigid under tensile loading and unable to sustain any compressive or bending loads. Springs can usually be assumed to be massless. These simplifications to a real system are in addition to any assumptions made about the kinematic behaviour of the system, e.g. that clearances at bearings, and flexibility, do not influence the geometric properties of the system. It is very important to bear in mind the assumptions made in setting up a mathematical model of a system when interpreting the results of an analysis. If the assumptions are invalid, then the results will also be invalid.

Once the mathematical model has been defined, a free body diagram must be drawn for *each* body in the system showing the forces which act *on* that body. The appropriate equations of motion (4.19), (4.30) or (4.43) are then applied. Next, any kinematic constraints must be identified and the resulting equations solved for the required variables.

When analysing a mathematical model of a system it is useful to check that the units of each term in any expressions obtained are consistent. If the units are consistent there is a chance that the answer is correct—if they are inconsistent there is *no* chance. Checking units can often point out algebraic errors and prevent them from being carried through a long calculation.

4.1 Turbo generator

Electric motors and generators, and steam and gas turbines are common examples of complete machines which may be considered as bodies in pure rotation. Let us consider a steam turbine driving a generator as shown in Fig. 4.36. If we know the torque-speed characteristics of the turbine and the generator then we can calculate the way in which the speed of the system will increase during start up.

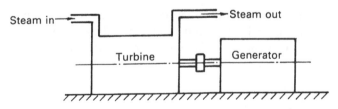

FIG. 4.36

Let the torque applied to the turbine rotor by the steam be M_s and let the electrical torque applied to the generator rotor be M_E. The load torque will oppose the motion of the generator and is therefore in the *opposite* direction from M_s. For mathematical simplicity let us assume that M_s is constant at all speeds and that the magnitude of M_E increases linearly with speed. Let the magnitude of M_E be given by

$$M_E = C\dot\theta, \tag{i}$$

where C is a constant and $\dot\theta$ is the angular velocity of the generator. The turbine and generator rotors are rigidly connected together so they may be considered dynamically as a single body in pure rotation. Let their combined moment of inertia about the axis of rotation be I. Now we draw the free body diagram of the rotor system and define the positive direction of the rotor displacement θ as shown in Fig. 4.37. M_s is in the positive direction so that the generator load torque M_E will be in the negative direction as shown. Since we are concerned only with angular motion, bearing reactions are not shown. The bearing forces act through the axis of rotation and therefore will not influence the angular motion of the system when we take moments about the axis of rotation.

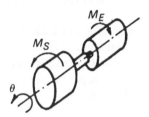

FIG. 4.37

Thus, using the moment equation (4.29),

$$\circlearrowleft \; M_s - M_E = I\ddot\theta \tag{ii}$$

so that, using eqn (i),

$$I\ddot\theta + C\dot\theta = M_s. \tag{iii}$$

This is a linear, second order, differential equation with constant coefficients, and with the constant term M_s on its right-hand side. It will have a solution for θ consisting of both a complementary function and a particular integral.

The solution of this equation is

$$\theta = A + Be^{-Ct/I} + \frac{M_s}{C}t - \frac{I}{C^2}M_s, \tag{iv}$$

where A and B are constants of integration.

Differentiating eqn (iv) with respect to time gives the angular velocity of the system as

$$\dot\theta = -\frac{C}{I}Be^{-Ct/I} + \frac{M_s}{C}. \tag{v}$$

If the system starts from rest so that when $t = 0$, $\theta = 0$ and $\dot{\theta} = 0$ we obtain from eqn (v)

$$B = \frac{M_s I}{C^2},$$

which on substitution into eqn (iv) gives $A = 0$. Thus the complete solution for the angular motion of the rotors is

$$\theta = \frac{M_s I}{C^2}[e^{-Ct/I} - 1] + \frac{M_s}{C}t,$$

$$\dot{\theta} = \frac{M_s}{C}[1 - e^{-Ct/I}],$$

and $\qquad \ddot{\theta} = \frac{M_s}{I}e^{-Ct/I}.$ \hfill (vi)

It is interesting to see how θ, $\dot{\theta}$ and $\ddot{\theta}$ vary with time.

As $t \to \infty$, $e^{-Ct/I} \to 0$. Hence the angular acceleration tends to zero, and the angular velocity approaches a constant value $\dot{\theta} = M_s/C$. At this speed therefore, the magnitudes of M_E and M_s are the same and there is no net torque on the system. θ, $\dot{\theta}$ and $\ddot{\theta}$ are plotted against time in Fig. 4.38.

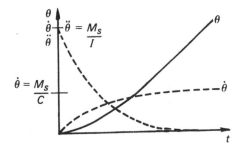

FIG. 4.38

The torque-speed characteristics of a real turbo generator would be more complex than those assumed here. However, the equation of motion, eqn (ii), would still apply.

4.2 Pendulum accelerometer

A rigid link supported on a horizontal pivot will, if the pivot is stationary, hang vertically downwards with its centre of mass directly below the pivot. If the pivot is given a *constant* horizontal acceleration, *a*, the link will assume a steady position at some angle θ to the vertical, as shown in Fig. 4.39. The angle θ of such a pendulum placed in a moving vehicle can be used as a measure of the acceleration of the vehicle. Sudden changes in acceleration will cause oscillation of the pendulum but this can be avoided over the required working range with careful design. By assuming that any oscillations have decayed to negligible amplitude, we can relate θ and a by a steady state analysis of the rigid link. Since in this steady state the angular position of the link is constant, the motion of the link is pure translation. The acceleration of all points in the link, including its centre of mass, will therefore be a. The free body diagram of the link is shown in Fig. 4.40 in which R_X, R_Y are the

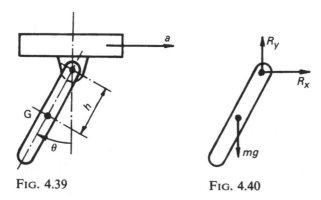

FIG. 4.39   FIG. 4.40

forces on the link at the pivot. We can now write the equations of motion, (eqn 4.19) for a body in pure translation as

$$\rightarrow R_x = ma, \tag{i}$$

$$\uparrow R_y - mg = 0. \tag{ii}$$

Also the resultant force must pass through the centre of mass so that the total moment about the centre of mass must be zero. Hence,

$$\circlearrowleft G - R_x h\cos\theta + R_y h\sin\theta = 0. \tag{iii}$$

From eqns (i), (ii) and (iii) we obtain

$$\tan\theta = \frac{R_x}{mg} = \frac{a}{g}. \tag{iv}$$

Thus, in the steady state condition, the acceleration of the pivot is directly proportional to $\tan\theta$. This type of accelerometer is frequently used where the acceleration being measured varies slowly with time.

4.3 Eccentric cam mechanism

The kinematics of the simple eccentric cam mechanism of Fig. 4.41 have been discussed on p. 41. Let us now consider its dynamic behaviour and

obtain expressions for the contact force between the cam and its flat faced
follower, the forces on the bearings supporting the cam at O, and the torque
required to drive the cam if its angular velocity ω is to be maintained constant.

This mechanism consists of two bodies: (i) the follower, which is in pure
translation, and (ii) the cam, which is a disc of radius r rotating about a fixed
point O. The position of the cam is defined by $\theta = \omega t$, and that of the follower
by y.

We must first draw the free body diagram for each of the bodies. Let the
mass of the follower be m_f. If we assume a coefficient of friction μ between
the cam and the follower, and also that the force between the follower and
its guide may be represented by the concentrated forces R_{x1} and R_{x2} the free
body diagram of the follower will be as shown in Fig. 4.42. We will neglect
any friction between the follower and its guide to simplify the algebra, but it
is suggested that the reader should investigate its influence.

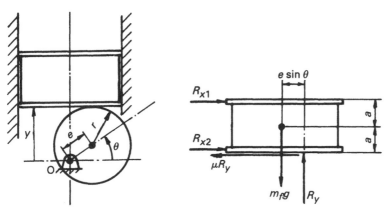

FIG. 4.41   FIG. 4.42

Using Fig. 4.42, the equations of motion, eqn (4.10), for the follower are

$$\rightarrow \quad R_{x1} + R_{x2} - \mu R_y = 0 \qquad (i)$$

$$\uparrow \quad R_y - m_f g = m_f \ddot{y} \qquad (ii)$$

and $\quad \circlearrowleft aR_{x2} - aR_{x1} + eR_y \sin \theta - a\mu R_y = 0. \qquad (iii)$

The free body diagram for the cam is shown in Fig. 4.43, in which F_x and
F_y are the forces at the bearings, and M is the external torque applied to the
cam shaft to maintain the motion. The contact forces are equal and opposite
to those on the follower in Fig. 4.42. Let the mass of the cam be m_c, and let
us assume that its centre of mass is at its geometric centre and that its moment
of inertia about O is I_O.

Since the body is rotating about a fixed point O, we can take moments about
that point, so that if the cam rotates at a constant angular velocity, eqns (4.30)

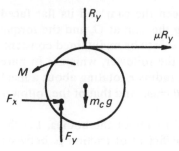

FIG. 4.43

give

$$\circlearrowleft \quad M - R_y e \cos \theta - \mu R_y (e \sin \theta + r) - m_c g e \cos \theta = I_0 \ddot{\theta} = 0, \quad \text{(iv)}$$

$$\rightarrow \quad F_x + \mu R_y = m_c \ddot{x}_G, \qquad \text{(v)}$$

and $\quad \uparrow \quad F_y - R_y - m_c g = m_c \ddot{y}_G. \qquad \text{(vi)}$

Equations (i) to (vi) are the equations of motion for the system.

However, the coordinates y, x_G, y_G and θ are not independent and we must now consider the kinematics of the system to establish their relationship.

From the geometry of the mechanism (p. 41),

$$y = r + e \sin \theta,$$

so that

$$\ddot{y} = -e\omega^2 \sin \theta. \qquad \text{(vii)}$$

Also $x_G = e \cos \theta, \qquad y_G = e \sin \theta,$

so that $\ddot{x}_G = -e\omega^2 \cos \theta$ and $\ddot{y}_G = -e\omega^2 \sin \theta.$

Thus y, x_G, y_G and their time derivatives can be expressed in terms of θ and ω, and by substituting eqn (vii) into eqn (ii) we obtain the contact force

$$R_y = m_f(g - e\omega^2 \sin \theta). \qquad \text{(viii)}$$

By substituting eqn (viii) into eqn (iv) and expressing $\sin^2 \theta$ and $\sin \theta \cos \theta$ in terms of 2θ,

$$M = \mu m_f \left[gr - \frac{e^2 \omega^2}{2} \right] + e\mu m_f(g - \omega^2 r) \sin \theta + ge(m_f + m_c) \cos \theta$$

$$- \frac{m_f e^2 \omega^2}{2} (\sin 2\theta + \mu \cos 2\theta). \qquad \text{(ix)}$$

The other variables such as F_x, F_y may now be determined if required.

The expression for the torque M required to drive the cam at a constant angular velocity ω contains a constant term, terms which depend upon $\sin \theta$

and cos θ, and terms which depend upon sin 2θ and cos 2θ. Thus the torque fluctuation during one revolution of the cam includes terms which vary at the cam frequency ω, (those in sin θ and cos θ) and others which vary at twice cam frequency 2ω, (those in sin 2θ and cos 2θ).

When interpreting these results we must be aware of the limitations of the analysis. In the physical system R_y is a contact force and can only act in the directions shown in the free body diagrams of Figs 4.42 and 4.43. R_y cannot be negative as this would require some adhesive force between the surfaces of the cam and follower. Hence eqn (viii) will only apply for positive R_y, i.e. when $g > e\omega^2 \sin \theta$. The kinematic relationships of eqn (vii) also require that contact be maintained.

The analysis is therefore valid over a complete cycle of operation only if $\omega^2 < g/e$. For higher speed operation a spring would be used to apply a downward force on the follower so that contact would be maintained to give correct kinematic behaviour. The free body diagram of the follower would then have to be modified to include the additional spring force.

4.4   Belt drive

We have already considered the kinematics of a belt drive in Chapter 1, application 1.2 and the static case of a belt and pulley in Chapter 3, application 3.5(b). Let us now consider the flat belt drive shown in Fig. 4.44 and take account of the dynamic effects when the drive runs at a constant speed. We will again assume that the belt is flexible and inextensible. The pulleys P_1 and P_2 will therefore have zero angular acceleration. Any element of belt in contact with a pulley will be moving on a circular path and its velocity will be constant in magnitude. It will therefore have an acceleration towards the centre of the pulley. Let us consider an element of belt on pulley P_1 as the pulley is just about to slip in the anti-clockwise direction relative to the belt. The forces on this element at the onset of slip are as shown in Fig. 4.45. As in the static case the element of belt subtends an angle $\delta\theta$ at the centre O_1 of the pulley and N is the normal force/unit length applied to the belt by the pulley. The

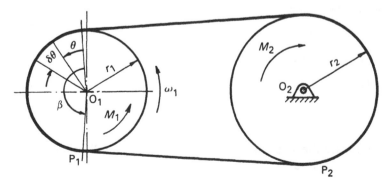

Fig. 4.44

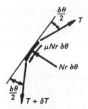

FIG. 4.45

acceleration a_r of the element is towards O_1 and has magnitude

$$a_r = \omega_1^2 r = \frac{v^2}{r},$$ (i)

where v is the linear velocity of the belt.

We can now apply Newton's Second Law to the element. In the radial and tangential directions we have

$$\nwarrow Nr\,\delta\theta - T\sin\frac{\delta\theta}{2} - (T + \delta T)\sin\frac{\delta\theta}{2} = -mr\,\delta\theta\,\frac{v^2}{r},$$ (ii)

and

$$\swarrow (T + \delta T)\cos\frac{\delta\theta}{2} - T\cos\frac{\delta\theta}{2} + \mu Nr\,\delta\theta = 0,$$ (iii)

where m is the mass/unit length of the belt.

Using the small angle approximations and taking the limit as $\delta\theta \to 0$ gives

$$Nr - T = -mv^2,$$ (iv)

and

$$\frac{\mathrm{d}T}{\mathrm{d}\theta} + \mu Nr = 0.$$ (v)

If we let

$$T_D = mv^2$$

we can express the tension as

$$T = T_S + T_D = T_S + mv^2.$$

Substituting this expression into eqns (iv) and (v) gives

$$Nr - T_s = 0,$$ (vi)

and

$$\frac{\mathrm{d}T_S}{\mathrm{d}\theta} + \mu Nr = 0.$$ (vii)

These equations are the same as eqns (iii) and (iv) in application 3.5(b), and their solution is therefore

$$T_S = T_{S1}e^{-\mu\theta}.$$ (viii)

At any value of θ the tension in the belt is given by

$$T = T_\text{S} + T_\text{D} = T_\text{S1}e^{-\mu\theta} + mv^2. \tag{ix}$$

If the tension in the tighter span is T_1 then

$$T_1 = T_\text{S1} + mv^2.$$

The tension T_2 in the slacker span will then be given by

$$T_2 = T_\text{S1}e^{-\mu\beta} + mv^2, \tag{x}$$

where β is the angle of wrap around the pulley P_1.

By comparing these results with those in application 3.5(b) we see that T_S can be considered as the static tension in the belt, and T_D as the additional tension due to dynamic effects. The dynamic tension T_D is the tension which would arise in the belt if it were moving around its path with a linear velocity v and the pulleys were removed.

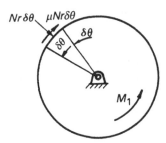

FIG. 4.46

Let us now draw the free body diagram of pulley P_1 as shown in Fig. 4.46 and determine the torque on the pulley which will cause slip. The forces at the bearings are not shown as these do not affect the moment equation for the pulley. A tangential frictional force is applied to the surface of the pulley by the belt over the arc from $\theta = 0$ to $\theta = \beta$. Over a small arc of the pulley which subtends an angle $\delta\theta$ at O_1 the tangential force is $Nr\,\delta\theta$ as shown. If the external moment applied to the pulley is M_1 then, by taking moments about O_1,

$$\overset{\curvearrowright}{(O_1)} \quad M_1 - \int_0^\beta \mu Nr^2 \, \text{d}\theta = 0.$$

Thus $\quad M_1 = \mu r^2 \displaystyle\int_0^\beta N \, \text{d}\theta, \tag{xi}$

which from eqn (vii) becomes

$$M_1 = \mu r \int_0^\beta T_S \, d\theta$$

$$= \mu r \int_0^\beta T_{S1} \, e^{-\mu\theta} \, d\theta$$

$$= T_{S1} r (1 - e^{-\mu\beta}). \tag{xii}$$

The torque which can be transmitted by the drive is therefore predicted to be independent of the belt velocity and to depend only upon the static tension T_S. This occurs because the normal force N, and hence the frictional force on the pulley, is independent of the belt speed. In a real drive this would not be the case as the belt would not be inextensible. Belt stretch would reduce both the angle of wrap and the normal force. As belt stretch is a function of total belt tension, which increases with speed, the torque transmitting capacity of the drive would reduce as the speed is increased.

4.5 Winches and hoists

(a) One of the simplest forms of lifting arrangement consists of a single pulley over which a rope is passed. One end of the rope is attached to the mass being lifted and a force T is applied to the free end of the rope as shown in Fig. 4.47. Let us assume that T is applied at some angle α to the horizontal. A mathematical model of this system can be set up, and if we assume that the mass of the rope is negligible the model will consist of two bodies:

  (i)   the pulley which is rotating about a fixed point O,
 (ii) the mass which is moving with pure translation.

In order to obtain expressions relating the force T to the acceleration of the mass we must first draw the free body diagrams for the bodies. The forces acting on the pulley are shown in Fig. 4.48 and are the tensions in the rope T and T_1, the pulley weight $m_P g$, and the force at the bearing at O which is

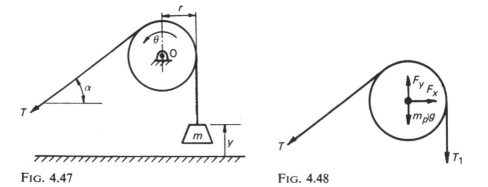

FIG. 4.47   FIG. 4.48

shown as components F_x and F_y. Friction at the bearings is neglected and we will assume that the centre of mass of the pulley is at O. We can now write the equations of motion for the pulley using eqns (4.30) as

$$\circlearrowleft \quad Tr - T_1 r = I_O \ddot{\theta}, \tag{i}$$

$$\rightarrow \quad F_x - T \cos \alpha = m_P \ddot{x}_G = 0, \tag{ii}$$

and $\quad \uparrow \quad F_y - T \sin \alpha - T_1 - m_P g = m_P \ddot{y}_G = 0. \tag{iii}$

The free body diagram for the mass is shown in Fig. 4.49. Only two forces, the tension T_1 in the rope, and the weight mg, act on the body and, since it is in pure translation, the resultant force must act through the centre of mass G. The weight acts through the centre of mass G and therefore the line of action of T_1 must also act through G.

We need only one equation of motion for the mass. For a body in pure translation eqns (4.19) give

$$\uparrow \quad T_1 - mg = m\ddot{y}. \tag{iv}$$

Provided that there is no slip between the rope and the pulley we can relate kinematically the angular velocity of the pulley and the linear velocity of the mass, giving, for the positive directions defined in eqns (i) and (iv),

$$r\dot{\theta} = \dot{y}. \tag{v}$$

From eqns (i), (iv) and (v) we obtain the acceleration of the mass as

$$\ddot{y} = \frac{T - mg}{m + I_O/r^2}. \tag{vi}$$

Values of F_x, F_y and $\ddot{y}$, for a given value of T can now be obtained if required. Note that F_x, F_y are the total forces on the pulley shaft in the x and y directions. A pulley is normally supported on two bearings which share the load. If the bearings are symmetrically placed as in Fig. 4.50 the load will be shared equally.

If the rope slips over the pulley $\dot{y}$ and $\dot{\theta}$ are not kinematically related so that eqn (v) is no longer valid. For this case we must relate the tensions T

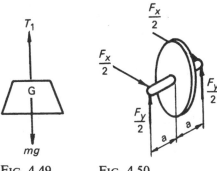

FIG. 4.49   FIG. 4.50

and T_1, using eqn (vii) of application 3.5(b), i.e.

$$T_1 = Te^{-\mu\beta},$$

where μ is the coefficient of friction between the belt and the pulley and β is the angle of wrap.

Equations (i) and (iv) then give

$$\ddot{\theta} = \frac{Tr(1 - e^{-\mu\beta})}{I_O} \tag{viii}$$

and $\qquad \ddot{y} = \dfrac{Te^{-\mu\beta} - mg}{m}.$ $\hfill$ (ix)

If the rope is released, T becomes zero and from eqn (ix)

$$\ddot{y} = -g,$$

i.e. the mass falls freely under gravity.

(b) Power operated lifting systems are usually more complicated than the simple pulley arrangment of (a), since it is difficult to apply a force to a free end of a rope over any appreciable distance. A hydraulic cylinder is sometimes used where only limited travel is needed, e.g. to raise the forks of a fork-lift truck. The usual method employed, where appreciable movement of the mass is required, is to wrap the free end of the rope around a cylindrical drum. The end of the rope is then securely fixed to the drum, so eliminating any possibility of slip. Rotation of the drum raises or lowers the load.

Lifting and lowering requires a high torque to be applied to the drum at a low speed. Most prime movers such as electric motors or I.C. engines, cannot economically provide this high torque. To match the motor characteristics with the output requirements a speed reduction, and hence a torque multiplication, (see kinematics p. 37, statics p. 129) must be introduced between the prime mover and the drum. A gear train is normally used for this purpose in electrically driven hoist units.

In the following we will consider the dynamics of an electrically driven winch used for hauling blocks, or trucks, up an incline. The example is also intended to demonstrate how geared systems may be analysed.

Figure 4.51 shows an electric motor which has a pinion, of pitch circle radius r_1, fixed to its output shaft. Let the moment of inertia of the motor armature and its pinion about the axis of rotation O_1 be I_1. The pinion meshes with a larger gear wheel of pitch circle radius r_2. This gear is fixed directly to a drum which is mounted on a shaft supported on bearings. Let the moment of inertia of this assembly about its axis O_2 be I_2. A rope, wrapped around the drum, is used to haul a block of mass m up an incline as shown. Let the coefficient of friction between the mass and the incline be μ. The duty cycle specifies

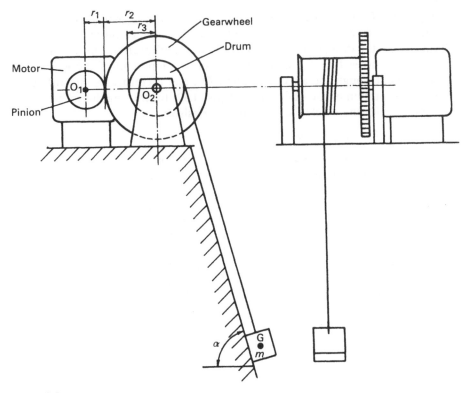

Fig. 4.51

that the winch must be able to raise the mass at velocity v over long distances. During acceleration of the mass from rest the motor will be required to exert an increased torque over a short period of time. The motor designer would need to know both the steady state torque required from the motor and the additional torque necessary for accelerating the system. To obtain expressions for these torques we must first write the equations of motion for each of the bodies in the system.

The system consists of three connected bodies:

(i) the motor armature and pinion assembly which rotates about the fixed point O_1,

(ii) the gear wheel and drum assembly which also rotates about the fixed point O_2, and

(iii) the mass m, which is in pure translation.

We can see, by inspection, that when the load is being raised the gear wheel and drum assembly will rotate in an anticlockwise direction and that the motor will rotate in a clockwise direction. These directions will be taken as positive when writing the equations of motion. We will consider first the armature and pinion. The forces and the electrical torque M acting on the body are shown in Fig. 4.52. P and S are the tangential and separating components of the tooth contact force (see p. 131 for a discussion of forces on gear teeth). We do not know the direction of P and so choose it arbitrarily. The weight m_1g, the bearing forces F_{1x}, F_{1y} and the separating component of the gear tooth force all act through the axis of rotation O_1, and therefore do not influence the angular motion of the body. Using eqns (4.30) for the angular motion only we obtain, with the clockwise direction as positive,

$$\text{①} \quad M - Pr_1 = I_1\ddot{\theta}_1. \tag{i}$$

Since we are only seeking an expression for M, the other two equations are not required. They are of course essential when finding the bearing forces.

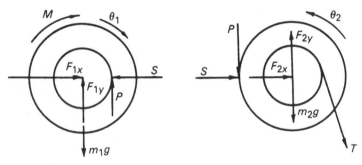

FIG. 4.52   FIG. 4.53

The free body diagram of the gear wheel and drum assembly is shown in Fig. 4.53. The gear tooth contact forces are *equal and opposite* to those on the pinion. Again we need only the equation for the angular motion of the system, and taking the anticlockwise direction as positive,

$$\text{②} \quad Pr_2 - Tr_3 = I_2\ddot{\theta}_2, \tag{ii}$$

where r_3 is the radius of the barrel and T is the tension in the rope.

The forces acting on the mass are shown in Fig. 4.54. Taking the positive directions indicated we obtain,

$$\nwarrow \quad T - mg \sin \alpha - \mu N = m\ddot{x}, \tag{iii}$$

$$\nearrow \quad N - mg \cos \alpha = 0. \tag{iv}$$

We have here assumed that the mass of the rope is small compared with m. In some cases, where a very long rope is needed, the mass of the rope is significant and cannot be neglected. Coring winches used for extracting sample cores of material from the sea bed provide a typical example where, in deep sea applications, the mass of the rope can be many times that of the load.

Equations (i) to (iv) are the dynamic equations for the system and contain six unknowns $\ddot{\theta}_1$, $\ddot{\theta}_2$, $\ddot{x}$, P, N and T. However θ_1, θ_2 and x are not independent,

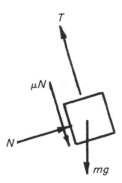

F<small>IG</small>. 4.54

and are related by the kinematic equations

$$\dot{\theta}_1 = \frac{r_2}{r_1} \dot{\theta}_2 \tag{v}$$

and $\dot{x} = r_3 \dot{\theta}_2.$ (vi)

If we now consider the condition when the mass has a constant velocity v up the plane, i.e. when $\ddot{\theta}_1 = \ddot{\theta}_2 = \ddot{x} = 0$, eqns (i) to (iv) give

$$N = mg \cos \alpha,$$

$$T = mg (\sin \alpha + \mu \cos \alpha),$$

$$P = mg \frac{r_3}{r_2} (\sin \alpha + \mu \cos \alpha),$$

and $M = M_s = mg \dfrac{r_1 r_3}{r_2} (\sin \alpha + \mu \cos \alpha).$ (vii)

where M_s is the motor torque required for constant speed lifting.

The signs in eqns (vii) show that the forces N, T, and P, act in the directions shown in the free body diagrams.

Equation (vii) for M_s gives the steady torque required to raise the load at a constant speed v, but some additional torque is required to accelerate the load from rest to v. Let us suppose that the design of the motor is such that it can produce a torque M_s over a long period of time but that, for short periods, it can exceed this value by an amount kM_s without overheating. We can now find the time that this overload has to be used when accelerating the system.

During acceleration the torque M available is given by

$$M = (1+k)M_s$$

$$= (1+k) mg \frac{r_1 r_3}{r_2} (\sin \alpha + \mu \cos \alpha). \tag{viii}$$

From eqns (iii) and (iv)

$$\ddot{x} = \frac{1}{m}[T - mg(\sin\alpha + \mu\cos\alpha)].\tag{ix}$$

By eliminating P from eqns (i) and (ii) and substituting for $\ddot{\theta}_1$, $\ddot{\theta}_2$ in terms of $\ddot{x}$ from eqns (v) and (vi) we obtain

$$T = -\frac{Mr_2}{r_ir_3} - \left(\frac{I_1r_2}{r_1^2r_3} + \frac{I_2}{r_3^2}\right)\ddot{x}.\tag{x}$$

Substituting T into eqn (ix) gives

$$\ddot{x} = \frac{kmg(\sin\alpha + \mu\cos\alpha)}{m + \dfrac{I_1r_2}{r_1^2r_3} + \dfrac{I_2}{r_3^2}}.\tag{xi}$$

Equation (xi) shows that $\ddot{x}$ is constant. Using the basic kinematics of Chapter 1 page 7, we can easily show that the time t^* taken for the mass to reach velocity v from rest will be

$$t^* = \frac{v\left(m + \dfrac{I_1r_1}{r_1^2r_3} + \dfrac{I_2}{r_3^2}\right)}{kmg(\sin\alpha + \mu\cos\alpha)}.\tag{xii}$$

Equation (xii) therefore gives the time for which the motor must operate in the overloaded condition.

Expressions for gear tooth force, rope tension, bearing reactions, etc., can also be extracted from the equations of motion of the system if required.

4.6   Balancing of rotating systems

Let us consider a simple machine which consists of a thin disc in pure rotation about its polar axis. No matter how simple the disc shape it is impossible to manufacture the disc such that its centre of mass coincides exactly with the axis of rotation. Let the distance between the axis of rotation and the centre of mass G be e, as shown in Fig. 4.55. The disc may be considered as a rigid body rotating about a fixed point and the force applied to the disc will be

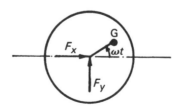

FIG. 4.55   FIG. 4.56

given by eqn (4.30), i.e.

$$F = ma_G. \tag{i}$$

The centre of mass G moves on a circular path of radius e. If the disc runs at a constant speed ω the centre of mass will have an acceleration $a_G = \omega^2 e$ towards the axis of rotation. The free body diagram of the disc is shown in Fig. 4.56. The force on the disc at the supporting bearings is shown in components F_x, F_y. Neglecting the weight of the disc, eqns (4.30) give

$$F_x = ma_{Gx} = -m\omega^2 e \cos \omega t,$$
$$F_y = ma_{Gy} = -m\omega^2 e \sin \omega t. \tag{ii}$$

The forces on the support structure will be equal and opposite to those on the disc so that the structure will experience forces

$$F_x = m\omega^2 e \cos \omega t,$$
$$F_y = m\omega^2 e \sin \omega t. \tag{iii}$$

These components may be considered as a single force vector of magnitude $m\omega^2 e$ rotating with angular velocity ω, as shown in Fig. 4.57. This force, which

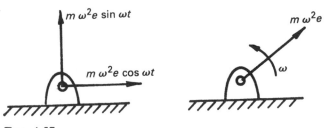

FIG. 4.57

is often referred to as the out of balance force, will shake the structure and could cause unacceptable vibration. To reduce this out of balance force the eccentricity e must be reduced by adding or removing material from the disc. The angular position of the centre of mass can be found by supporting the disc in very low friction bearings, or by allowing it to roll along a pair of parallel horizontal knife edges so that the disc will come to rest with G vertically below its axis. Material is then added or removed until the disc will remain at rest in any position. Balancing machines are also available in which the disc is run at some speed and the direction and magnitude of the out of balance force measured. The amount of material to be removed, and its position, are then calculated and displayed. The use of such a machine achieves the same results as could be obtained statically. This process is therefore called *static* balancing, even though the disc may be rotating during the procedure.

The foregoing refers only to a *thin* disc. In most machines, e.g. electric motors, the rotating parts cannot be considered thin, and the procedure for balancing such rotors is more complex. The full procedure will not be considered in detail here but a simplified system will be used to illustrate the additional features.

Consider the system shown in Fig. 4.58 which consists of two identical discs mounted on a massless, but rigid, shaft AB. Let each disc have mass eccentricity *e*, and let them be mounted such that the eccentricities are on diametrically opposite sides of the shaft centre line. The centre of mass of the *total* system will therefore be on the axis of rotation and the system would therefore be *statically* balanced.

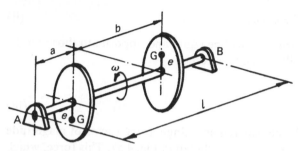

FIG. 4.58

Let us now consider the system rotating at some constant speed ω. The centre of mass of each disc will have an acceleration $\omega^2 e$ towards the axis of rotation. Each disc will therefore apply a force to the shaft of magnitude $m\omega^2 e$. The free body diagram of the shaft is shown in Fig. 4.59(a) where the eccentricities are considered instantaneously to be in the plane of the paper. F_{SA} and F_{SB} are the forces *on* the shaft at the bearings. For equilibrium of the shaft,

$$\uparrow \, \Sigma F = 0, \text{ and } \Sigma M = 0.$$

Thus $\uparrow \, F_{SA} + F_{SB} + m\omega^2 e - m\omega^2 e = 0,$

so that $F_{SA} = -F_{SB}.$

Also by taking moments about end A

$$\text{Ⓐ} - m\omega^2 e a + F_{SB} l + m\omega^2 e (b + a) = 0,$$

so that $F_{SB} = -\dfrac{m\omega^2 e\, b}{l}.$ (iv)

Thus the equal and opposite bearing forces provide a couple on the shaft, and by Newton's Third Law an equal and opposite couple will be applied to the frame. Whilst there is no *resultant* out of balance force on the frame, the forces on the frame at A and B will produce an out of balance couple about an axis perpendicular to AB which rotates with the shaft at frequency ω.

These out of balance forces at the bearings arise because of the *moments* of the out of balance forces caused by the individual discs. To achieve balance

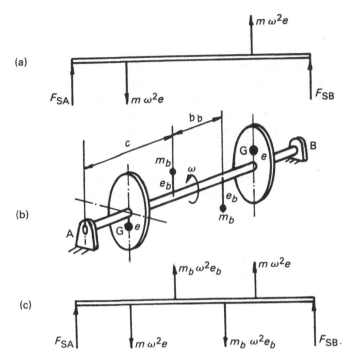

FIG. 4.59

mass must be added or removed to create an equal and opposite *moment*, whilst still maintaining static balance. Full balance can only be achieved by adding *two* balancing masses in *two* planes separated by a distance b_b as shown in Fig. 4.59(b). The angular position of these balance masses must be such that the centre of mass of the system remains on the axis of rotation, i.e. for our particular case they must be 180° apart. For simplicity, we will choose both balance masses to have mass m_b and mass eccentricity e_b.

The free body diagram of the balanced shaft is shown in Fig. 4.56(c) and by taking moments about end A we have

$$\circled{A} \quad -m\omega^2 ea - m_b\omega^2 e_b(c + b_b) + F_{SB}l + m_b\omega^2 e_b c + m\omega^2 e(a + b) = 0,$$

so that $m\omega^2 eb - m_b\omega^2 e_b b_b + F_{SB}l = 0.$ (v)

Thus if we choose $meb = m_b e_b b_b$, then $F_{SB} = 0$, and since $F_{SB} = -F_{SA}$, it follows that there will be no out of balance forces at the bearings.

In considering the forces F_{SA}, F_{SB} we have neglected the weight of the system which will cause constant downward forces on the frame at the support bearings. Such forces are not a problem. Only the rotating out of balance effects give rise to periodic forces which cause noise and vibration.

The foregoing shows that to produce balance in a rotating system which cannot be considered as a single thin disc, material must be added or removed in *two* planes if the out of balance moment is to be eliminated. Two plane balancing is called *dynamic* balancing since the effects of the out of balance moments can only be detected when the system is rotating.

Rotors in real machines come in all shapes and sizes. By extending the above arguments we can show that any rotor can be *dynamically* balanced by the addition, or subtraction, of mass in any *two* separated planes.

An interesting comparison can be made by comparing the balancing of a cycle wheel which is thin and motor car wheel which is comparatively thick. The cycle wheel can be satisfactorily balanced by the addition of a single mass to achieve *static* balance. The motor car wheel should be balanced using a dynamic balancing machine with balancing masses added to *both* sides of the wheel. If a garage uses a dynamic balancing machine but adds masses only to one side of the wheel, then only *static* balance can be achieved, and out of balance moments will still be present.

4.7 The slider crank mechanism

The dynamics of the slider crank mechanism of Fig. 4.60 will now be analysed.

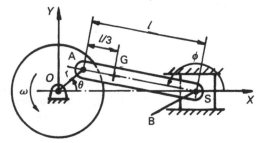

FIG. 4.60

Other linkage mechanisms, e.g. the four bar linkage may be considered in a similar manner. Such mechanisms are used widely in packaging and other production machinery, and the slider crank mechanism forms the basis of most internal combustion engines. The slider S may be considered as a rigid body in pure translation, the connecting rod AB as a rigid body in general plane motion and the crank OA as a rigid body which rotates about a fixed point O. If the crank OA rotates at a constant angular velocity ω the motion of the connecting rod and slider may befound from a kinematic analysis. This has been carried out in Chapter 1, application 1.6. We can now apply the equations of motion to each individual body in the mechanism to find the forces and torques necessary to maintain the motion.

We will consider the mechanism in the general position shown in Fig. 4.60 where θ and ϕ define the angular positions, measured anticlockwise from OX, of the crank and the connecting rod. The position of the slider is defined by its distance x_s along OX from O, and the position of the centre of mass G_c of the connecting rod is given by coordinates x_{Gc}, y_{Gc}.

Let the slider have mass m_s and let its centre of mass G_s be coincident with the pivot at B. The connecting rod has mass m_c and moment of inertia I_G about an axis through its centre of mass G_c. The crank and flywheel assembly has mass m_f and moment of inertia I_O about the fixed point O. We will assume that the centre of mass of the crank and flywheel is at O.

We now draw a free body diagram for each of the bodies in the mechanism.

The free body diagram of the slider is shown in Fig. 4.61. Friction has been neglected so that the only forces acting on the slider are the forces from the connecting rod and the guide.

R_{y1} and R_{y2} are the contact forces on the slider from the guide and are assumed to act at the corners of the slider. F_{xB} and F_{yB} are the components of the pin force on the slider at B applied by the connecting rod. The directions of these forces are not known so they have been assumed to act in the directions shown.

Applying the equations of motion, eqns (4.19), for a body in pure translation gives

$$\rightarrow \quad F_{xB} = m_s \ddot{x}_s, \tag{i}$$

$$\uparrow \quad F_{yB} + R_{y1} + R_{y2} - m_s g = m_s \ddot{y}_s = 0, \tag{ii}$$

and $\quad \text{G}_s) \, c(R_{y2} - R_{y1}) = 0. \tag{iii}$

The only forces acting on the connecting rod are its weight and the contact forces at the connections to the crank and to the slider. These are shown in the free body diagram of Fig. 4.62.

The components of the force on the connecting rod at B are equal and opposite to those acting on the slider at B and the force at A is represented by components F_{xA} and F_{yA} as shown.

Applying eqns (4.43) for a body in general plane motion gives

$$\rightarrow \quad F_{xA} - F_{xB} = m_c \ddot{x}_{Gc} \tag{iv}$$

$$\uparrow \quad F_{yA} - F_{yB} - m_c g = m_c \ddot{y}_{Gc} \tag{v}$$

and $\quad \text{G}_c) - F_{xA} \dfrac{l}{3} \sin \phi + F_{yA} \dfrac{l}{3} \cos \phi - F_{xB} \dfrac{2l}{3} \sin \phi + F_{yB} \dfrac{2l}{3} \cos \phi = I_G \ddot{\phi}.$

$$\tag{vi}$$

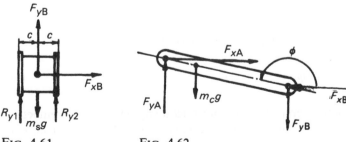

FIG. 4.61   FIG. 4.62

The forces acting on the crank are shown in the free body diagram of Fig. 4.63, where M_O is an external torque applied to the crank by the driving motor. The centre of mass of the crank will have zero acceleration since it is at O. The angular acceleration of the crank is also zero because we have assumed that it rotates at a constant angular velocity.

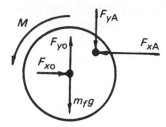

FIG. 4.63

The forces on the crank at A are equal and opposite to the forces on the connecting rod at A. F_{xO} and F_{yO} are the forces on the crank at its bearings.

We can now write the equations of motion for the crank using eqns (4.30) as

$$\rightarrow F_{xO} - F_{xA} = 0, \tag{vii}$$

$$\uparrow F_{yO} - F_{yA} - m_f g = 0, \tag{viii}$$

and $\quad \circlearrowright M_O - F_{yA} r \cos \theta + F_{xA} r \sin \theta = 0. \tag{ix}$

In eqns (i) to (viii) all of the acceleration terms can be found, as functions of ω, from the kinematics of the mechanism. This leaves 8 unknown forces and the unknown torque M_O to be found.

In the general case these unknown forces are best found by writing a small computer program.

If we take a *special case* in which the mass of the connecting rod is small compared with that of the slider, its effect may be neglected and eqns (iv) and (v) reduce to

$$F_{xA} - F_{xB} = 0,$$

and $\quad F_{yA} - F_{yB} = 0. \tag{x}$

Substituting these values into eqn (vi), and putting $I_G = 0$, we obtain

$$\frac{F_{yA}}{F_{xA}} = \frac{F_{yB}}{F_{xB}} = \tan \phi, \tag{xi}$$

so that the forces on the connecting rod must act *along* AB, i.e. the connecting rod will be in pure tension or compression. (Remember that this is true *only* for a *massless* link.)

From eqns (i) and (x)

$$F_{xA} = m_s \ddot{x}_s,$$ (xii)

and from eqn (xi)

$$F_{yA} = m_s \ddot{x}_s \tan \phi.$$

The driving torque in eqn (ix), can therefore be expressed as

$$M_O = m_s \ddot{x}_s \tan \phi (r \cos \theta) - m_s \ddot{x}_s r \sin \theta$$

$$= m_s \ddot{x}_s r (\tan \phi \cos \theta - \sin \theta).$$ (xiv)

In eqn (xiv) $\ddot{x}_s$ and ϕ are both functions of θ, and by using the results of the kinematic analysis of page 47, M_O can be expressed directly in terms of ω and θ.

Even for this much simplified case the algebra involved in seeking an analytical solution for M_O is tedious. It is again more convenient to use a small computer program to obtain a numerical solution of the kinematic equations at a series of crank angles. Substitution of the numerical values of $\ddot{x}_s$, ϕ and θ into eqn (xiv) will give values of M_O over the complete cycle of operation.

It is interesting to consider the behaviour of the system when the crank radius is small, i.e. when $r \ll l$. This assumption greatly simplifies the algebra, but still illustrates some of the main characteristics of the slider crank mechanism.

With the assumption that $r \ll l$, $\tan \phi$ may be assumed small and eqn (xiv) may be approximated as

$$M_O = -m_s \ddot{x}_s r \sin \theta.$$ (xv)

From eqn (vi) p. 47 the kinematic analysis of the mechanism gives

$$\ddot{x}_s = -r\dot{\theta}^2 \cos \theta + l\ddot{\phi} \sin \phi + l\dot{\phi}^2 \cos \phi.$$

When ϕ is assumed small this may be approximated as

$$\ddot{x}_s = -r\dot{\theta}^2 \cos \theta,$$ (xvi)

which on substitution into eqn (xv) gives

$$M_O = m_s r^2 \dot{\theta}^2 \sin \theta \cos \theta = \frac{m_s r^2 \dot{\theta}^2}{2} \sin 2\theta.$$ (xvii)

Since $\theta = \omega t$ we have

$$M_O = \frac{m_s \omega^2 r^2}{2} \sin 2\omega t.$$ (xviii)

Thus the torque M_O which has to be applied by the drive motor to maintain a constant crank speed varies at a frequency of 2ω, i.e. at *twice* the running

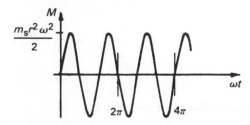

FIG. 4.64

speed as shown in Fig. 4.64. In practice this is difficult to achieve so that speed
fluctuations would occur. These can be reduced by using a flywheel.

Let us now consider the forces required in the holding down bolts which
secure the machine to its foundations. Figure 4.65 shows the freebody diagram
of the frame of the machine in which H_1, H_2 and V are the forces applied to

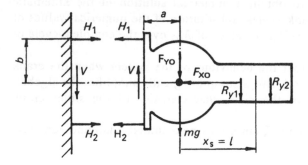

FIG. 4.65

it by the holding down bolts and mg is the weight of the frame. It is assumed
that the centre of mass of the frame is at the crank shaft bearing, again to
simplify the algebra. The forces applied to the foundation on which the machine
is mounted are equal and opposite to H_1, H_2 and V. Also shown are the forces
F_{xO}, F_{yO} applied to the frame at the crank shaft bearing and R_{y1}, R_{y2} at the
sliding bearing. If we make the above assumptions that the mass of the
connecting rod is small and that $r \ll l$, we can show, using eqns (i) to (xii) that

$$F_{xO} \approx -m_s\omega^2 r \cos \theta$$

and $$F_{yO} \approx m_s g - (R_{y1} + R_{y2}) + m_f g = m_f g, \qquad (xix)$$

The total force in the holding down bolts must maintain static equilibrium of
the machine frame. Thus

$$\uparrow \ V - F_{yO} - R_{y1} - R_{y2} - mg = 0,$$

$$\rightarrow \ -H_1 - H_2 - F_{xO} = 0,$$

and taking moments about the upper bolt

$$① - (F_{yO} + mg)a - 2H_2b - R_{y1}(a + l - c) - R_{y2}(a + l + c)$$
$$- F_{xO}b = 0. \qquad \text{(xx)}$$

The forces R_{y1}, R_{y2} will, in fact, change their position along the axis of the machine as the slider moves. However since we have assumed that $r \ll l$, the variation in position is small and can be neglected in eqns (xx).

From eqns (xx) and (xix)

$$H_1 \approx \frac{(m_f + m + m_s)ga + m_sgl}{2b} + \frac{m_s\omega^2 r \cos\theta}{2},$$

and

$$H_2 \approx -\frac{(m_f + m + m_s)ga + m_sgl}{2b} + \frac{m_s\omega^2 r \cos\theta}{2}. \qquad \text{(xxi)}$$

The forces H_1 and H_2 in the bolts consist, therefore, of a constant term and a periodic term. The constant terms are due to the moments produced by the weights of the components and are equal in magnitude but opposite in direction. Both periodic terms have the same sign and together represent a horizontal fluctuating force

$$H = m_s\omega^2 r \cos\omega t. \qquad \text{(xxii)}$$

Thus a harmonic out of balance force $m_s\omega^2 r \cos\omega t$ along the slider axis is applied to the foundations supporting the machine. The magnitude and frequency of this force varies with the running speed of the machine. Using the dimensions of a typical engine the reader should compare the values of M_O and H obtained using correct values of l and r with those obtained using the assumption $r \ll l$ as above. A short computer program will be required to obtain numerical values of M_O and H.

4.8 Balancing of reciprocating machines

We have seen that it is possible to eliminate, or reduce to an acceptable level, the out of balance forces transmitted to the foundations of a rotating machine. The problem is much more difficult to overcome with a reciprocating machine.

A simple mass equal to that of the slider, located at the end of the crank as shown in Fig. 4.66(a), would produce out of balance forces on the frame as shown in Fig. 4.66(b). The horizontal component of this force is the same as that produced by the slider crank mechanism as given by eqn (xxii). Thus if we arrange a balancing mass m_s at radius r *opposite* the crank, as shown in Fig. 4.66(c), the force on the frame due to the reciprocating mass will be completely balanced by the equal and opposite force $m_s\omega^2 r \cos\omega t$ from the balancing mass. Unfortunately, we have now an out of balance force $m_s\omega^2 r$ $\sin\omega t$ in the vertical direction. All that has been achieved is to change the out of balance horizontal force on the frame into a vertical force of the same magnitude. It is however possible to share the out of balance force between

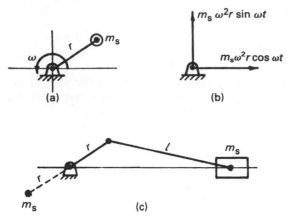

FIG. 4.66

the vertical and horizontal directions. If for example we choose the balancing mass m_b to be half of m_s, or the product of m_b and its radial position r_b to be half of $m_s r$, then the horizontal out of balance force will be halved, but a force of equal magnitude will be introduced in the vertical direction. By choosing $m_b r_b$ appropriately it is possible to transfer a proportion of the out of balance force from the direction of the reciprocating motion to a perpendicular direction. The benefits of such partial balancing depend upon the types of mounting used for the machine.

It is possible to achieve balance by introducing two *counter-rotating* balance masses each of mass $m_s/2$ at radius r, or of equivalent $m_b r_b$, as shown in Fig. 4.67. The horizontal components of the forces due to the balance masses will cancel the $m_s \omega^2 r \cos \omega t$ term from the reciprocating mass and their vertical components will be equal and opposite. This approach involves considerable complication of the system.

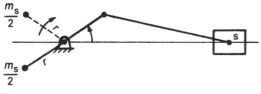

FIG. 4.67

It must be remembered that in the above we have assumed that the mass of the connecting rod is negligible and that $r \ll l$. If these approximations are not valid (they would not be in most modern I.C. engines) the situation is more complex and all the terms in eqns (i) to (ix) would have to be considered. The out of balance forces would include higher harmonics of the running speed and the effect of the connecting rod would be to generate out of balance moments as well as additional out of balance forces. These effects are discussed in detail in more advanced texts.

4.9 Robot arm

Figure 4.68 shows a simple two degree of freedom robot arm. The arm CD carries the robot gripper hand at the point C and is driven by means of the electric torque motor EM_2. The armature and stator of this motor are rigidly connected to arms CD and AB respectively and its output axis passes through the point G_2 as shown. A counterbalance mass C_2 is added to the arm CD at D so that the centre of mass of the arm also lies at G_2.

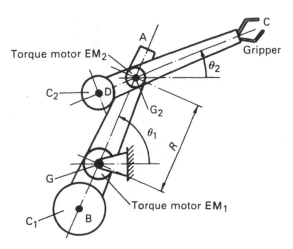

FIG. 4.68

A second torque motor EM_1 is connected to the arm AB and has its output axis passing through the point G. The stator of this motor is rigidly connected to ground and thereby supports the whole robot arm. A second counterbalance mass C_1 is added to arm AB at B and is chosen so that the centre of mass of the *complete* robot arm *always* lies at G. (The reader should verify that this can be achieved provided that the centre of mass of arm CD lies at the point G_2.)

We shall assume that the masses of arms AB and CD are m_1 and m_2 respectively and that I_1 and I_2 represent their moments of inertia about axes through G and G_2 respectively. The angular positions of the arms, at any instant, are given by θ_1 and θ_2 as shown, and the distance $GG_2 = R$.

The robot consists of two rigid bodies AB and CD. Body AB is in pure rotation about the fixed point G and body CD is in general plane motion. To obtain the equations of motion for the robot we first have to draw the free body diagrams for each of the bodies in the system.

Let us start by considering arm CD. Its free body diagram is shown in Fig. 4.69. M_2 represents the torque applied to CD by the armature of the torque motor EM_2. The components F_{x2}, F_{y2} represent the forces applied to link CD at G_2 from the motor bearings, and m_2g is the weight of CD.

Using eqns (4.43) for a body in general plane motion we obtain, for link CD

$$\rightarrow \quad F_{x2} = m_2\ddot{x}_{G2}, \tag{i}$$

$$\uparrow \quad F_{y2} - m_2g = m_2\ddot{y}_{G2}, \tag{ii}$$

and $\quad \widehat{G_2} \quad M_2 = I_2\ddot{\theta}_2. \tag{iii}$

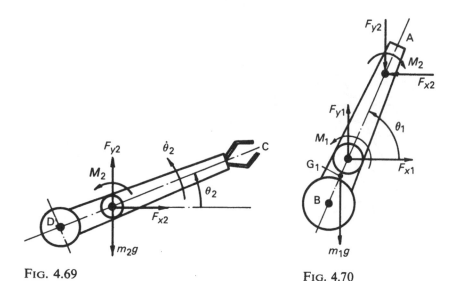

FIG. 4.69   FIG. 4.70

The free body diagram of arm AB is shown in Fig. 4.70. This arm is in pure rotation about the fixed point G. The forces and the moment acting on arm AB at the connection G_2 are equal and opposite to those acting on arm CD. Components F_{x1} and F_{y1} represent the forces on the arm at the fixed pivot and the weight of the arm m_1g acts through the centre of mass of the arm G_1. Motor EM_1 exerts a torque M_1 on the arm as shown. To ensure that the centre of mass of the whole robot lies at G, the centre of mass G_1 of arm AB is positioned a distance r from G where

$$r = \frac{m_2R}{m_1}. \tag{iv}$$

Using eqns (4.30) for a body in pure rotation

$$\rightarrow \quad F_{x1} - F_{x2} = m_1\ddot{x}_{G1}, \tag{v}$$

$$\uparrow \quad F_{y1} - F_{y2} - m_1g = m_1\ddot{y}_{G1}, \tag{vi}$$

and ⑤ $M_1 - M_2 + F_{x2}R \sin \theta_1 - F_{y2}R \cos \theta_1 + m_1 gr \cos \theta_1 = I_1 \ddot{\theta}_1 .$

$$\text{(vii)}$$

The kinematic constraints of the system allow us to express x_{G1}, y_{G1} and x_{G2}, y_{G2} in terms of the angular positions of the arms. For this case

$$x_{G1} = -r \cos \theta_1 ,$$

$$y_{G1} = -r \sin \theta_1 , \tag{viii}$$

and $x_{G2} = R \cos \theta_1 ,$

$$y_{G2} = R \sin \theta_1 . \tag{ix}$$

Differentiating eqns (viii) and (ix) twice with respect to time gives

$$\ddot{x}_{G1} = r\ddot{\theta}_1 \sin \theta_1 + r\dot{\theta}_1^2 \cos \theta_1$$

$$\ddot{y}_{G1} = -r\ddot{\theta}_1 \sin \theta_1 + r\dot{\theta}_1^2 \sin \theta_1 , \tag{x}$$

and $\ddot{x}_{G2} = -R\ddot{\theta}_1 \sin \theta_1 - R\dot{\theta}_1^2 \cos \theta_1$

$$\ddot{y}_{G2} = R\ddot{\theta}_1 \cos \theta_1 - R\dot{\theta}_1^2 \sin \theta_1 . \tag{xi}$$

By substituting eqns (i) to (xi) into eqn (vii) we obtain

$$M_1 - M_2 = (I_1 + m_2 R^2)\ddot{\theta}_1 . \tag{xii}$$

Equations (iii) and (xii) are the equations of motion of the system. If the motor torques are known the angular accelerations can be obtained. The angular velocity and position of the arms can be computed by integrating the angular accelerations. M_1 and M_2 would not normally be constant so that a numerical procedure would have to be used to integrate eqns (iii) and (xii).

The velocity and position of a point on the hand could then be obtained from the kinematic constraint that

$$x_H = R \cos \theta_1 + l \cos \theta_2$$

and $y_H = R \sin \theta_1 + l \sin \theta_2 ,$

where l is the distance of the point from G_2.

It can be seen from eqn (iii) that the motion of arm CD can be controlled by motor torque M_2. The motion of arm AB however depends upon the value of $(M_1 - M_2)$. This shows that we must consider the motion of arm CD when considering how to control the motion of arm AB.

By placing the centre of mass of arm CD on the axis of motor EM$_2$ and the centre of mass of the *whole* robot arm on the axis of motor EM$_1$ we can see that the control of the robot can be made to be independent of the weights $m_1 g$ and $m_2 g$. This arrangement greatly simplifies the computer control algorithm used to generate the motor torques M_1 and M_2. The disadvantage of this balanced design is that the counterbalance masses at B and D may give rise to large values for I_1, I_2 and m_2. For given torque motors this would tend to reduce the speed of response of the robot.

4.10   Vibration and vibration isolation

We have shown that machines in which the major components are in pure rotation, such as in steam or gas turbines, can be balanced by the addition or removal of mass in two separated planes. In practice perfect balance can never be achieved but it is possible to reduce the out of balance forces in such machines to an acceptably low level at an acceptable cost. To achieve the same result with a reciprocating machine is much more difficult and requires the addition of counter rotating shafts.

Out of balance forces in any machine will cause periodically varying forces to be transmitted into the supporting structure or foundations. If the machine is firmly bolted down the out of balance will be directly transmitted by the fixing bolts. This is of little importance if the foundation is a large rigid block, but in the more usual case where the foundation is not rigid the periodic forces can produce unacceptable levels of noise and vibration. For example an air conditioning plant with a reciprocating air compressor mounted in its usual position on the roof of a building can cause annoying vibration of the building.

Periodic forces can also arise from other sources, for example from the cutting tool in a lathe or from aerodynamic forces caused by turbulence. To reduce the transmission of such periodic forces into the supporting structure flexible elements can be introduced between the source of the excitation and the support structure or foundations.

The detailed analysis of vibrating systems is very complex, but we can illustrate some of the principal effects by using a simple mathematical model in which the flexible elements are assumed massless. Let us consider the case of a machine supported on springs and initially we will assume that there are no externally applied forces. A machine would normally be supported on at

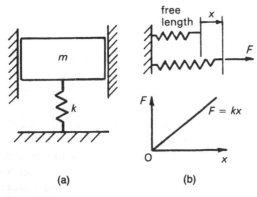

(a)   (b)

FIG. 4.71

least four springs but we will, in our model, combine them into a single equivalent spring. We will also assume that the machine can move only with pure translation in the vertical direction so that horizontal and rotational motions are not considered.

We will assume that the support consists of linear spring of stiffness k and that the machine can be considered as a rigid body of mass m. The idealised system is shown in Fig. 4.71(a). A linear spring of stiffness k produces a force which is directly proportional to its extension x such that the force in the spring

$$F = kx, \qquad\qquad\qquad\qquad\qquad\text{(i)}$$

as shown in Fig. 4.71(b).

Metallic springs behave in this manner but rubber springs (often used in anti-vibration mountings) can deviate from this assumption.

In the SI system of units k will have dimensions of N/m, i.e. force/unit extension.

Let us now consider the equilibrium of the system when it is stationary. When the machine is lowered gently onto the springs they will be compressed by an amount y_s called the static deflection, as shown in Fig. 4.72. The machine is then in equilibrium under the action of its weight and the force applied to it by the spring, as shown in the free body diagram of Fig. 4.73. Hence for vertical equilibrium of the mass we have

$$\downarrow mg - ky_s = 0,$$

so that $y_s = mg/k.$ $\qquad\qquad\qquad\qquad\qquad\text{(ii)}$

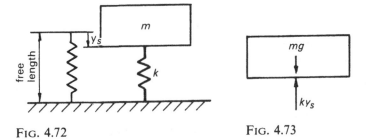

FIG. 4.72   FIG. 4.73

The springs must be designed such that the stresses caused by such a deflection are within the allowable levels for the spring material.

If the machine is disturbed from this equilibrium position its resulting motion can be predicted by obtaining and solving the equations of motion for the mass. First we need to draw the free body diagram of the system for some general position of the mass. Let us consider the forces acting on the mass as it passes through some general position defined by the displacement y, measured from the equilibrium position, as shown in Fig. 4.74. Vertically downwards is taken as the positive direction. The spring will now be compressed by a further amount y so that the force in the spring will now be $k(y_s + y)$. This force will act vertically upwards on the body as shown in the free body diagram of Fig. 4.75.

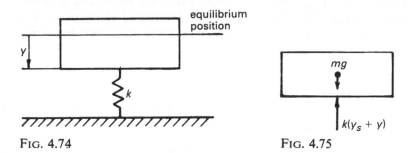

FIG. 4.74   FIG. 4.75

The body is in pure translation so that from eqns (4.19)

$$\downarrow mg - k(y_s + y) = m\ddot{y}.$$

But $mg = ky_s$ from eqn (ii), so that

$$m\ddot{y} + ky = 0. \tag{iii}$$

Equation (iii) is a second order linear differential equation in the displacement and is the equation of motion of a single degree of freedom vibrating system. If we rewrite eqn (iii) as $\ddot{y} = -(k/m)y$ we see that the direction of the acceleration $\ddot{y}$ of the mass is always opposite from that of the deflection y.

The solution to eqn (iii) is well known and can be written

$$y = A \sin p_n t + B \cos p_n t, \tag{iv}$$

where A and B are constants and $p_n = \sqrt{k/m}$.

Equation (iv) may also be expressed as

$$y = C \sin(p_n t + \phi) \tag{v}$$

where $C = \sqrt{A^2 + B^2}$ and $\tan \phi = B/A$.

That eqns (iv) and (v) are solutions of eqn (ii) can be checked by substituting them back into eqn (iii).

The constants A and B, and hence C and ϕ can be obtained if values of y or $\dot{y}$ are known at particular values of t. Two values are required since we need to obtain two unknown constants. Usually we know initial conditions, i.e. those at $t = 0$.

For example if the machine is displaced a distance y_0 from the equilibrium position and released from rest, we have

$$y = y_0 \quad \text{and} \quad \dot{y} = 0 \qquad \qquad \text{at } t = 0.$$

Substituting for y into eqn (iv) with $t = 0$ gives

$$y_0 = \text{B}.$$

By differentiating eqn (iv) with respect to time and substituting the second condition we obtain

$$\dot{y}_0 = Ap_n = 0.$$

Hence the resulting free motion of the mass is described by

$$y = y_0 \cos p_n t. \tag{vi}$$

Equation (vi) represents simple harmonic motion which has already been discussed in Chapter 1 (see p. 5).

Our simple mathematical model therefore predicts a free vibration as shown in Fig. 4.76 of amplitude equal to the initial displacement from the equilibrium position of the machine. This motion will continue for all t without decay.

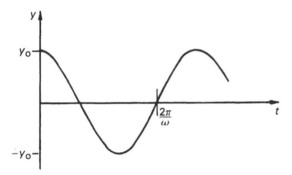

FIG. 4.76

However, we know that such free vibration will decay with time because of the energy losses inherent in any real system due to friction, air resistance and hysteresis in the spring materials. Since more advanced mathematical models are needed to cope with these effects they will not be considered here.

The resulting vibration occurs at a frequency p_n which is called the natural frequency of the system. Its value depends upon the mass of the machine and the stiffness of its supports, but is *independent* of the initial conditions. (The reader should investigate the effect of changing the initial conditions to $y = 0$, $\dot{y} = \dot{y}_0$, at $t = 0$, i.e. zero displacement from equilibrium but some initial velocity).

Let us now consider the effect of some external periodic force $F \cos \omega t$ applied to the machine. To obtain the equation of motion under these conditions we again draw the free body diagrams of the mass and the spring as the machine passes through some general position defined by y from the equilibrium position. From the free body diagram of the mass shown in Fig. 4.77

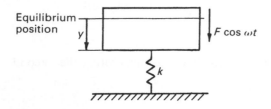

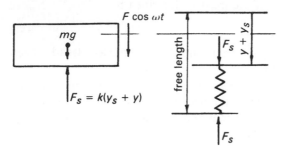

FIG. 4.77

we obtain

$$mg - k(y_s + y) + F \cos \omega t = m\ddot{y},$$

so that $m\ddot{y} + ky = F \cos \omega t.$ (vii)

The left hand side of eqn (vii) is the same as that of eqn (iii) but a forcing term has been added to the right-hand side. Its solution is given by the sum of the complementary function and the particular integral.

The complementary function is the solution of eqn (vii) with the right-hand side set to zero, so that

$$y = A \sin p_n t + B \cos p_n t,$$

as before. This again represents free vibration at the natural frequency p_n.

The particular integral can be obtained by any of the standard methods of solution, e.g. D operator, Laplace transform etc. The reader should confirm that the solution is

$$y = \frac{F \cos \omega t}{k - m\omega^2},$$ (viii)

so that the full solution for y is given by

$$y = A \sin p_n t + B \cos p_n t + \frac{F \cos \omega t.}{k - m\omega^2}.$$ (ix)

The motion, or response, of the mass is therefore given by a forced vibration at the frequency ω upon which is superimposed a free vibration at the natural frequency p_n. The amplitude of the free motion at the natural frequency will depend upon the initial conditions and would, in a real system, gradually decay with time because of energy losses. The motion at the forcing frequency will persist as long as the force $F \cos \omega t$ is applied. These two components are referred to as the *transient* and *steady state* parts of the response.

Thus immediately after switching on the machine the motion would be represented by eqn (ix). However as the terms corresponding to the free motion die away only the forced motion would remain. The response would therefore approach the steady state solution given by eqn (viii). Typical responses are shown in Fig. 4.78.

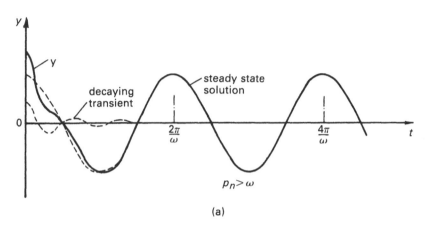

(a)

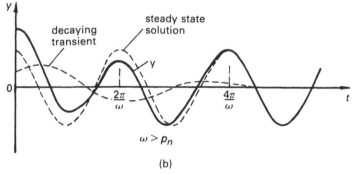

(b)

FIG. 4.78

If the amplitude during the initial few cycles of vibration is not such as to overstress the support springs, only the steady state behaviour is of interest.

Let us now examine how the amplitude of the steady state vibration varies as the forcing frequency ω changes.

The steady state component of eqn (iii) may be expressed as

$$y = Y_0 \cos \omega t,$$

where Y_0 is the amplitude of vibration.

From eqn (ix), this amplitude may be expressed as

$$Y_0 = \frac{F}{k} \frac{1}{(1 - \omega^2/p_n^2)}.$$ (x)

The amplitude of vibration is therefore directly proportional to the amplitude of the applied force, but it also depends upon the ratio of excitation frequency to the natural frequency, i.e. ω/p_n.

Let us first consider the effect on the amplitude of vibration of varying ω whilst keeping the magnitude F of the applied force constant. If ω is zero the amplitude is F/k. In this case the applied force is a steady force and the deflection is constant.

As ω increases the denominator initially decreases so that the amplitude of the vibration increases. When ω is equal to the natural frequency p_n, the denominator becomes zero and the amplitude becomes very large. Under these conditions, with the forcing frequency equal to the natural frequency, the mass-spring system is said to be in *resonance*. Up to this frequency the amplitude has been positive so that the displacement has been in phase with the force F. If we now consider ω to be very slightly greater than p_n the denominator remains very small but has changed sign. This means that as soon as the exciting frequency exceeds the natural frequency the displacement becomes 180° out of phase with the force. As ω is increased further the displacement remains 180° out of phase and its magnitude decreases. When the forcing frequency is very high the denominator becomes very large and Y_0 approaches zero.

The effect on Y_0 of varying ω is shown in Fig. 4.79. Often the modulus of Y_0 is plotted as in Fig. 4.80.

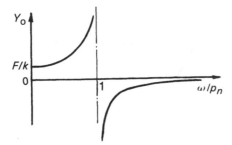

FIG. 4.79

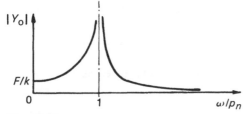

FIG. 4.80

It is important to note that in producing these results the *magnitude* of the force F was kept constant and only the forcing *frequency* was varied. The relationship between Y_0 and F at a given frequency is linear so that doubling F would result in a doubling of the amplitude. Even if F is very small an infinite amplitude is predicted at resonance when $\omega = p_n$.

In any real system there will be effects which we have not included in our simple mathematical model. These effects would limit the amplitude of vibration. For example there would be some energy losses in the system. These losses could be due to hysteresis in the materials of the flexible elements or may be introduced to control the vibration by using dampers. The forces which cause energy losses are collectively known as damping forces. If the damping forces were included in the model the response would be as shown in Fig. 4.81. The maximum amplitude and the sharpness of the peak will be determined by the damping forces. However, our simple model shows that the ratio of the forcing frequency to the natural frequency of the system has a major effect on the amplitude of vibration of the system and that it is important not to excite a machine at a frequency near the natural frequency of the machine on its supports.

The force transmitted into the foundations depends upon the extension of the springs as shown by the free body diagrams of Fig. 4.82. This force will

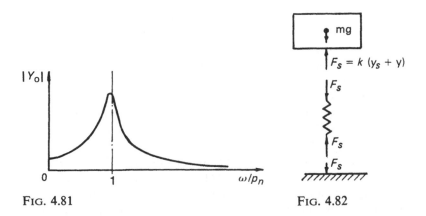

FIG. 4.81   FIG. 4.82

consist of a steady component equal to the weight, mg, of the machine plus a fluctuating component due to the vibration. Only the fluctuating component F_t will cause noise and vibration in the supporting structure. In the steady state this force is given by

$$F_t = ky = \frac{F}{1 - \omega^2/p_n^2} \cos \omega t. \qquad \text{(xi)}$$

Thus as the forcing frequency varies the force transmitted to the supporting structure is influenced in the same way as the amplitude Y_0. The ratio F_t/F is plotted against ω/p_n in Fig. 4.83.

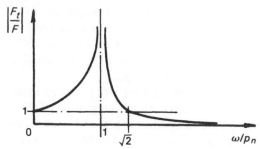

FIG. 4.83

If the machine were solidly bolted to a very stiff support structure, k would be very large. The natural frequency p_n would also be very large and, from eqn (xi), the transmitted force would be equal to the disturbing force. The addition of flexible elements between the machine and its support structure will only be beneficial if the ratio $F_t/F < 1$.

From eqn (xi)

$$\frac{F_t}{F} = \frac{1}{1 - \omega^2/p_n^2} \tag{xii}$$

which gives $|F_t/F| = 1$ when $\omega/p_n = 0$ or $\omega/p_n = \sqrt{2}$. Hence $|F_t/F| < 1$ only when $\omega > p_n\sqrt{2}$. Usually the forcing frequency corresponds to the running speed of the machine and is fixed. The flexible elements must therefore have as low a stiffness as possible to ensure that the natural freuency of the system is low. The ratio ω/p_n will then be large and the greater its value the smaller will be the force transmitted. As can be seen, flexible mountings are not always beneficial. If it is not possible to achieve a natural frequency less than $\omega/\sqrt{2}$ the introduction of flexible elements will *increase* the force transmitted and it would be better for the machine to be solidly bolted down.

In cases where the periodic exciting force arises from out of balance forces in the machine the magnitude of F will not be constant but will vary with running speed. Application 4.6 showed that the out of balance force H produced by a reciprocating machine could, with some simplfying assumptions, be represented as

$$H = m_s\omega^2 r \cos \omega t = F \cos \omega t.$$

Thus $$F = m_s\omega^2 r \tag{xiii}$$

and is proportional to the square of the running speed. By substituting eqn (xiii) into eqn (x) the amplitude of vibration

$$Y_0 = \frac{\omega^2 r m_s}{k} \frac{1}{(1 - \omega^2/p_n^2)},$$

which, since $k = p_n^2 m$, may be rewritten

$$Y_0 = \frac{rm_s}{m} \frac{\omega^2/p_n^2}{(1 - \omega^2/p_n^2)}. \qquad \text{(xiv)}$$

The magnitude of Y_0 varies with ω as shown in Fig. 4.84. As ω increases the amplitude of the vibration builds up until, at resonance when $\omega = p_n$, very large amplitudes are predicted. The amplitude then decreases until it approaches a value of rm_s/m when $\omega/p_n \gg 1$.

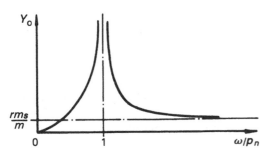

FIG. 4.84

From eqn (xi) the magnitude of the force F_t transmitted to the foundation can be expressed as

$$F_t = \frac{\omega^2 rm_s}{(1 - \omega^2/p_n^2)}. \qquad \text{(xv)}$$

Thus $\quad \dfrac{F_t}{F} = \dfrac{1}{(1 - \omega^2/p_n^2)},\qquad \text{(xvi)}$

which is the same as eqn (xii).

Again the running speed ω must be greater than $p_n\sqrt{2}$ if the flexible mountings are to be beneficial.

Also when $\omega \gg p_n$ the force transmitted to the foundation

$$F_t = kg = \frac{krm_s}{m},$$

and does not vary with speed. It is proportional to the spring stiffness, so again low stiffness supports are beneficial. The ratio F_t/F, however, continues to decrease because although F_t remains substantially constant F continues to increase with the square of ω.

Machines have, of course, to be started and stopped. A machine on properly designed, flexible supports must therefore run through the speed range where large amplitudes are likely to occur. Our simple analysis assumes a constant value of ω and so cannot predict the effect of accelerating the machine up to its normal running speed. However, if the machine is accelerated rapidly through the region in which ω is near to p_n the amplitude should not have time to build up to a dangerous level. Care must be taken to ensure that damage does not occur.

Normally vibration is undesirable and steps have to be taken, as in the above case, to minimise its effects. In some cases, such as in vibrating conveyers, we need to maximise the vibration. Such machines would therefore be designed to operate in a resonant manner, i.e. the natural frequency of the machine on its supporting springs would be made equal to the excitation frequency.

Exercises

1   A uniform bar AB of length l is accelerated over a rough horizontal plane (coefficient friction μ) in the direction BA by a force P applied at A. Find the tension in the bar at a section R where RA $= a$.

2   Fig. 4.85 shows a simple pendulum of mass m and length l. If the support point O is given a constant acceleration $a_x = a_0$ and $a_y = ka_0$, as shown, determine the constant angle θ made by the pendulum, and the tension in its wire.

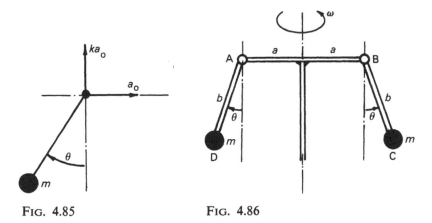

FIG. 4.85   FIG. 4.86

3   Derive an expression for the steady angle θ in the mechanism shown in Fig. 4.86 in terms of the lengths a, b and the constant angular velocity about the vertical axis. All masses except those at C and D may be neglected, and the hinges at A and B may be assumed frictionless.

4   Determine the range of ω for which the slider A in Fig. 4.87 will remain in the position shown if the coefficient of friction between slider and rod is 0.3.

5   Fig. 4.88 shows a block of mass m^* resting on a smooth horizontal plane. A uniform cantilever AB of mass m and length l is rigidly clamped to

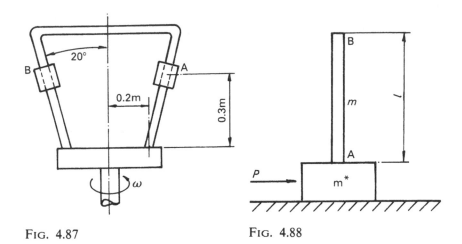

FIG. 4.87   FIG. 4.88

the block at end A. Determine the bending moment applied to the cantilever by the block if a horizontal force *P* is applied to the block as shown.

6   In the Scotch-yoke mechanism shown in plan in Fig. 4.89 the crank OA rotates counter-clockwise with an angular velocity of 10 rad/sec. A small block A slides along the slot CD in the reciprocating member CDB. Show that CDB moves with simple harmonic motion in the line OE, and find its maximum velocity. OA = 50 mm.

   If the motion of block A in the slot CD is opposed by dry friction of coefficient 0.3, and if friction elsewhere is negligible, find as a function of the crank angle θ shown in Fig. 4.89 the torque required at the crank to drive the mechanism when θ lies between 0° and 90°. The mass of CDB is 5 kg and the mass of all other parts may be neglected.

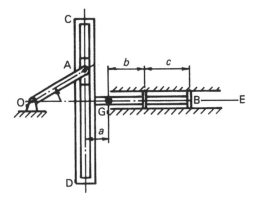

FIG. 4.89

7   A monorail car with front-wheel drive of mass 5 tonnes is suspended from
two wheels 10 m apart as shown in Fig. 4.90. Its centre of mass G is
mid-way between the wheels and 4 m below the rail. If it is designed to
achieve a maximum speed of 200 km/h with uniform acceleration in 40
seconds from rest on a level rail, determine the minimum required
coefficient of friction between the front wheel and the rail. Neglect the
mass of the wheels.

   Discuss whether a higher acceleration could be achieved with rear wheel
drive, assuming that the coefficient of friction is the same. **Hint:** draw the
free body diagrams of the wheels.

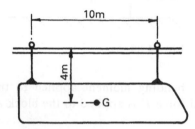

FIG. 4.90

8   Fig. 4.91 shows a block of mass m supported on two idlers. The rollers
are driven at high angular velocities in the directions shown such that
slip always occurs at points A and B.
   If the centre of mass G is offset from the centre line PQ by an amount
x, show that the block acceleration is given by

$$\ddot{x} = -\frac{g\mu}{b - \mu a} x$$

where μ is the coefficient of friction between the block and the rollers.

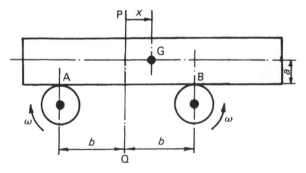

FIG. 4.91

9 The mass of a uniform circular disc of radius R and centre O is m. A hole of radius $R/2$ and centre C is drilled out of the disc where OC $= R/4$. Find the moment of inertia of the drilled disc about an axis through C perpendicular to the plane of the disc in terms of m^* and R, where m^* is the mass of the drilled disc.

10 Fig. 4.92 shows a wire loop which is in the shape of a sector of a circle of radius R. If the mass/unit length of the wire is ρ, determine from first principles the moment of inertia of the loop about axes OX and OY.

Use the perpendicular axis theorem to determine the moment of inertia of the loop about an axis through O perpendicular to the plane of the loop.

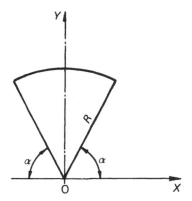

FIG. 4.92

11 Fig. 4.93 shows a rectangular prism of mass m with dimensions a, b and c. Determine the moment of inertia I_Z about an axis OZ which passes through the geometric centre O.

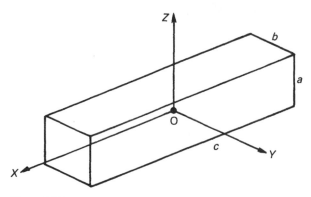

FIG. 4.93

12 A symmetric disc can rotate freely in bearings at its centre C. A torque
$T = 8 \sin 2t$ Nm is applied to the disc when it is initially at rest. Neglecting
friction, derive an expression for θ, the subsequent rotation of the disc,
as a function of time t. The moment of inertia of the disc about a polar
axis through C is $0.4 \, \text{kgm}^2$.

13 A homogenous rectangular plate ABCD shown in Fig. 4.94 measuring
3m × 2m and of mass 60 kg can rotate in a vertical plane about a fixed
frictionless hinge O attached to corner A. If it is released from rest with
AG horizontal find the force at the hinge immediately after release.

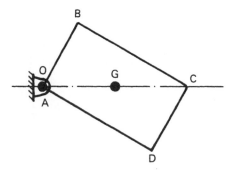

FIG. 4.94

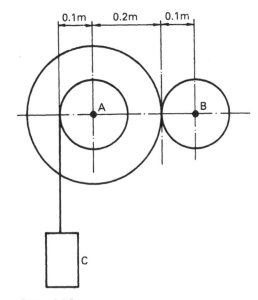

FIG. 4.95

14 A mass m is fastened to one end of a light rope. The rope is coiled round a circular pulley of radius r which is left free to rotate about its axis. Prove that the mass descends with acceleration $g/(1 + I/mr^2)$, where I is the moment of inertia of the pulley about its axis of rotation.

15 The two gear wheels shown in Fig. 4.95 may rotate with negligible friction about their fixed centres at A and B. An integral pulley fixed to the left-hand wheel carries a light cable which supports a body C of mass 4 kg. If the assembly is released from rest, find the acceleration of C and the tension in the cable. The moment of inertia about an axis through A of the left-hand wheel and pulley assembly is 0.2 kg m^2 and the moment of inertia about an axis through B of the right-hand wheel is 0.04 kg m^2.

16 A uniform circular disc of mass m and centre C is mounted with eccentricity OC $= e$ at mid-span of a rigid shaft AB (Fig. 4.96). The shaft is supported at A and B in identical short bearings: it rotates at constant speed ω under the action of a constant driving torque T applied at end A and a resisting torque developed by friction in the bearings. Gravity and friction elsewhere may be neglected.

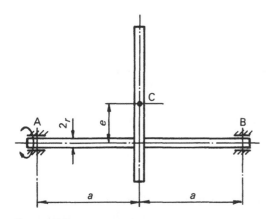

FIG. 4.96

If the friction between the bearing and the shaft is dry and has a coefficient μ, derive expressions for the normal and tangential components of the force at each bearing. Hence find the power that must be supplied by the torque T to maintain the given motion. The effective radius of each bearing is $r = 10$ mm, $e = 50$ mm, $\mu = 0.2$, $\omega = 50$ rad/s and the disc has mass 20 kgm.

17 A parallelogram four bar linkage consists of three uniform bars AB $= a$, BC $= 2a$, CD $= a$, hinged together at B and C and to a fixed frame at A and D. If a torque T is applied about A to AB when BC is perpendicular to AB and CD, find the instantaneous angular acceleration of AB starting from rest. Neglect gravity and friction. The masses of the links are m, $2m$, m respectively.

18   Fig. 4.97 shows a two-stage gear reduction in which the three gears turn on the fixed centres O_1, O_2 and O_3, about which their respective moments of inertia are 0.03 kg m^2, 0.30 kg m^2 and 4.00 kg m^2. A constant torque of 20 Nm is applied to the driving gear about O_1 and a load torque of magnitude $0.2 \omega_3^2$ Nm opposes the rotation of the output gear about O_3, where ω_3 is the angular velocity of the output gear about O_3 in units of rads/sec.

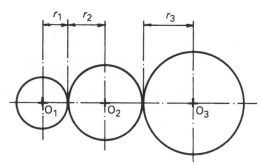

FIG. 4.97

Find (a) the initial angular acceleration of the driving gear starting from rest, and (b) the angular velocity of the driving gear when steady speed conditions have been reached.

The effective pitch-circle radii of the gears are

$$r_1 = 0.05 \text{ m}, \qquad r_2 = 0.10 \text{ m} \quad \text{and} \quad r_3 = 0.20 \text{ m}.$$

Neglect friction.

19   A slider crank mechanism is shown in Fig. 4.98. In the position shown the mechanism is at rest. The slider is required to have an initial acceler-

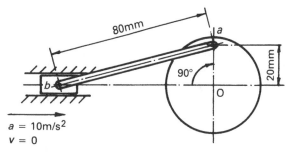

$a = 10\text{m/s}^2$
$v = 0$

FIG. 4.98

ation of 10 m s^{-2}. Calculate the magnitude of the force F which must be applied to the slider to give it the required acceleration, and the force in the connecting rod ab.

The mass of the slider is 0.1 kg, the polar moment of inertia of the flywheel to which the coupler is pinned is $2 \times 10^{-4} \text{ kg m}^2$ and the mass of the coupler may be neglected. Gravity and friction may also be neglected. **Hint** Use a graphical method to obtain the accelerations of the links.

20 A heavy uniform bar AB rests in a horizontal position on two identical pulleys with its centre of mass G midway between the centres of the pulleys, as shown in Fig. 4.99. Each pulley may turn freely and without friction about its centre. If a constant torque T is applied to the left-hand pulley, derive an expression for the initial acceleration $\ddot{x}$ of the bar, assuming no slip between the bar and either pulley.

The mass of the bar is m. Each pulley has radius r and moment of inertia I about an axis through its centre.

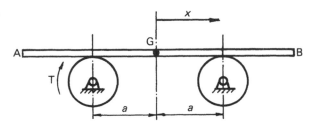

FIG. 4.99

21 The wheel shown in Fig. 4.100 is supported by a cable and has mass m and moment of inertia I_O about an axis through 0. Determine the initial angular acceleration of the wheel and the tension in the cable when the wheel is released from rest under gravity, in the position shown.

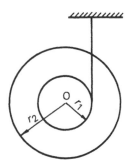

FIG. 4.100

22   A uniform circular disc of mass m and radius r is released from rest on a rough plane inclined to α to the horizontal, as shown in Fig. 4.101. The coefficient of friction between disc and plane is μ. Write down the equations of motion of the disc, and hence derive an expression for the maximum allowable value of inclination α if the disc is not to slip on the plane. Discuss the motion if this value of α is exceeded.

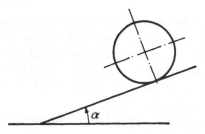

Fig. 4.101

23   If the rollers in Fig. 4.102 are uniform and identical, and if there is no slip between *either* rollers and ground or rollers and slab, derive an expression in terms of the given quantities for the initial acceleration of the slab. The mass of the slab is M and each roller has mass m.

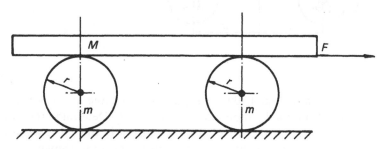

Fig. 4.102

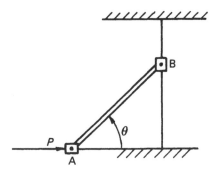

Fig. 4.103

24 Fig. 4.103 shows the plan view of a heavy steel door AB which closes an access way in the bulkhead of a ship. The edges of the door at A and B slide along perpendicular guides under the action of a hydraulic ram that provides a constant force P at A. Friction may be neglected, the door is of width $AB = 2a$, its mass is m and its moment of inertia about the vertical axis through its centre of mass G is $I = mk^2$ where G is at the mid-point of AB.

   Show that the equation of motion of the door can be expressed as $\ddot{\theta} = f(\theta)$ and find the function $f(\theta)$.

25 An epicyclic gear, shown in Fig. 4.104 consists of an arm OA, a fixed sun wheel S of centre O, and a planet wheel P attached to OA by a bearing

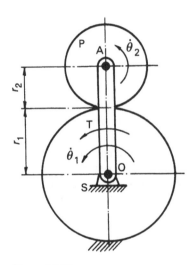

FIG. 4.104

at A. When the arm OA rotates, P rolls without slip on S. Show that the angular velocity $\dot{\theta}_2$ of P is related to the angular velocity $\dot{\theta}_1$ of OA by the expression

$$r_2\dot{\theta}_2 = (r_1 + r_2)\dot{\theta}_1$$

where r_1 and r_2 are the effective radii of S and P respectively.

   If a torque T is applied to OA about O, derive the equations of motion for the arm OA and for the planet wheel P. Hence find an expression for the angular acceleration of OA in terms of T. The moment of inertia of OA about an axis through O is I_1 and the moment of inertia of P about an axis through A is I_2. The mass of P is m. Neglect gravity and friction at the bearings.

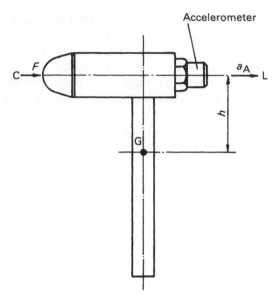

Fig. 4.105

26 Fig. 4.105 shows a hammer used in vibration testing of structures. A small accelerometer is mounted in the head of the hammer and produces an output corresponding to the acceleration along CL. This is used to calculate the force F that the hammer applies to the structure.

If the force applied by the hand can be neglected determine the relationship between F and the acceleration a_A recorded by the accelerometer.

The hammer has mass m and its centre of mass G is a distance h below the line of action of F. The moment of inertia of the hammer about an axis through G is I_G.

Neglect the angular velocity of the hammer during the impact.

5

Work and energy

Basic Theory

We have shown in Chapter 4 how the behaviour of systems can be analysed by the direct application of Newton's Laws of Motion. An alternative approach, based on the First Law of Thermodynamics, will now be considered which can, in certain circumstances, provide a simpler and quicker analysis.

5.1 Work done by a force

Figure 5.1 shows a force F applied to a point P in a mechanical system S.

We will suppose that when the force F is applied P undergoes a displacement δr and moves to P'. It is assumed that the direction of F and δr are inclined to one another at an arbitrary angle θ, as shown in Fig. 5.2.

Let us assume that the displacement from P to P' is given by a displacement $\delta r \cos \theta$ along the line of action of F from P to Q, followed by a displacement $\delta r \sin \theta$, perpendicular to the line of action of F, from Q to P'. Because the point P has been displaced from P to P' by the force F we say that *work* has

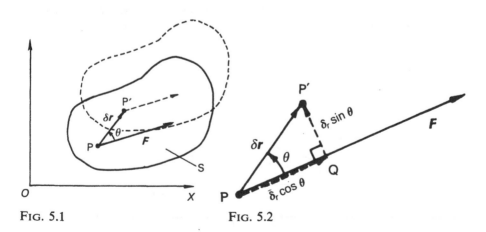

FIG. 5.1   FIG. 5.2

been done *on* the system by the externally applied force. The amount of work, δW, done by the force during this displacement is a scalar quantity defined by

$$\delta W = F\,\delta r \cos\theta, \tag{5.1}$$

where F is the magnitude of the force and $\delta r \cos\theta$ is the displacement of the point of application of $\mathbf{F}$ measured in the direction of $\mathbf{F}$.

From this definition and from Fig. 5.2 it can be seen that no work is done by the force when its point of application is displaced along a direction which is perpendicular to its line of action, i.e. no work is done during the displacement from Q to P'.

It also follows from eqn (5.1) that the work done will be positive provided the displacement $\delta r \cos\theta$ is in the same direction as the force $\mathbf{F}$. If the displacement of P along the line of action of $\mathbf{F}$ is in the opposite direction from that of the force, as shown in Fig. 5.3, the work done is negative. In this case we say that work has been done by the system S on the system which is applying the force.

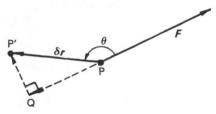

FIG. 5.3

It can be seen from eqn 5.1 that δW will be positive when $\cos\theta > 0$, which requires $-\pi/2 < \theta < \pi/2$.

It is possible to rewrite eqn 5.1 in terms of the components of $\mathbf{F}$ and δr along OX and OY. Suppose that the lines of action of $\mathbf{F}$ and δr are inclined with respect to the axis OX by angles α and β respectively, as shown in Fig. 5.4. Equation (5.1) can therefore be rewritten as

$$\delta W = F\,\delta r \cos(\beta - \alpha),$$

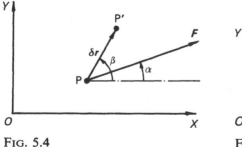

FIG. 5.4   FIG. 5.5

since, in this case, $\theta = (\beta - \alpha)$. Thus by expanding $\cos(\beta - \alpha)$, we have

$$\delta W = F\, \delta r(\cos \beta \cos \alpha + \sin \beta \sin \alpha). \tag{5.2}$$

The components of F and δr along OX and OY are given by

$$F_x = F \cos \alpha, \qquad F_y = F \sin \alpha, \qquad \delta x = \delta r \cos \beta$$

and $\delta y = \delta r \sin \beta$.

Thus, by direction substitution into eqn (5.2) the work done can also be expressed as

$$\delta W = F_x\, \delta x + F_y\, \delta y. \tag{5.3}$$

The total work done is therefore the sum of the work done by each component of the force.

If we now consider the special case shown in Fig. 5.5, where the force is directed along the axis OY, i.e. $F_x = 0$ and $F_y = F$, and the point P is displaced only in the direction OX, i.e. $\delta y = 0$, it can be seen from eqn (5.3) that the work done is $\delta W = 0$. Thus, no work is done by a force when its point of application is moved in a direction which is perpendicular to its line of action.

When the directions of both F and δr lie along the axis OX, i.e. $F_x = F$, $\delta r = \delta x$ and $F_y = \delta y = 0$, the work done is given by

$$\delta W = F\, \delta x. \tag{5.4}$$

For a finite displacement of P in the direction OX, from a position x_1 to a position x_2, the total work done on the system by the force is given by taking the limit as $\delta x \to 0$ and integrating eqn (5.4),

i.e. $$W = \int_{x_1}^{x_2} F\, dx. \tag{5.5}$$

If F is a constant eqn (5.5) can be integrated directly to give

$$W = F(x_2 - x_1).$$

In the case, where F varies with x, i.e. $F = F(x)$, we need to know how F varies with the position x. Suppose, for example, that $F(x)$ varies as shown in Fig. 5.6. The work done in moving its point of application from position

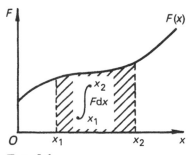

FIG. 5.6

x_1 to position x_2 is, from eqn (5.5), the area bounded by $F(x)$ between $x = x_1$ and $x = x_2$. The work done by a force can therefore be obtained by evaluating, either graphically or numerically, the area under the force-displacement curve.

From its definition, work has units of Nm, or in terms of the fundamental units kg m^2 s^{-2}. The Nm is often called the Joule(J) so that $1\,J = 1\,Nm = 1$ kg m^2 s^{-2}.

Example 5.1   Let us consider the system S to be a simple spring fixed at one end and let the point P be located at the free end of the spring, as shown in Fig. 5.7(a).

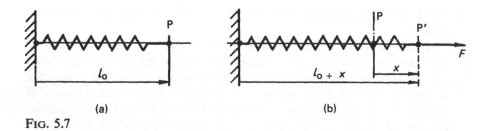

(a)   (b)

FIG. 5.7

We will calculate the work done by the force F when the point P is displaced an amount x from its free position.

The force required to produce this displacement is shown in Fig. 5.8 and is given by the relationship

$$F = kx, \tag{i}$$

where k is the stiffness of the spring.

The work done by the force F is given by eqn 5.5 and for this case is

$$W = \int_0^x F \, dx = \int_0^x kx \, dx = \tfrac{1}{2}kx^2. \tag{ii}$$

Equation (ii) shows that the work done is proportional to the square of the displacement. The graph of F against x is the straight line shown in Fig. 5.8.

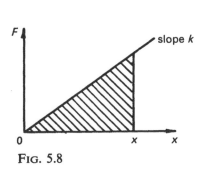

FIG. 5.8

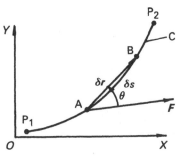

FIG. 5.9

The area of the triangle formed between $x = 0$ and x is $\frac{1}{2}kx^2$ which is equal to the work done by F.

Work done by a force moving along a curved path

Suppose we wish to calculate the work done by a force F when its point of application is moved from P_1 to P_2 along some path C in the XY plane, as shown in Fig. 5.9. We can see that if the point of application of the force moves from a point A on the curve to a neighbouring point B then the displacement of the point of application is approximately $\delta r = \delta s$, where δs is the arc length from A to B. The work done between A and B is therefore given by

$$\delta W = (F \cos \theta)\, \delta s. \tag{5.6}$$

Since, in the limit as $\delta s \rightarrow 0$, θ can be interpreted as the angle between the line of action of F and the tangent to the curve at A, the quantity F cos θ is the component of F measured along the direction of the tangent.

   Thus by setting

$$F_t = F \cos \theta,$$

the work done by the force F can be written as

$$\delta W = F_t\, \delta s. \tag{5.8}$$

This work is done by F_t the tangential component of F. As before, the component which is perpendicular to the displacement at A does no work. Therefore, as the point of application of the force is displaced along the curve from position P_1 to P_2, the total work done is found by integrating eqn (5.8)

i.e. $$W = \int_{s_1}^{s_2} F_t(s)\, \mathrm{d}s. \tag{5.9}$$

To evaluate this integral it is necessary to define the curve C and express the tangential component F_t as a function of the arc length s. If eqns (5.5) and (5.9) are compared it can be seen that eqn (5.5) is in fact a special case of eqn (5.9). The result given by eqn (5.5) is for the case where the curve C is given by a straight line parallel to the axis OX.

Example 5.2   To show how eqn (5.9) may be used let us determine the work done by a force of constant magnitude when its point of application P is moved around the circumference of a circle of radius R. We will first assume that the line of action of the force always remains tangential to the circle, as shown in Fig. 5.10. The work done in moving the point of application of the force from point A to a neighbouring point B is given by eqn (5.8) as

$$\delta W = F_t\, \delta s. \tag{i}$$

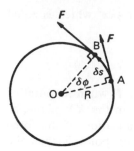

FIG. 5.10

For this case $F_t = F$ and $\delta s = R\,\delta\phi$, so that

$$\delta W = FR\delta\phi. \tag{ii}$$

Since the quantity $FR = M$, the moment of the force F about the centre O of the circle eqn (ii) can also be written in the form,

$$\delta W = M\,\delta\phi. \tag{iii}$$

This shows that the work done by the moment of a force is given by multiplying the magnitude of the moment by its angular displacement measured in the direction of the moment. If the total angular displacement is from $\phi = 0$ to $\phi = \phi_0$ the total work done is

$$W = \int_0^{\phi_0} M\,d\phi. \tag{iv}$$

Since, for this case, M is constant, eqn (iv) can be integrated to give

$$W = M\phi_0 = FR\phi_0. \tag{v}$$

For one complete revolution $\phi_0 = 2\pi$ so that

$$W = (2\pi R)F. \tag{vi}$$

If we now consider the case where the force F acts radially outwards, as shown in Fig. 5.11, then $F_t = 0$. The force F therefore does no work as the point of application of the force moves around the circumference of the circle.

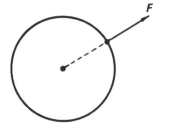

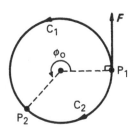

FIG. 5.11   FIG. 5.12

5.2 Conservative and non-conservative forces

Let us reconsider the case of Fig. 5.10 where a force of constant magnitude
F is applied tangentially to the circumference of a circle of radius R and
centre O. Suppose that the point of application of the force is initially at P_1
and is then displaced around the circumference, following the arc C_1, until it
reaches P_2, as shown in Fig. 5.12. The work done by the force in going from
P_1 to P_2, is given by

$$W_{C_1} = FR\phi_0, \tag{5.10}$$

where ϕ_0 is the obtuse angle $P_1\hat{O}P_2$ as shown. Let us now repeat this exercise
but, instead of following arc C_1 to reach P_2, let us move the point of application
of the force along arc C_2. For this case the work done by the force is

$$W_{C_2} = -FR(2\pi - \phi_0) \tag{5.11}$$

so that

$$W_{C_2} = W_{C_1} - 2\pi FR. \tag{5.12}$$

Thus $W_{C_2} \neq W_{C_1}$, which shows that the work done by the force when its point
of application moves from P_1 to P_2 depends upon the path taken.

  Although we have only proved this result for the special case shown in Fig.
5.12, the result is generally true.

  The work done by a force moving between two points P_1 and P_2 therefore
depends upon the path taken by the point of application of the force. In other
words, if the force F in Fig. 5.13 moves from P_1 to P_2 along the path C_A the

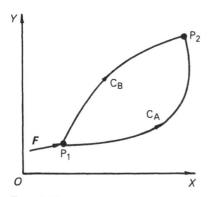

FIG. 5.13

work done by the force will be different from that when it moves from P_1 to
P_2 along path C_B. The work W_A done by the force moving along C_A is given
by

$$W_A = \int_{C_A} F_{t_A} \, ds_A, \tag{5.13}$$

and similarly, along C_B

$$W_B = \int_{C_B} F_{t_B} \, ds_B, \tag{5.14}$$

where F_{t_A} and F_{t_B} are the components of F tangential to the paths C_A and C_B. Generally $W_A \neq W_B$.

If the work done by the force in going from P_1 to P_2 *depends* upon the path taken, the force F is said to be *non-conservative*.

When the work done is *independent* of the path taken, i.e. when $W_A = W_B$, the force F has special properties and is said to be *conservative*.

For this case the work done depends only on a function, V, which is evaluated at points P_1 and P_2. The work done by a conservative force can therefore be expressed as

$$W = -(V(x_2, y_2) - V(x_1, y_1)), \tag{5.15}$$

where (x_1, y_1) and (x_2, y_2) are the coordinates of P_1 and P_2 respectively.

The quantity V is called the potential function of the system which *applies* the conservative force and is a function of *position* only. The negative sign in eqn (5.15) is purposely included to ensure that the potential function of a conservative system decreases as the forces produced by the system do positive work. It will be shown later that the function V represents the potential energy stored in a conservative system.

The relationship between V and a *conservative* force F can be found by considering the work done by F when its point of application is displaced from a position (x, y) to a neighbouring position $(x + \delta x, y + \delta y)$. Using eqn (5.3), the work done by the components of F is given by

$$\delta W = F_x \, \delta x + F_y \, \delta y. \tag{5.16}$$

From eqn (5.15) the work done is also

$$\delta W = -(V(x + \delta x, y + \delta y) - V(x, y)). \tag{5.17}$$

If we now use the rules of partial differentiation, the value of V at $(x + \delta x, y + \delta y)$ can be written as

$$V(x + \delta x, y + \delta y) = V(x, y) + \frac{\partial V}{\partial x} \delta x + \frac{\partial V}{\partial y} \delta y. \tag{5.18}$$

When eqn (5.18) is substituted into eqn (5.17) and the result equated with eqn (5.16) we have

$$F_x \, \delta x + F_y \, \delta y = -\frac{\partial V}{\partial x} \delta x - \frac{\partial V}{\partial y} \delta y. \tag{5.19}$$

This result must be true for all values of δx and δy.

Thus for **F** to be conservative

$$F_x = -\frac{\partial V}{\partial x},$$

and $$F_y = -\frac{\partial V}{\partial y}.$$ (5.20)

Example 5.3 The force due to gravity is an example of a conservative force. Consider the case shown in Fig. 5.14. A particle of mass m is shown located at a point $P_1(x_1, y_1)$ in the vertical plane and acting on it is a vertical gravitational force

$$F_y = -mg.$$ (i)

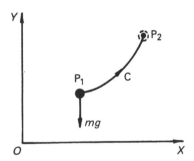

FIG. 5.14

Let us suppose that while this force acts upon it the particle moves to a new position $P_2(x_2, y_2)$ along some arbitrary path C. Since we have chosen the direction of the axis OX to be horizontal, the horizontal component of the gravitational force is zero.

The work done by the force mg is given by integrating eqn (5.3) as

$$W = \int_{y_1}^{y_2} F_y \, dy = -mg \int_{y_1}^{y_2} dy,$$ (ii)

i.e. $$W = -mg(y_2 - y_1).$$ (iii)

The work done is therefore a function *only* of the final and initial positions of P and is independent of the path C.

If we now compare eqn (iii) with eqn (5.15),

$$V(x_2, y_2) = mgy_2 \quad \text{and} \quad V(x_1, y_1) = mgy_1.$$

The potential function V is in this case independent of x and may therefore be written as

$$V = mgy.$$ (iv)

We can now extend this result for a particle to a rigid body of total mass m. Let the particle P_i shown in Fig. 5.15 be displaced vertically from y_{i1} to y_{i2} along some path.

The work δW_i done by the gravitational force acting on the particle of mass m_i is therefore

$$\delta W_i = -m_i g(y_{i2} - y_{i1}).$$

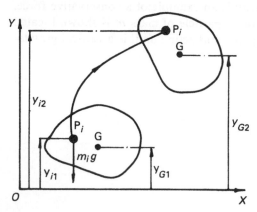

FIG. 5.15

If we sum the work done by the gravitational force on each of the particles in the body, the total work done is

$$W = -g \sum m_i(y_{i2} - y_{i1}) = -g\{\sum m_i y_{i2} - \sum m_i y_{i1}\}, \tag{v}$$

where the values of $(y_{i2} - y_{i1})$ may be different for each of the particles in the body.

Since the vertical positions y_{G1} and y_{G2} of the centre of mass of the body are defined by

$$m y_{G1} = \sum m_i y_{i1} \quad \text{and} \quad m y_{G2} = \sum m_i y_{i2},$$

eqn (v) gives

$$W = -mg(y_{G2} - y_{G1}). \tag{vi}$$

By comparison with eqn (5.15) the potential function $V = mgy_G$. If the vertical displacement of the centre of mass of the body is expressed as

$$h_G = y_{G2} - y_{G1},$$

then

$$W = -mgh_G. \tag{vii}$$

The work done by gravity is determined by the vertical displacement of the centre of mass of the body. Since the gravitational force always acts vertically downwards negative work is done by the force during upward displacements of G. Positive work is only produced when the displacement of G is downwards.

The potential function

$$V = mgh_G \qquad\qquad\qquad \text{(viii)}$$

and will be increased when the centre of mass of the body is raised and be reduced when the centre of mass is lowered.

5.3   Power

When the point of application of a force F is displaced an amount δr eqn (5.1) gives the work done by the force as

$$\delta W = F \delta r \cos \theta \qquad\qquad\qquad (5.21)$$

where θ is the angle between the directions of F and δr.

If this work δW is carried out in a small time interval δt, then the rate P at which the work is done is

$$P = \lim_{\delta t \to 0} \frac{\delta W}{\delta t} = \frac{dW}{dt}. \qquad\qquad\qquad (5.22)$$

P is the rate at which work is transferred from one system to another by a force and is called the *power* of the system which is applying the force. For example, the power of an engine is the rate at which the engine can do work on another system.

The units of power are Js^{-1}, Nms^{-1} or in fundamental units, kgm^2s^{-3}. Power is often quoted in Watts (W) so that $1W = 1\ Js^{-1} = 1\ Nms^{-1} = 1\ kgm^2s^{-3}$.

Using eqn (5.21) power may also be expressed as

$$P = \lim_{\delta t \to 0} F \frac{\delta r}{\delta t} \cos \theta. \qquad\qquad\qquad (5.23)$$

The term $\delta r / \delta t$ is the magnitude of the velocity vector v of the point of application of the force.

Thus $P = Fv \cos \theta,$ $\qquad\qquad\qquad (5.24)$

in which θ can be interpreted as the angle between the force vector F and the velocity vector v of the point of application of the force, as shown in Fig. 5.16.

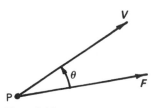

FIG. 5.16

When F and v are given in terms of their components F_x, F_y and v_x, v_y eqn (5.3) shows that the power P can also be written in the form

$$P = F_x \dot{x} + F_y \dot{y}$$

$$= F_x v_x + F_y v_y. \tag{5.25}$$

Similarly if the point of application of the force moves with $v = ds/dt$ along a curve C as shown in Fig. 5.17 then eqn (5.8) gives the power P as

$$P = F_t \frac{ds}{dt} = F_t v. \tag{5.26}$$

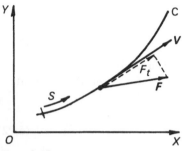

FIG. 5.17

The power is therefore determined by the tangential component of the force F_t, and the velocity v of the point of application of the force along its curve.

From example 5.2 the work δW done by a moment, or torque, M can be written as

$$\delta W = M \, \delta \phi.$$

The rate at which work is done by the moment is therefore

$$P = \lim_{\delta t \to 0} M \frac{\delta \phi}{\delta t} = M \frac{d\phi}{dt} = M \dot{\phi}. \tag{5.27}$$

For example, in a case where the moment M is applied to the end of a shaft, $\dot{\phi}$ will be the angular velocity of the shaft.

Example 5.4   Figure 5.18(a) shows the torque characteristic of a motor in which the torque M varies with its rotational speed ω such that

$$M = M_s \left(1 - \frac{\omega}{\omega_0} \right). \tag{i}$$

Let us find how the power output of the motor varies with speed.
From eqn (5.27)

$$P = M\omega = M_s \left(1 - \frac{\omega}{\omega_0} \right) \omega. \tag{ii}$$

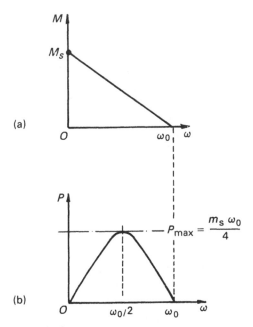

FIG. 5.18

The speed at which maximum power is developed can be obtained by differentiating eqn (ii) with respect to ω and equating the result to zero, i.e.

$$\frac{\mathrm{d}P}{\mathrm{d}\omega} = M_s(1 - 2\omega/\omega_0) = 0. \qquad \text{(iii)}$$

Thus, at maximum power,

$$\omega = \omega_0/2 \quad \text{and} \quad P_{max} = M_s\omega_0/4.$$

The graph of P against ω is as shown in Fig. 5.18(b).

When a motor with such a torque characteristic is used to drive a steady load, the size of the motor can be minimised by ensuring that the motor runs at a speed of $\omega_0/2$, i.e. at its maximum power condition. If the required steady angular velocity ω_L of the load is different from $\omega_0/2$ it would be desirable to connect the motor and load together via a gearbox with speed ratio $\omega_0/2 : \omega_L$.

5.4 Internal energy

Let us consider the *closed* system, shown in Fig. 5.19(a), with external forces F_i applied at points P_i. This system will contain a fixed amount of matter and will be separated from its surroundings by a boundary S which is free to move. Matter is not permitted to leave or enter the system by crossing the boundary S. The system could, for example, be a collection of rigid bodies and the boundary could be defined by the surfaces of the bodies, as shown in Fig.

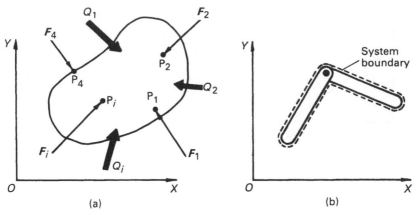

FIG. 5.19

5.19(b). As the system moves the points of application of the forces will move so that work will be done on the system by the forces. If work W is done on the system by the externally applied forces, and if a total amount of heat Q is transferred across the system boundary, then the state of the system will change. For example, the heat transferred into the system will cause its temperature to rise and could also produce displacements within the system due to thermal expansion. Since work occurs because of the displacements of the points of application of the external forces there will be accompanying changes to the positions and velocities of other points within the system. The temperature, displacement and velocity are the properties at a point within the system and are referred to as system states.

A measure of the total change in the state of the system can be calculated from the First Law of Thermodynamics. This law states that

$$W + Q = \Delta U, \tag{5.28}$$

where the quantity ΔU is called the change in the *internal energy* of the system. The internal energy, U, is a function of the system states and eqn 5.28 gives a simple method of determining the additional energy stored in the system due to the total work done on the system and the total heat transferred across its boundary.

In eqn (5.28) the units of Q and U must be the same as those for W, i.e. J, Nm or $kg\,m^2\,s^{-2}$. Since W and Q are scalar quantities, which have the same value in all reference frames then, from eqn (5.28), the internal energy U is also a scalar quantity.

In many machines the heat flow Q may be neglected. The machine is then said to behave *adiabatically* since $Q = 0$, and eqn (5.28) reduces to

$$W = \Delta U. \tag{5.29}$$

If, in addition, no work is done then

$$\Delta U = 0. \tag{5.30}$$

The internal energy of the machine is then fixed at some constant value which is independent of the configuration of the machine,

i.e. $\qquad U_1 = U_2,$ $\hfill$ (5.31)

where the suffixes 1 and 2 refer to separate configurations of the machine.

Equation (5.31) is often referred to as the principle of conservation of energy, and means that when no work is done on, or by, a system, and there is no heat flow into or out of the system, the internal energy of the system remains constant.

The conditions represented by eqns (5.29) and (5.30) do not, however, represent the situation which arises in a mechanical system where friction is present. In such a machine the friction forces generated at the sliding surfaces cause the machine's temperature to rise and heat will be lost to the surroundings. However, during continuous operation the temperature of the machine would reach a steady value and would not change significantly during one cycle of operation. In this case the machine behaves isothermally and there would be a steady heat loss from the machine which must be included in eqn (5.28). A typical example is a gearbox in which heat is generated by the relative sliding motion of the gear teeth. The temperature of the gearbox and its lubricating oil will rise until a steady temperature is reached at which the heat loss from the hot surface of the gearbox is equal to the net work done on the gearbox by the drive and load torques.

To derive an expression for U which is appropriate to the system under consideration we must first consider the various independent forms of internal energy. These can each be calculated from a knowledge of the states of the system once the mass content and the constitutive properties of the system, such as the material specific heats and elasticities are known. For the special case of mechanical systems, i.e. systems built from simple machine elements such as rigid bodies and springs, the internal energy will include contributions from:

(i)   potential energy due to the positions of the bodies in a gravitational field,
(ii) kinetic energy due to the velocities of the particles making up the bodies,
(iii) strain energy in the bodies due to relative displacement of points in the bodies,
(iv) thermal energy due to the temperatures of the bodies.

When applying the First Law of Thermodynamics to a system, the *system boundary* must first be clearly defined. There is usually a number of possible choices for the boundary, all correct, which will lead to the required solution by different routes. We will see that the system enclosed by a boundary can be as simple as a single particle or as complex as a complete machine.

5.5 Potential Energy

Particle

Let us consider a particle P of mass m_P positioned a distance h above the surface of the Earth. For this position Newton's Law of Gravitation predicts that there will be an attractive force F_g acting on both the Earth and the particle given by

$$F_g = \frac{G_c m_e m_P}{(R_e + h)^2},$$
(5.32)

where G_c is the universal gravitational constant and m_e and R_e are the mass and radius of the Earth respectively. To maintain the particle and the Earth in equilibrium, an external force F_O must be applied to each body as shown in the free body diagrams of Fig. 5.20(a).

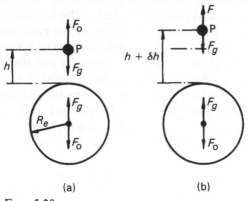

(a)   (b)

FIG. 5.20

Thus $F_O - F_g = 0$,

so that $F_O = \dfrac{G_c m_P m_e}{(R_e + h)^2}.$
(5.33)

Let us now increase the separation h by applying a total force $F = F_O + F_P$ to the particle and a force F_O to the Earth, as shown in the free body diagram of Fig. 5.20(b).

Since the force applied to the Earth exactly balances the gravitational attraction caused by the particle the Earth will remain in equilibrium as the particle moves.

Applying Newton's Second Law of Motion to the particle gives

$$F - F_g = m_P \frac{dv_P}{dt},$$
(5.34)

where v_P is the velocity of the particle.

we can write

$$\frac{dv_P}{dt} = \frac{dv_P}{dh} \cdot \frac{dh}{dt} = v_P \frac{dv_P}{dh}$$

and since $F = F_O + F_P$ and $F_O = F_g$, eqn (5.34) can be rewritten,

$$F_P = mv_P \frac{dv_P}{dh}. \tag{5.35}$$

Let us now draw a system boundary around the particle and the Earth, as shown in Fig. 5.21 and apply the First Law of Thermodynamics to the system, i.e.

$$W + Q = \Delta U. \tag{5.36}$$

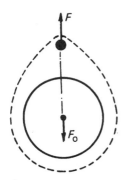

Fig. 5.21

No heat is transferred across the boundary of the system and the work done on the system by the external forces F on the particle and F_O on the Earth is given by

$$W = F \, dh = (F_O + F_P) \, dh. \tag{5.37}$$

The force F_0 applied to the Earth is not displaced and therefore does no work. Using eqns (5.33) and (5.35) we can write

$$W = \int_{h_1}^{h_2} \frac{G_c m_e m_P}{(R_e + h)^2} \, dh + \int_{h_1}^{h_2} m_P v_P \frac{dv_P}{dh} \, dh,$$

$$= \int_{h_1}^{h_2} \frac{G_c m_e m_P}{(R_e + h)^2} \, dh + \int_{v_1}^{v_2} m_P v_P \, dv_P, \tag{5.38}$$

where h_1 and h_2 represent the initial and final positions of the particle. We shall now arrange that the particle has zero velocity at positions h_1 and h_2,

i.e. $v_1 = v_2 = 0.$

The integral of the second term on the right-hand side of eqn (5.38) is therefore zero

Thus $$W = \int_{h_1}^{h_2} \frac{G_c m_e m_P}{(R_e + h)^2} \, dh = G_c m_e m_P \frac{(h_2 - h_1)}{(R_e + h_1)(R_e + h_2)}. \tag{5.39}$$

The change in the internal energy of the system caused by the displacement of the particle from a point h_1 to a point h_2 above the Earth is therefore given by substituting eqn (5.39) into eqn (5.36) so that

$$\Delta U = G_c \frac{m_e m_P (h_2 - h_1)}{(R_e + h_1)(R_e + h_2)}. \tag{5.40}$$

Suppose we now restrict the use of this equation to values of h which are much less than the radius of the Earth, i.e. $h \ll R_e$. Equation (5.40) can now be rewritten as

$$\Delta U \approx \frac{G m_e m_P}{R_e^2} (h_2 - h_1) = \frac{G m_e m_P}{R_e^2} \Delta h. \tag{5.41}$$

The constant $G m_e / R_e^2 = g$ and is equal to 9.81 m s^{-2}. This constant is the well-known acceleration due to gravity.

Hence if we let $V_1 = m_P g h_1$ and $V_2 = m_P g h_2$ then

$$\Delta U = m_P g (h_2 - h_1) = (V_2 - V_1). \tag{5.42}$$

The internal energy of the system, i.e. of the particle *and* the Earth, is therefore increased by an amount proportional to the displacement $\Delta h = (h_2 - h_1)$. The term, $V = m_P g h$, used in eqn (5.42) is often referred to as the potential energy of the particle due to its position in a gravitational field.

Rigid body

If we now consider a number of particles of which the typical particle P_i of mass m_i moves from a height h_{i1} above the surface of the Earth to a height h_{i2}, then the change in total internal energy of the system is obtained by summing the eqn (5.42) for all the particles in the body,

i.e. $$\Delta U = \sum m_i g (h_{i2} - h_{i1}). \tag{5.43}$$

This result applies to any collection of particles whether they are connected or not. If the particles constitute a rigid body of total mass m, then from the definition of centre of mass

$$mh_G = \sum m_i h_i,$$

where h_G is the height of the centre of mass of the body above the Earth's surface.

Equation (5.43) may therefore be written as

$$\Delta U = mg(h_{G2} - h_{G1}) = mg\Delta h_G. \tag{5.44}$$

The change in internal (potential) energy of the system is therefore determined by the vertical displacement of the centre of mass of the system relative to the Earth.

The potential energy of a system consisting of a rigid body and the Earth is thus given, to a very good approximation, as

$$V = mgh_G. \tag{5.45}$$

This is usually referred to as the potential energy of the body although, strictly speaking, we should refer to the complete system of the body and the Earth.

If we now compare eqn (5.44) with eqn (viii) in example 5.3, it can be seen that the internal energy stored in a rigid body due to its position in a gravitational field is the same as the potential function V associated with the gravitational force mg acting on the body.

The foregoing and example 5.3 show that gravitational forces can be treated either as externally applied forces acting on the body, in which case the Earth is excluded from the system boundary, or as a form of internal energy, in which case the Earth must be included within the boundary as part of the system.

5.6   Kinetic Energy

Particle

Figure 5.22 shows the components F_x and F_y of an external force F acting on a system which consists of a single particle of mass m. If the particle is free to move in the OXY plane then, from Newton's Second Law,

$$F_x = m \frac{dv_x}{dt} \tag{5.46}$$

and $\quad F_y = m \dfrac{dv_y}{dt},$

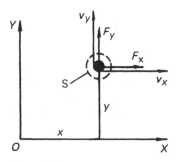

FIG. 5.22

where v_x and v_y are the components along OX and OY of the particle's absolute velocity, v. The work done on the system by the forces when the particle is displaced by differential amounts dx along OX and dy along OY is given by

$$dW = F_x\, dx + F_y\, dy. \tag{5.47}$$

If we now substitute for F_x and F_y using eqns (5.46), dW can be written as

$$dW = m\left(\frac{dv_x}{dt} dx + \frac{dv_y}{dt} dy\right).$$

$$= m(v_x\, dv_x + v_y\, dv_y)$$

$$= \tfrac{1}{2}m\, d(v_x^2 + v_y^2), \tag{5.48}$$

since $v_x\, dv_x = \dfrac{d(v_x^2)}{2}.$

By writing $v^2 = v_x^2 + v_y^2$, where v is the magnitude of the velocity of the particle, eqn (5.48) shows that the work dW, done by F on the system during the displacement, is

$$dW = d(\tfrac{1}{2}mv^2) \tag{5.49}$$

and is equal to a change in the quantity $\tfrac{1}{2}mv^2$. If we now apply the First Law of Thermodynamics, as given by eqn (5.28), to our system we obtain

$$dW = \Delta U = d(\tfrac{1}{2}mv^2).$$

The change in the quantity $\tfrac{1}{2}mv^2$ can therefore be interpreted as a change in internal energy. The quantity $\tfrac{1}{2}mv^2$ is called the *kinetic energy*, T, of the particle so that

$$T = \tfrac{1}{2}mv^2. \tag{5.50}$$

As would be expected, kinetic energy has units of $kg\, m^2\, s^{-2}$.

For finite displacements eqn (5.49) can be integrated to show that the work done on the particle is equal to the change in its kinetic energy,

i.e. $$W = \int_{v_1}^{v_2} d\left(\frac{mv^2}{2}\right) = \tfrac{1}{2}mv_2^2 - \tfrac{1}{2}mv_1^2. \tag{5.51}$$

Example 5.5 Let us consider the motion of a particle which is thrown vertically into the air. We will assume that it is thrown upwards from a height y_1 above the ground with an initial velocity $v1$. We will obtain an expression for the velocity of the particle during its subsequent motion. The system boundary can be defined in two ways,

(i)   around the particle alone,
(ii)   around the particle and the Earth.

We will consider both cases and develop parallel solutions. In both cases Q will be assumed to be zero and air drag on the particle will be neglected.

The system boundary diagrams are shown in Figs. 5.23(a) and (b).

(i)   In this case the system boundary is drawn around the particle so that the gravitational force is an *external* force which does work on the system. Now

$$W = \Delta U. \tag{i}$$

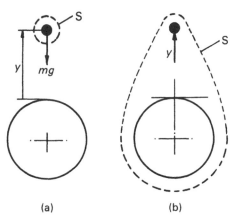

(a)   (b)

FIG. 5.23

The work done on the system by the external force mg is

$$W = -mg(y_2 - y_1). \tag{ii}$$

The internal energy of the system i.e. the particle, is due solely to its kinetic energy so that

$$\Delta U = \tfrac{1}{2}m(v_2^2 - v_1^2). \tag{iii}$$

The substitution of eqns (ii) and (iii) into eqn (i) yields

$$-mg(y_2 - y_1) = \tfrac{1}{2}m(v_2^2 - v_1^2)$$

or $v_2^2 = v_1^2 - 2g(y_2 - y_1).$ \hfill (iv)

(ii)   In Fig. 5.23(b) the gravitational forces are *internal* to the system. No external forces act so that no work is done on the system. Therefore

$$W = \Delta U = 0. \tag{v}$$

The internal energy of the system is composed of both potential and kinetic energy. Thus

$$\Delta U = mg(y_2 - y_1) + \tfrac{1}{2}m(v_2^2 - v_1^2). \tag{vi}$$

From eqns (v) and (vi)

$$0 = mg(y_2 - y_1) + \tfrac{1}{2}m(v_2^2 - v_1^2) \tag{vii}$$

or $\quad v_2^2 = v_1^2 - 2g(y_2 - y_1).$ $\hfill$ (viii)

Both approaches give the same result and show that as the particle moves upwards its velocity reduces and will be zero when

$$y_2 - y_1 = \frac{v_1^2}{2g}. \tag{ix}$$

Also when the particle passes through its original position as it falls back to Earth $y_2 = y_1$ so that

$$v_2 = v_1. \tag{x}$$

Note that because energy is not a vector quantity eqns (iv) and (viii) give only the magnitude of the velocity of the particle.

Rigid body

Having defined the kinetic energy of a particle we are now able to consider the kinetic energy of a rigid body. Let us consider a typical particle P_i of mass m_i in the rigid body shown in Fig. 5.24. If the particle has a velocity v_i its kinetic energy will be

$$T_i = \tfrac{1}{2}m_i v_i^2.$$

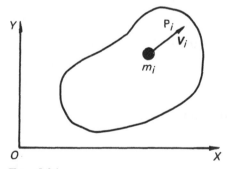

FIG. 5.24

The body can be considered as a large number of individual particles rigidly connected together. Therefore the total kinetic energy T of the body can be obtained by summing the kinetic energies of all the particles in the body so that

$$T = \sum T_i = \tfrac{1}{2}\sum m_i v_i^2. \tag{5.52}$$

As we saw in Chapters 1 and 4 the velocities v_i of the particles in the rigid body are not independent of one another and their relationship depends upon the type of motion of the body, i.e. whether the body is in pure translation, pure rotation or general plane motion. We will follow the previous pattern by considering each case separately.

Pure translation

When a rigid body moves with pure translation, as shown in Fig. 5.25, the angular velocity ω of the body is zero and all points in the body have the *same* velocity v. (For details see Chap. 1, p. 24.)

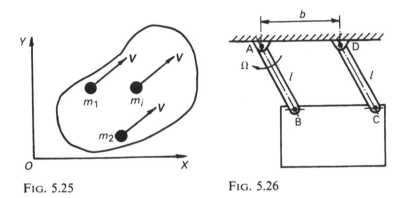

FIG. 5.25   FIG. 5.26

Thus since $v_i = v$ for all the points in the body, v^2 may be taken outside the summation of eqn (5.52) which then reduces to

$$T = \tfrac{1}{2}v^2(\sum m_i). \tag{5.53}$$

The quantity $\sum m_i$ is equal to the total mass m of the body. Equation (5.53) therefore becomes

$$T = \tfrac{1}{2}mv^2. \tag{5.54}$$

Thus a rigid body moving with pure translation has kinetic energy which is the same as that of an equivalent particle of the same mass moving with the same velocity.

Example 5.6 Figure 5.26 shows two identical rods AB and DC of equal length l supporting a heavy plate of mass m at the points B and C. If the connections at A, B, C and D are simple pin joints and are positioned so that AD = BC = b determine the kinetic energy of the plate when rod AB rotates with angular velocity Ω.

The first step in the solution of this problem is to consider the motion of the plate. We know from our results on the kinematics of rigid bodies that the velocities of the points B and C on the plate are related by the vector equation

$$v_C = v_B + v_{CB},$$ (i)

where v_{CB} is the velocity of point C relative to point B. The direction of v_{CB} must be perpendicular to the line CB since the body is rigid, and its magnitude is given by

$$v_{CB} = \omega b,$$ (ii)

where ω is the angular velocity of the plate. Point B moves on a circular path of radius l so that its velocity

$$v_B = \Omega l$$ (iii)

and is in a direction perpendicular to AB as shown in Fig. 5.27.

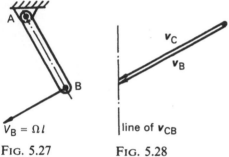

$$V_B = \Omega l$$ line of v_{CB}

FIG. 5.27   FIG. 5.28

The velocity of point C is unknown but its direction must be perpendicular to link CD. Because the points ABCD form a parallelogram the direction of v_c is also perpendicular to AB and is therefore parallel to v_B.

A graphical solution of the vector eqn (i) can be obtained by the method used in Chapter 1. It can be seen from Fig. 5.28 that $v_C = v_B$ and $v_{CB} = 0$. This shows, from eqn (ii), that the angular velocity of the plate is zero. The plate therefore moves in pure translation and all points in the plate have velocity $v = v_B$.

Hence, by using eqn (5.54), the kinetic energy of the plate is given by

$$T = \tfrac{1}{2}mv_B^2$$
$$= \tfrac{1}{2}ml^2\Omega^2.$$ (iv)

The most difficult part of this solution was to prove that the plate moves with pure translation. This result could have been arrived at by inspection since, because of the parallelogram construction of the mechanism, v_B must be identical with v_C for all positions of the mechanism. The plate must therefore be in pure translation.

Pure rotation

Figure 5.29 shows a rigid body which is supported at the point O by means of a pin joint. The body is therefore free to rotate in the OXY plane about an axis which passes through O.

In this case the kinematic constraints associated with pure rotation apply to v_i in eqn (5.52). Let our typical particle P_i be located at a fixed distance r_i

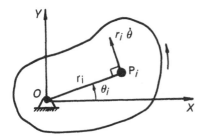

FIG. 5.29

from O. The particle will move in a circular path of radius r_i and centre O. From eqn (1.21) its velocity

$$v_i = r_i \dot{\theta}_i$$

and is in a direction perpendicular to OP_i as shown in Fig. 5.29 where $\dot{\theta}_i$ is the angular velocity of the line OP_i.

Thus, substituting for v_i in eqn (5.52),

$$T = \tfrac{1}{2} \sum m_i (\dot{\theta}_i r_i)^2.$$

Since the body is assumed rigid all lines in the body will have the same angular velocity $\dot{\theta}$. We can now remove the $\dot{\theta}_i^2$ term from the summation to give

$$T = \tfrac{1}{2} \dot{\theta}^2 \sum m_i r_i^2. \tag{5.55}$$

The term $\sum m_i r_i^2$ should now be familiar to the reader as the definition of the moment of inertia, I_O, of the body about an axis through O. (See Section 4.4), so that eqn (5.55) may be rewritten as

$$T = \tfrac{1}{2} I_O \dot{\theta}^2. \tag{5.56}$$

Thus the kinetic energy of a rigid body in pure rotation can be expressed in terms of its moment of inertia about an axis through the fixed point and its angular velocity.

Example 5.7   A uniform link of mass m and length $2l$ which can rotate, without friction, about a fixed horizontal axis through O is shown in Fig. 5.30. Let us suppose that the link is released from rest in the upright vertical position, (i.e. $\theta = 0$) and is allowed to fall freely under the action of gravity.

Let us obtain an expression for the angular velocity of the link as a function of the angle θ.

If we choose the external surfaces of the link to be our system then the external forces acting on the system will be the weight mg of the link and the forces R_H and R_V at the pivot, as shown in Fig. 5.31.

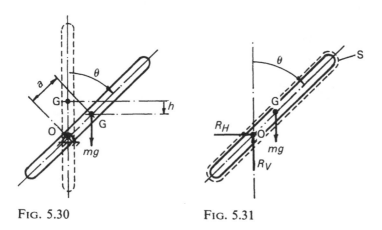

FIG. 5.30   FIG. 5.31

We can apply the First Law of Thermodynamics to the system, i.e.

$$W + Q = \Delta U. \tag{i}$$

In this case $Q = 0$ and ΔU represents the change in kinetic energy of the link. Initially the link is at rest and so has zero kinetic energy. Suppose that when the link has rotated through an angle θ its angular velocity is $\dot{\theta}$. Its kinetic energy T, from eqn. (5.56), is

$$T = \tfrac{1}{2}I_O\dot{\theta}^2, \tag{ii}$$

where I_O is the moment of inertia of the link about an axis through O. Equation (ii) also represents the change in internal energy ΔU of the system since the initial kinetic energy is zero.

It can be seen from Fig. 5.31 that the weight mg is the only external force which does work during the rotation θ. Forces R_H and R_V do not move and so do no work.

The force mg may be considered to act at the centre of mass G of the body which is a distance a from O as shown.

During the rotation θ the force mg moves a distance

$$h = a(1 - \cos\theta) \tag{iii}$$

along its line of action, as shown in Fig. 5.30. The work done *on* the system by the force *mg* is therefore

$$W = mga(1 - \cos \theta), \tag{iv}$$

and is positive since the vertical displacement of G is in the same direction as *mg*.

Substituting eqns (ii) and (iv) into eqn (i) gives

$$mga(1 - \cos \theta) = \tfrac{1}{2} I_O \dot{\theta}^2, \tag{v}$$

so that

$$\dot{\theta}^2 = \frac{2mga}{I_O}(1 - \cos \theta). \tag{vi}$$

If we now set $a = l$, it can be seen that eqn (v) becomes identical to eqn (vii) of example 4.4. Equation (vi) therefore represents the first integral of the equation of motion of the link. The foregoing shows that an expression for the angular velocity of the link can be obtained immediately from the energy eqn (i) without reference to the governing equations of motion.

By applying the parallel axis theorem (see p. 177) to the link its moment of inertia about an axis through O can be written as

$$I_O = I_G + ma^2 \tag{vii}$$

$$= m(k_G^2 + a^2) \tag{viii}$$

where k_G is the radius of gyration of the link about an axis through G.

When eqn (viii) is substituted into eqn (vi) the expression for the angular velocity becomes

$$\dot{\theta}^2 = \frac{2ga}{(k_G^2 + a^2)}(1 - \cos \theta). \tag{ix}$$

By differentiating eqn (ix) with respect to a and setting $d\dot{\theta}^2/da = 0$ it can be seen that, for any angle θ, the angular velocity $\dot{\theta}$ will have a maximum value when

$$a = k_G. \tag{x}$$

Thus, to achieve maximum angular velocity of the link, O must be distant from G by an amount equal to the radius of gyration of the link about G. In this case

$$\dot{\theta}^2_{\max} = \frac{g}{k_G}(1 - \cos \theta). \tag{xi}$$

General plane motion

Let us consider the particle P_i of mass m_i to be part of a rigid body which is moving with general plane motion. Let the particle have a velocity v_i as shown in Fig. 5.32. In this case the kinematic constraints for a rigid body in general plane motion apply (Chapter 1, p. 26). These constraints show that the velocity of any point P_i can be expressed in terms of the velocity, v_Q, of an arbitrary point Q in the body and the angular velocity $\dot{\theta}$ of the body by the vector equation

$$v_i = v_Q + v_{iQ}. \tag{5.57}$$

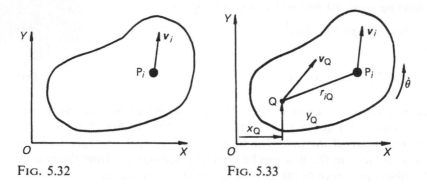

FIG. 5.32   FIG. 5.33

The vector v_{iQ} is the velocity of P_i relative to Q, and its magnitude is given by

$$v_{iQ} = r_{iQ}\dot{\theta},$$

where r_{iQ} is the length of the line joining P_i and Q as shown in Fig. 5.33. The direction of v_{iQ} is perpendicular to the line QP_i as shown in Fig. 5.34. If the line QP_i makes an angle θ_i with OX it can be seen from Fig. 5.34 that the components of v_{iQ}, measured in the directions of OX and OY, are given by

$$\dot{x}_{iQ} = -\dot{\theta}r_{iQ}\sin\theta_i = -\dot{\theta}(y_i - y_Q)$$

and $$y_{iQ} = \dot{\theta}r_{iQ}\cos\theta_i = \dot{\theta}(x_i - x_Q). \tag{5.58}$$

The velocity of point P_i can now be expressed in component form using eqns (5.57) and (5.58) as

$$\dot{x}_i = \dot{x}_Q - \dot{\theta}(y_i - y_Q), \tag{5.59}$$

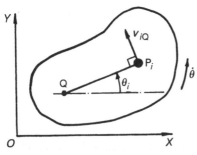

FIG. 5.34

and $\quad \dot{y}_i = \dot{y}_Q + \dot{\theta}(x_i - x_Q),$

where $\dot{x}_Q$ and $\dot{y}_Q$ are the components of v_Q along OX and OY.

The kinetic energy of particle P_i is thus given by

$$T_i = \tfrac{1}{2}m_i v_i^2$$
$$= \tfrac{1}{2}m_i(\dot{x}_i^2 + \dot{y}_i^2). \tag{5.60}$$

When eqns (5.59) are substituted into eqn (5.60) the expression for T_i can be rewritten as

$$T_i = \tfrac{1}{2}m_i\{(\dot{x}_Q^2 + \dot{y}_Q^2) + 2\dot{\theta}\dot{y}_Q(x_i - x_Q) - 2\dot{\theta}\dot{x}_Q(y_i - y_Q)$$
$$+ \dot{\theta}^2(x_i - x_Q)^2 + \dot{\theta}^2(y_i - y_Q)^2\}. \tag{5.61}$$

The kinetic energy T for the whole body is now given by summing the values of T_i for all the particles in the body,

i.e. $\quad T = \sum \tfrac{1}{2}m_i(\dot{x}_Q^2 + \dot{y}_Q^2) + \sum \dot{\theta}\dot{y}_Q m_i(x_i - x_Q)$

$$- \sum \dot{\theta}\dot{x}_Q m_i(y_i - y_Q) + \tfrac{1}{2}\sum \dot{\theta}^2 m_i[(x_i - x_Q)^2 + (y_i - y_Q)^2]. \tag{5.62}$$

Since the factors in eqn (5.62) involving $\dot{x}_Q$, $\dot{y}_Q$ and $\dot{\theta}$ are independent of the suffix i they may be brought out of the summation sign to give

$$T = \tfrac{1}{2}(\dot{x}_Q^2 + \dot{y}_Q^2) \sum m_i - \dot{\theta}\dot{y}_Q x_Q \sum m_i + \dot{\theta}\dot{x}_Q y_Q \sum m_i$$
$$+ \dot{\theta}\dot{y}_Q \sum m_i x_i - \dot{\theta}\dot{x}_Q \sum m_i y_i$$
$$+ \tfrac{1}{2}\dot{\theta}^2 \sum m_i[(x_i - x_Q)^2 + (y_i - y_Q)^2]. \tag{5.63}$$

Let us now examine eqn (5.63) term by term. Since $m = \sum m_i$ is the total mass of the body and $v_Q^2 = \dot{x}_Q^2 + \dot{y}_Q^2$ we can see that

$$\tfrac{1}{2}(\dot{x}_Q^2 + \dot{y}_Q^2) \sum m_i = \tfrac{1}{2}mv_Q^2,$$

$$\dot{\theta}\dot{y}_Q x_Q \sum m_i = m\dot{\theta}\dot{y}_Q x_Q,$$

and $\quad \dot{\theta}\dot{x}_Q y_Q \sum m_i = m\dot{\theta}\dot{x}_Q y_Q. \tag{5.64}$

Using the definition of the position of the centre of mass of a body i.e.

$$mx_G = \sum m_i x_i \quad \text{and} \quad my_G = \sum m_i y_i,$$

$$\dot{\theta}\dot{y}_Q \sum m_i x_i = m\dot{\theta}\dot{y}_Q x_G$$

and $\quad \dot{\theta}\dot{x}_Q \sum m_i y_i = m\dot{\theta}\dot{x}_Q y_G.$

The final term in eqn (5.63)

$$\sum m_i[(x_i - x_Q)^2 + (y_i - y_Q)^2] = \sum m_i r_{iQ}^2.$$

This term defines the moment of inertia I_Q of the body about an axis through the point Q.

Substituting these expressions into eqn (5.57) then gives

$$T = \tfrac{1}{2}mv_Q^2 + m\dot{\theta}\dot{y}_Q(x_G - x_Q) - m\dot{\theta}\dot{x}_Q(y_G - y_Q) + \tfrac{1}{2}I_Q\dot{\theta}^2. \tag{5.65}$$

Equation (5.65) is a general expression for the kinetic of a rigid body, in terms of the angular velocity $\dot{\theta}$ of the body and the motion of an arbitrary point Q in the body. Fortunately this expression can be greatly simplified if we choose the point Q to be at the centre of mass G.

In this case

$$T = \tfrac{1}{2}mv_G^2 + \tfrac{1}{2}I_G\dot{\theta}^2. \tag{5.66}$$

We can see from eqn (5.66) that the kinetic energy is made up from two separate terms. The first term is related to the translation of the body and is the kinetic energy of an equivalent particle of mass m located at G. The second term determines the contribution made by the rotation of the body and is given by considering the body to be in pure rotation about an axis passing through G.

The expression given by eqn (5.66) again shows the importance of the position of the centre of mass G when considering the motion of rigid bodies moving with general plane motion.

Example 5.8   Figure 5.35 shows a uniform disc of mass m and radius R rolling without slip on a rough horizontal surface. Let us determine the kinetic energy of the disc when its centre O has a velocity v_O as shown.

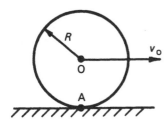

FIG. 5.35

The disc is moving with general plane motion since it is capable of simultaneous translation and rotation. Thus using eqn (5.66) the kinetic energy of the disc is given by

$$T = \tfrac{1}{2}mv_G^2 + \tfrac{1}{2}I_G\omega^2. \tag{i}$$

Since the disc is uniform its centre O coincides with its centre of mass G and

$$v_G = v_O. \tag{ii}$$

Before we can calculate T we must determine the angular velocity ω of the disc. If the disc is rolling on the plane without slip ω and v_G are not independent. For the disc to roll without slip there must be no relative velocity between the point A on the disc and a coincident point on the ground. The ground is stationary and therefore

$$v_A = 0. \tag{iii}$$

We can now use the vector relationship between the velocities of points O and A to write

$$v_O = v_A + v_{OA} = v_{OA}. \tag{iv}$$

Now $\quad v_{OA} = \omega R$

in a direction perpendicular to AO so that, from eqn (iv)

$$v_O = \omega R. \tag{v}$$

in the same direction.

If we now substitute eqns (ii) and (v) into eqn (i) the kinetic energy of the disc can be written

$$T = \tfrac{1}{2}(m + I_G/R^2)v_O^2. \tag{vi}$$

We have already shown in example 4.8 that the moment of inertia of a uniform disc about its polar axis is given by

$$I = \frac{mR^2}{2}. \tag{vii}$$

Using this expression eqn (vi) gives the kinetic energy of the disc as

$$T = \tfrac{3}{4}mv_O^2. \tag{viii}$$

Let us now consider the effect of adding a particle of mass m_P to a point P on the circumference of the disc. We will calculate the kinetic energy of the combined body when the particle is in the position shown in Fig. 5.36. It will again be assumed that the disc rolls without slip and that its centre O moves with velocity v_O.

The kinetic energy T_P of the particle is

$$T_P = \tfrac{1}{2}m_P v_P^2. \tag{ix}$$

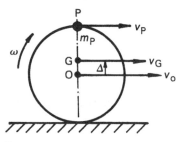

FIG. 5.36

The velocity v_P of point P is found using the velocity relationship

$$v_P = v_O + v_{PO}.$$

For the position shown the directions of the vectors v_P, v_O and v_O are all horizontal so that

$$v_P = v_O + \omega R = 2v_O. \tag{x}$$

Thus

$$T_P = 2m_P v_O^2. \tag{xi}$$

The kinetic energy of the combined system is now found by adding eqns (viii) and (xi) to give

$$T = \tfrac{3}{4}mv_O^2 + 2m_P v_O^2$$
$$= \frac{(3m + 8m_P)v_O^2}{4}. \tag{xii}$$

This result can also be found by direct application of eqn (5.66) once v_G and I_G have been found.

The distance Δ of the centre of mass G of the combined system from the point O is determined from

$$(m + m_P)\Delta = m_P R,$$

i.e.

$$\Delta = \frac{m_P}{m + m_P} R. \tag{xiii}$$

The velocity of G is given by the vector equation

$$v_G = v_O + v_{GO},$$

and since the directions of v_O and v_{GO} are parallel we can write

$$v_G = v_O + \omega \Delta$$
$$= \left(1 + \frac{\Delta}{R}\right) v_O.$$

Thus

$$v_G = \left(\frac{m + 2m_P}{m + m_P}\right) v_O. \tag{xiv}$$

The moment of inertia of the combined body about an axis through O is

$$I_O = \left(\frac{mR^2}{2} + m_P R^2\right). \tag{xv}$$

Using the parallel axis theorem, eqn (4.35) the moment of inertia I_G of the combined system about an axis through G can be found as

$$I_G = I_O - (m + m_P)\Delta^2$$

$$= \left(\frac{mR^2}{2} + m_P R^2\right) - (m + m_P)\Delta^2$$

$$= \tfrac{1}{2}mR^2(m + 3m_P)/(m + m_P)). \qquad \text{(xvi)}$$

The kinetic energy of the combined system is therefore given by

$$T = \tfrac{1}{2}(m + m_P)v_G^2 + \tfrac{1}{2}I_G\omega^2$$

$$= \tfrac{1}{2}(m + m_P)\left(\frac{m + 2m_P}{m + m_P}\right)^2 v_O^2 + \tfrac{1}{4}mR^2\left(\frac{m + 3m_P}{m + m_P}\right)\frac{v_O^2}{R^2}$$

$$= \left(\frac{3m + 8m_P}{4}\right)v_O^2, \qquad \text{(xvii)}$$

which agrees with the result of eqn (xii).

Suppose we now recalculate the kinetic energy of the system for the case where the particle at P has moved to a position shown in Fig. 5.37. Since the disc rolls without slip, the velocity v_P of the particle in this position will be zero. The particle is therefore instantaneously at rest and has no kinetic energy. The kinetic energy of the combined system is therefore determined by the kinetic energy of the disc, which, from eqn (viii) is given by

$$T = \tfrac{3}{4}mv_O^2.$$

Let us repeat this calculation using the general result given by eqn (5.66), i.e. from

$$T = \tfrac{1}{2}(m + m_P)v_G^2 + \tfrac{1}{2}I_G\omega^2.$$

From the vector equation $v_G = v_O + v_{GO}$ the magnitude of v_G is

$$v_G = v_O - \omega\Delta$$

$$= v_O(1 - \Delta/R)$$

$$= \frac{mv_O}{m + m_p}. \qquad \text{(xviii)}$$

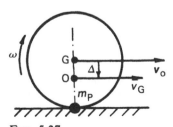

FIG. 5.37

By using the value of I_G given by eqn (xvi) and substituting eqn (xviii) into eqn (5.66) we find that

$$T = \tfrac{3}{4}mv_O^2,$$

as expected.

5.7 Strain Energy

When a force is applied to a system which is elastic the force will cause the system to deform. The dimensions and geometry of the system will change but in most cases these changes will be small compared with the gross dimensions of the system. For example, a gear tooth will deflect under load by some very small amount. In general the displacements produced in an elastic system will vary throughout the system. The whole system becomes strained and its internal energy will increase in proportion to the amount of work done *on* the system by the forces causing the strains.

Some machine elements are, however, designed to be flexible so that under the action of forces they can experience considerable dimensional changes. A typical example is a coil spring.

The relationship between the deflection of the spring and the applied force is an important characteristic of the spring. For example, a linear elastic spring is one in which the force F required to produce a deflection in the spring is directly proportional to the deflection x. We will only consider simple springs where x is measured at the point of application of the force along its line of action. For this type of spring

$$F = kx. \tag{5.67}$$

The constant k is called the stiffness of the spring and has units of $N\,m^{-1}$. If we plot a graph of F against x we obtain the straight line of slope k as shown in Fig. 5.38.

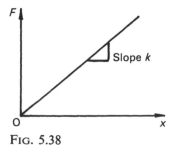

Slope k

FIG. 5.38

Let us consider the spring of stiffness k, shown diagramatically in Fig. 5.39(a), with a force F applied to its free end. For equilibrium of the spring there must be an equal and opposite force applied to its fixed end as shown in the free body diagram of Fig. 5.39(b). We shall now measure the displace-

(a)

(b)

(c)

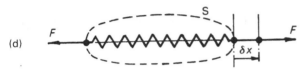

(d)

FIG. 5.39

ment, x, of the point on the spring at which the force is applied. The datum position corresponds to the case where $F = 0$ and the force required to maintain a displacement x is, from eqn (5.67)

$$F = kx.$$

Let us now consider the deflected spring as a system with boundary S, as shown in Fig. 5.39(d). We can apply the First Law of Thermodynamics,

$$W + Q = U_2 - U_1$$

to this system as the free end of the spring is displaced an additional amount δx.

We will assume that $Q = 0$. Work is done on the system by the external force F as the free end moves through the displacement δx, so that

$$\delta W = F \, \delta x. \tag{5.68}$$

The force F at the fixed end of the spring does no work on the system since its point of application is stationary. The total work done on the system during a displacement from x_1 to x_2 is obtained by taking the limit as δW and $\delta x \to 0$ and integrating eqn (5.68) between the limits x_1 and x_2.

Thus

$$W = \int_{x_1}^{x_2} F \, dx = \int_{x_1}^{x_2} kx \, dx$$

$$= \tfrac{1}{2}kx_2^2 - \tfrac{1}{2}kx_1^2. \tag{5.69}$$

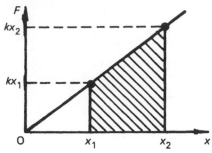

FIG. 5.40

This is equal to the area under the force-deflection graph between x_1 and x_2 as shown in Fig. 5.40.

Thus, if $x_1 = 0$ and $x_2 = x$, the work done by the force when its point of application is displaced an amount x from the unstretched position of the spring is given by

$$W = \tfrac{1}{2}kx^2.\tag{5.70}$$

Substituting eqn (5.69) into the First Law of Thermodynamics we obtain

$$\tfrac{1}{2}kx_2^2 - \tfrac{1}{2}kx_1^2 = U_2 - U_1.\tag{5.71}$$

Equation (5.71) shows that the spring experiences a change in internal energy when work is done on it by an external force. This internal energy is called the *strain energy*, V_s, of the spring where

$$V_s = \tfrac{1}{2}kx^2.\tag{5.72}$$

The strain energy stored in the spring is therefore determined by its extension x, i.e. by the difference between its stretched length and its free length.

We should note that if the spring is compressed an amount x (i.e. $x < 0$) the strain energy stored in the spring is still given by eqn (5.72). The reader should confirm that this is the case by repeating the above analysis for $x < 0$.

The units of V_s are, from eqn (5.71),

$$\frac{N}{m}m^2 = \frac{kg\,m^2}{s^2}$$

which agrees with our original definition of energy.

Equation (5.72) is a general result for simple linear springs. Let us now consider some typical springs and obtain expressions for the strain energy stored in them when they are deflected.

Example 5.9 Helical spring　A *close* coiled helical spring is shown in Fig. 5.41. If the wire diameter d is small compared with the mean radius R of the spring,

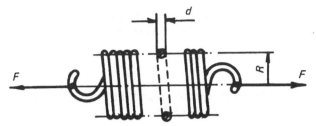

FIG. 5.41

then the extension of the spring under a given load F is given by*

$$F = \left(\frac{Gd^4}{64R^3n}\right)x, \tag{i}$$

where n is the number of coils in the spring and G is the modulus of rigidity of the spring material. If d is not small compared with R, or if the spring is not close coiled, additional terms arise in the expression.

The stiffness k of such a spring is, from eqn (5.67),

$$k = \frac{F}{x} = \frac{Gd^4}{64R^3n}. \tag{ii}$$

The strain energy stored in this spring is determined by its extension x and, from eqns (5.72) and (ii), is given by

$$V_s = \frac{1}{2}\left(\frac{Gd^4}{64R^3b}\right)x^2. \tag{iii}$$

Example 5.10 Cantilevered leaf spring Figure 5.42 shows a flexible cantilevered beam of length l with a transverse force F applied to its free end. Due to the force the free end of the cantilever is displaced an amount x.

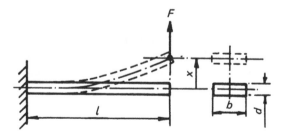

FIG. 5.42

It can be shown that F and x are related by the expression*

$$F = \frac{Ebd^3}{4l^3}x, \tag{i}$$

* See elementary text on strength of materials.

where E is the Young's modulus of the beam material, b is the width of the beam and d is its depth. The stiffness of the beam is therefore

$$k = \frac{F}{x} = \frac{Ebd^3}{4l^3}. \tag{ii}$$

The strain energy V_s stored in the beam due to an end force which produces an end displacement x is therefore, from eqn (5.72),

$$V_s = \frac{1}{2}\left(\frac{Ebd^3}{4l^3}\right)x^2. \tag{iii}$$

Beams with other end fixings, such as simple supports at both ends, can be treated in a similar manner *provided that x* is taken to mean the deflection of the point at which the force is applied, measured in the direction of the force.

Example 5.11 Torsional spring Figure 5.43 shows a circular shaft of radius R and length l, rigidly clamped at one end A. A twisting moment M is applied to the free end of the shaft at B and causes the section at B to twist an amount θ, as shown.

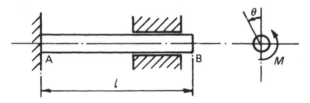

FIG. 5.43

The relationship between the applied moment and the twist in the shaft at B is given by*

$$M = \frac{GJ}{l}\theta, \tag{i}$$

where G is the modulus of rigidity of the shaft material, and $J = \pi R^4/2$ is the polar second moment of area of the shaft cross-section. If the torsional stiffness, k_θ, of the shaft is defined to be the twisting moment required to produce unit *angular* deflection, i.e. M/θ, then

$$k_\theta = \frac{GJ}{l}. \tag{ii}$$

Equation (i) can be written as

$$M = k_\theta\theta \tag{iii}$$

which, by analogy, is of the same form as the eqn (5.67). The units of torsional stiffness will therefore be Nm rad^{-1}.

*See elementary text on Strength of Materials.

The work done by the moment M during a small angular displacement $\delta\theta$ has been shown (Example 5.2, eqn (iii)) to be given by

$$\delta W = M\,\delta\theta. \tag{iv}$$

For a total twist θ the work done on the bar by the twisting moment is obtained by integrating eqn (iv) over the range 0 to θ, i.e.

$$\begin{aligned}
W &= \int_0^\theta M\,\mathrm{d}\theta \\
&= \int_0^\theta k_\theta\theta\,\mathrm{d}\theta \\
&= \tfrac{1}{2}k_\theta\theta^2.
\end{aligned} \tag{v}$$

The strain energy V_s stored in the shaft due to a twist θ is therefore given by

$$V_s = \tfrac{1}{2}k_\theta\theta^2. \tag{vi}$$

Example 5.12 Mass-spring system   When a spring functions as an actuator and is used to drive a system, such as a mass, the force provided by the spring is *conservative*. To show this let us consider the mass-spring arrangement shown in Fig. 5.44. Any friction between the mass and the ground will be neglected.

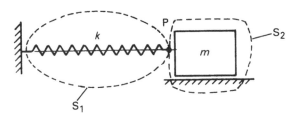

FIG. 5.44

Let the spring be system S_1 and the mass be system S_2, and let us propose that system S_1 applies a force F to the system S_2 at the point of connection P between their boundaries.

Suppose that, due to the action of this force, the point P on the boundary of system S_2 becomes displaced from an initial position $x = x_1$ to a final position $x = x_2$, as shown in Fig. 5.45(a). The displacement x, of P, is measured from a datum given by the unstretched position of the spring.

The work done by the force F on system S_2 during this displacement is thus given by

$$W = -\int_{x_1}^{x_2} F\,\mathrm{d}x. \tag{i}$$

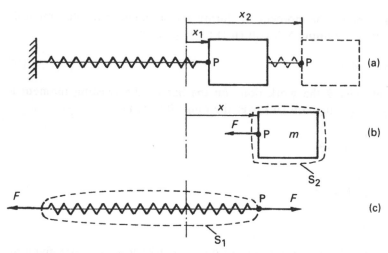

FIG. 5.45

If the end of the spring is displaced to the *right* by an amount x, as shown in Fig. 5.45(c), then using eqn (5.67) we can write

$$F = kx, \qquad (ii)$$

where k is the stiffness of the spring.

When eqn (ii) is substituted into eqn (i) the integration of eqn (i) between the limits x_1 and x_2 gives the work done on the mass by the force F as

$$W = -(\tfrac{1}{2}kx_2^2 - \tfrac{1}{2}kx_1^2). \qquad (iii)$$

It can be seen that the work done by F depends only on the initial and final positions of the point P. By comparing eqn (iii) with eqn (5.15) we can see that the potential energy function of the spring is given by

$$V = \tfrac{1}{2}kx^2 \qquad (iv)$$

which, from eqn (5.72), is identical to the strain energy V_s.

The foregoing shows that the force exerted by the spring on the mass is conservative and that positive work can only be done on the mass by the spring at the expense of the potential (strain) energy stored in the spring. For example, suppose that the spring of Fig. 5.44 is initially *compressed* by an amount x_0. When released from rest the mass will start to move to the right due to the action of the spring force. The amount of compression in the spring is reduced and the strain energy stored in the spring falls. When the spring becomes unstrained, i.e. as point P passes through its datum position, $x = 0$ and the potential energy stored in the spring reduces to zero. The work done on the mass by the spring is then given by setting $x_1 = -x_0$ and $x_2 = 0$ in eqn (iii) and is

$$W = \tfrac{1}{2}kx_0^2. \qquad (v)$$

If we now refer to our results on the kinetic energy of a particle we can see that the work done on the mass increases its kinetic energy. Since the mass starts from rest its velocity v when it passes through the datum position can be calculated directly by applying the First Law of Thermodynamics, i.e.

$$W + Q = \Delta U, \tag{vi}$$

to system S_2.

In this case $Q = 0$, $W = \frac{1}{2}kx^2$ and ΔU is the change in kinetic energy of the mass. Since the mass is initially at rest eqn (vi) gives

$$\tfrac{1}{2}kx_0^2 = \tfrac{1}{2}mv^2. \tag{vii}$$

Hence

$$v = x_0\sqrt{k/m}. \tag{viii}$$

If we consider the mass and the spring together to be a *complete* system the result of eqn (viii) can also be derived by applying the First Law of Thermodynamics directly to the combined system as shown in Fig. 5.46.

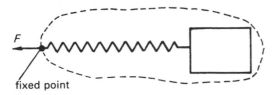

F

fixed point

FIG. 5.46

The only external force applied to this total system is the force at the fixed end of the spring. This force, since it does not move along its line of action, does no work on the system. From the First Law of Thermodynamics

$$W + Q = \Delta U.$$

For our system $W = 0$ since no external forces which do work are applied to the system boundary. We will also assume that no heat is transferred across the boundary so that $Q = 0$. Thus

$$\Delta U = U_2 - U_1 = 0, \tag{ix}$$

where U_1 and U_2 are the initial and final internal energies of the system.

In the initial state the spring is compressed by an amount x_0 and the mass is at rest. The initial internal energy of the system is therefore due to the strain energy in the spring, i.e.

$$U_1 = \tfrac{1}{2}kx_0^2. \tag{x}$$

In the final state the mass reaches a position where the spring is unstretched, i.e. $x = 0$. The internal energy of the system in the final state is thus due to the kinetic energy of the mass, i.e.

$$U_2 = \tfrac{1}{2}mv^2. \tag{xi}$$

Substituting for U_1 and U_2 in eqn (ix) gives

$$\tfrac{1}{2}mv^2 - \tfrac{1}{2}kx_0^2 = 0,$$

so that

$$v = x_0\sqrt{k/m}. \tag{xii}$$

The foregoing has shown that the same result has been obtained by

(i)   drawing our system boundary around the mass alone and considering the work done on this system, i.e. the mass, by the external forces which are *applied* to the system, and by

(ii) drawing our system boundary around both mass and spring and considering the work done on the system, i.e. the mass *and* spring, by the external forces *applied* to the system.

In eqn (xii) the term $\sqrt{k/m}$ is the natural frequency, p_n, of free vibration of the mass on the spring (application 4.9, eqn (iv)). The maximum velocity v of the mass occurs as it passes through the position $x = 0$ and from eqn (xii) can be written as

$$v = p_n x_0. \tag{xiii}$$

From the results of applying Newton's Second Law to the system in engineering application 4.8 we have shown that the velocity of the mass at any instant during its cycle of vibration may be expressed as

$$v = x_0 p_n \sin p_n t. \tag{xiv}$$

Hence the maximum velocity of the mass is, from eqn (xiv)

$$v = x_0 p_n$$

which is the same as predicted by eqn (xiii).

5.8 Thermal Energy

When heat is transferred *into* a system the temperature of the system will always increase provided no transformation of phase, such as from solid to liquid, occurs. The change in the internal energy corresponding to this increase in temperature is equal to the change in the thermal energy of the system. This change in the internal energy is given by

$$\Delta U = mC_v \Delta T_s, \tag{5.73}$$

where m is the mass of the system,

 C_v is the specific heat of the substance within the system

and ΔT_s is the change in temperature.

In most machines the changes in temperature during steady state running are small. Friction at sliding connections, such as bearings, will cause an initial temperature rise. The temperature will then reach some steady value where the heat generated by friction is equal to the heat lost from the machine because of the temperature difference between the machine and its surroundings.

Thus, in the steady state condition, the internal energy due to thermal effects remains constant.

Thermal energy must however be taken into account when considering transient behaviour. A typical example is the friction brake. Such brakes heat up rapidly when applied and the change in internal energy of the system due to the rise in temperature is very significant.

Engineering Applications

When applying the results of this section to an engineering problem we will again use the earlier assumptions and idealisations to produce a mathematical model of the system. It is an *essential* first step to define, and sketch in on the diagram of the system, the boundary of the system to which the First Law of Thermodynamics is to be applied. There is usually a number of possibilities for this boundary. The consequence of using different boundaries is illustrated in some of the following examples by presenting parallel solutions.

5.1 Child's toy

Figure 5.47 shows a child's toy loop-the-loop game. A small ball bearing is placed on the ramp, as shown, and is then allowed to roll down under the

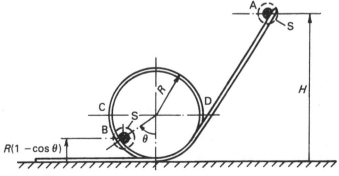

FIG. 5.47

action of gravity. The ball enters the loop at the bottom of the ramp and will then travel along a circular path of radius R, until it either falls off the track or exits again at the bottom.

Let us consider the motion of the ball around the loop when it is released from rest at a position A, which is a vertical height H above the bottom of the ramp.

We will assume that the ball can be treated as a particle and that friction can be neglected.

In Fig. 5.47 a system boundary S has been drawn around the ball and the gravitational force mg will be treated as an externally applied force which does work on the system within S.

If we consider the state of the ball at position A and at a general position B on the loop, then the First Law of Thermodynamics, when applied to the system, gives

$$W_{AB} = U_B - U_A \qquad\qquad\qquad\qquad \text{(i)}$$

if there is no heat flow into the system.

W_{AB} is the work done on the ball during the displacement from A to B, and U_A and U_B are the internal energies of the system at A and B.

In the general position B the external forces acting on the ball, and hence the system, are shown in the free body diagram of Fig. 5.48. N represents the contact force between the ball and the track and, since friction is neglected, it always acts in a direction which is perpendicular to the track. The weight mg of the ball acts vertically downwards.

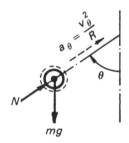

FIG. 5.48

The work done on the system is determined by the vertical displacement of the point of application of the external force mg and from Fig. 5.47 is given by

$$W_{AB} = mg(H - R(1 - \cos\theta)). \qquad\qquad\qquad \text{(ii)}$$

The contact force N does *no work* since its line of action is always perpendicular to the displacement of the ball along the track.

The internal energy of the system is determined by the kinetic energy of the ball and since the ball is initially at rest we have

$$U_A = 0. \qquad\qquad\qquad\qquad\qquad\qquad \text{(iii)}$$

When the position B is reached we shall assume that the velocity of the particle is v_θ. Hence

$$U_B = \tfrac{1}{2}mv_\theta^2. \tag{iv}$$

Substitution of equations (ii), (iii) and (iv) into eqn (i) gives

$$mg(H - R(1 - \cos\theta)) = \tfrac{1}{2}mv_\theta^2, \tag{v}$$

from which v_θ is given by

$$v_\theta^2 = 2g(H - R(1 - \cos\theta)). \tag{vi}$$

If the ball remains in contact with the track at all points eqn (vi) shows that its velocity will be a maximum at $\theta = 0$ and a minimum at $\theta = \pi$, i.e.

$$(v_\theta)^2_{\text{max}} = 2gH, \tag{vii}$$

and $\quad (v_\theta)^2_{\text{min}} = 2g(H - 2R). \tag{viii}$

Since N represents the contact force between the ball and the track then the ball will remain in contact with the track provided that

$$N > 0. \tag{ix}$$

Since the ball moves on a circular track of radius R it will have an acceleration component

$$a_\theta = \frac{v_\theta^2}{R} \tag{x}$$

towards the centre of the track, as shown.
Applying Newton's Second Law to the ball we have, in the radial direction,

$$N - mg\cos\theta = \frac{mv_\theta^2}{R},$$

i.e. $\quad N = mg\cos\theta + \dfrac{mv_\theta^2}{R}. \tag{xi}$

This equation shows that loss of contact with the track can only occur if $\cos\theta < 0$, i.e. when the point B lies on the upper portion of the loop between the points C and D. However, eqn (vi) shows that the ball will only go beyond the point C if $v_\theta^2 > 0$ at $\theta = \pi/2$, i.e. when

$$H > R. \tag{xii}$$

If we now substitute eqn (vi) into eqn (xi) the contact force N becomes

$$N = \frac{2mg}{R}(H - R(1 - \tfrac{3}{2}\cos\theta)). \tag{xiii}$$

Thus, for the ball to remain in contact at the top of the loop, where $\theta = \pi$, eqn (xiii) shows that the height H must be chosen such that

$$H > \tfrac{5}{2}R. \tag{xiv}$$

For values of H satisfying inequality (xiv) the ball will pass smoothly around the loop and will exit at $\theta = 2\pi$, according to eqn (vi), with a velocity given by

$$v_\theta^2 = 2gH, \tag{xv}$$

which is the same as the velocity at entry to the loop. At the top of the loop the velocity is given by eqn (viii).

Inequality (xiv) shows that $(v_\theta)_{min}^2$ must be greater than gR to reach this position. The limiting condition when $(v_\theta)_{min}^2 / R = g$ shows that for contact to be maintained at $\theta = 180°$ the downward acceleration v_θ^2 / R of the particle must be greater than g.

When $5/2\, R > H > R$, eqn (xiii) shows that the ball will lose contact with the track, after point C, at a poisition given by

$$\cos\theta = \frac{2}{3}\left(1 - \frac{H}{R}\right). \tag{xvi}$$

If $H < R$ the ball will remain in contact with the track but will not pass the point C. Equation (vi) shows that the ball will come to rest on the loop when its vertical displacement is equal to H, i.e. when

$$R(1 - \cos\theta) = H.$$

Thereafter it will roll back down the loop and will return to its original position. This motion will continue indefinitely if no energy is lost due to friction.

The problem could also have been approached by drawing the system boundary around the particle and the Earth. In this case no external forces would act on the system, but with the particle at A the system would have internal potential energy mgH. The reader should check that this alternative approach gives the same results.

5.2 Lift system

The mine cage of mass m shown in Fig. 5.49 is installed in a mine shaft and is raised and lowered by means of a wire rope, one end of which is firmly attached to the top of the cage. At the top of the shaft the other end of the rope is wrapped around a drum which is driven by an electric motor through a single stage gearbox as shown. To raise the cage the drum is rotated in an anticlockwise direction by the motor and the cable is wound up.

The drum and its integral gear G_1 are free to rotate, with angular velocity ω_1, on bearings about an axis through O_1 and have a combined moment of

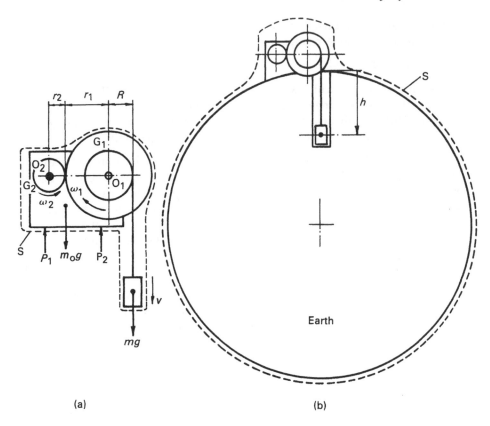

(a)   (b)

F<small>IG</small>. 5.49

inertia I_1 about their axis through O_1. Similarly the drive motor and the gear G_2 rotate, with angular velocity ω_2, on bearings about an axis O_2 and have a combined moment of inertia I_2 about this axis. The pitch circle diameters of the gears G_1 and G_2 are $2r_1$ and $2r_2$ respectively, and the drum is assumed to have a diameter $2R$, as shown.

Let us assume that the cage is stationary at the surface of the mine shaft and a catastrophic failure occurs which cuts off power to the motor at a time when the safety brakes on the drum gave failed. We will calculate the velocity v with which the cage impacts the base of the shaft a distance h below the surface.

The problem can be approached in two ways by drawing the system boundary S

(i)   around the motor, gearbox and cage so that external forces mg, m_0g, P_1 and P_2 are applied as shown in Fig. 5.49(a),

(ii) around the motor, gearbox, cage and the Earth, so that there are no external forces applied to the system as in Fig. 5.49(b).

We will apply the First Law of Thermodynamics to both systems to show that both give the same results.

Now　　$W + Q = \Delta U.$　　(i)

For both systems we will assume that no heat flows across the boundary S and that there is no change in temperature of the parts as the cage drops down to shaft. This is equivalent to assuming that there is no friction in the system. We now have to consider each of our two systems separately.

(i)　The forces P_1 and P_2 are provided by the holding down bolts and maintain the drive assembly (i.e. the motor, the gears and the drum) in vertical equilibrium. These forces together with the weight, $m_0 g$, of the drive assembly do no work on the system since their points of application remain stationary.

In this case only the external force mg, due to the weight of the cage, does work on the system as it moves a distance h down the shaft. Thus

$$W = mgh.$$　　(ii)

The initial internal energy U_1 will be zero since the system is assumed to be initially at rest. The internal energy of the system when the cage reaches the bottom of the shaft will be equal to the kinetic energy of the system since we have assumed no rise in temperature. The cage is in pure translation and the gears and drum are in pure rotation so that

$$\Delta U = U_2 - U_1$$
$$= \tfrac{1}{2}mv^2 + \tfrac{1}{2}I_1\omega_1^2 + \tfrac{1}{2}I_2\omega_2^2. \qquad \text{(iii)}$$

Substitution of eqns (ii) and (iii) into eqn (i) gives

$$mgh = \tfrac{1}{2}mv^2 + \tfrac{1}{2}I_1,\,\omega_1^2 + \tfrac{1}{2}I_2\omega_2^2 \qquad \text{(iv)}$$

(ii)　Here there are no external forces applied to the system so that

$$W = 0 \qquad \text{(v)}$$

Suppose we now define the surface of the Earth as our datum from which we will calculate the internal energy of the system due to its position in a gravitational field.

Initially the system will have zero internal energy, i.e. $U_1 = 0$ since the system is at rest and the cage is at the surface.

As the cage falls the system will gain kinetic energy but will lose potential energy due to its position h below the surface. When the case contacts the base of the shaft the system will have kinetic energy

$$T = \tfrac{1}{2}mv^2 + \tfrac{1}{2}I_1\omega_1^2 + \tfrac{1}{2}I_2\omega_2^2$$

and potential energy $V = -mgh.$ Thus

$$U_2 = T + V$$

and $\Delta U = U_2 - U_1$

$$= \tfrac{1}{2}mv^2 + \tfrac{1}{2}I_1\omega_1^2 + \tfrac{1}{2}I_2\omega_2^2 - mgh. \tag{vi}$$

Substituting eqns (v) and (vi) into eqn (i) gives

$$0 = \tfrac{1}{2}mv^2 + \tfrac{1}{2}I_1\omega_1^2 + \tfrac{1}{2}I_2\omega_2^2 - mgh. \tag{vii}$$

Eqns (iv) and (vii) give the same result.

The angular velocities of the gears G_1 and G_2 are not independent of v and from the kinematics of the system (see engineering application 1.3)

$$v = \omega_1 R,$$

and $$\omega_2 = \omega_1 \frac{r_1}{r_2} = \frac{v}{R}\frac{r_1}{r_2}.$$

Substituting for ω_1 and ω_2 in either of eqn (iv) or eqn (viii) gives

$$v^2 = \frac{2gh}{1 + \dfrac{1}{mR^2}\left(I_1 + \left(\dfrac{r_1}{r_2}\right)^2 I_2\right)}. \tag{viii}$$

Eqn (viii) gives the magnitude of the velocity of the cage as it reaches the base of the shaft. By inspection of the denominator in eqn (viii) we see that if mR^2 is increased for given values of I_1, I_2, r_1 and r_2, this velocity will increase. If the values of I_1 or I_2 are increased the maximum velocity will be reduced. It is important to note that a change in I_2, since it is multiplied by the *square* of the gear ratio, has a much greater effect than a change in I_1.

In many winch systems the kinetic energy of the high speed motor is much greater than that of the load and the drum. With large gear ratios the inertia of the motor is often the dominant effect during acceleration and braking.

5.3 Flywheel press

Figure 5.50 shows the principal components of a flywheel press which is used to stamp out solenoid laminations from steel strip. The press is driven by an electric motor which is directly coupled to the flywheel. A crank-slider mechanism is driven by the flywheel and the tool used to punch out the laminations is rigidly fixed to the slider. The tool makes contact with the strip before the slider approaches its lower extreme, or bottom dead centre (BDC), position. In this position the bearings of the crank-slider mechanism at O, A and B are almost in line and large forces can be developed at the punch tool without causing excessive side forces on the slider or large torques on the flywheel. When the slider passes through the BDC position the punching operation is finished. The tool thereafter retracts, the metal strip is indexed forward and the cycle begins again.

The crank and flywheel are in pure rotation, the connecting rod is in general plane motion and the punch is in pure translation.

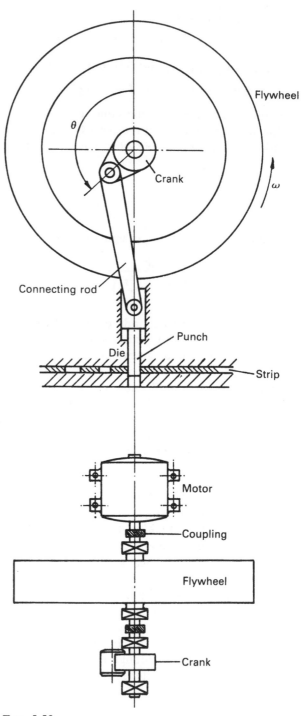

Fig. 5.50

During the stamping of a lamination the forces generated at the tool will decelerate the press. This causes the angular velocity of the flywheel to fall and a minimum value ω_{min} is reached at the end of the stroke. Once the stamping operation is complete the motor torque will accelerate the flywheel. Its angular velocity will then increase to a maximum value ω_{max}, which is reached just before the next stamping operation occurs. The variation of the angular velocity of the flywheel over one cycle is therefore of the form sketched in Fig. 5.51.

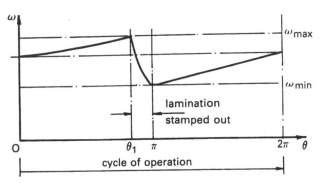

Fig. 5.51

For smooth continuous operation of the press it is necessary to (i) select a motor which has sufficient drive torque to maintain the cycle, and (ii) reduce the speed fluctuations of the flywheel to an acceptable level. We will now carry out an energy analysis of the press to find out how these requirements can be achieved.

Suppose that the drive torque acting on the flywheel due to the action of the drive motor is constant and has a value M_O. Let the rotation of the crank, and hence that of the flywheel, be defined by the angle θ from the vertical corresponding to the upper extreme or top dead centre (TDC) position, as shown by the line diagram of the crank-slider mechanism in Fig. 5.52.

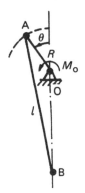

Fig. 5.52

When the punch tool touches the strip, let the crank angle be θ_1, as shown in Fig. 5.51. A further rotation, which advances θ to 180°, will cause the punch tool to penetrate the strip and to move through a distance h. For intermediate positions let the penetration of the tool into the sheet be defined by the displacement x measured from the top surface of the sheet, as shown in Fig. 5.53. If the crank radius is much less than the length of the connecting rod ($R \ll l$), which would be the case in most presses, the tool penetration into the strip can be determined from the geometry of Fig. 5.50 and shown to be given by

$$x \approx R(\cos \theta_1 - \cos \theta). \qquad (i)$$

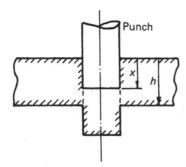

Fig. 5.53

Let us now consider the force on the tool during the stamping of the lamination. Experiment has shown that the force F rises rapidly to a value F_0 and then reduces in an almost linear manner to zero such that, to a good approximation,

$$F = F_0(h - x)/h. \qquad (ii)$$

Figure 5.54 shows how the tool force varies with slider position. The force is zero until contact is made with the sheet. Immediately thereafter it rises to

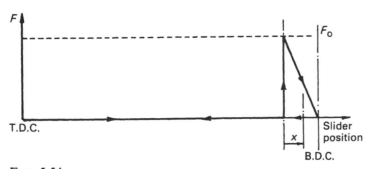

Fig. 5.54

its maximum value $F = F_0$. As the tool penetrates the sheet the force falls linearly from its maximum value and is zero at the end of the stroke.

The arrows on the diagram indicate how the force varies as the slider moves from TDC to BDC and back to TDC.

Let us now consider the crank and flywheel and the connecting rod and punch as a system with an external torque M_O and external forces F and N applied as shown in Fig. 5.55. The force N is the normal force on the slider from the guide. Friction at the bearings at O, A and B, and at the guide will be neglected. The forces at the bearing O are not shown since they do no work on the system.

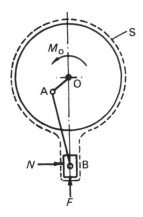

FIG. 5.55

In most presses the flywheel is large and the mass of the connecting rod and punch are small compared with that of the flywheel. We will therefore neglect these small masses to simplify the analysis.

To find the motor torque required to maintain a constant cycle time we can apply the First Law of Thermodynamics to the system over a complete cycle. Let us start from TDC on one cycle and consider the state of the system at TDC on the next cycle.

Now if we assume no heat flow to the system, $Q = 0$, so that

$$W_c = \Delta U, \tag{iii}$$

where W_c is the total work done on the system in one complete cycle. (Note that this assumes that the machine is at the same temperature as its surroundings and conforms with our assumptions of no friction in the system).

In eqn (iii) ΔU represents the change in the kinetic energy of the system and since we require the angular velocity of the flywheel, and hence the velocities of all points in the system, to be the same each time TDC is reached, $\Delta U = 0$. Thus the total work, W_c, done on the system over one cycle must be zero. The force N has no displacement along its line of action and therefore does no work. Only the torque M_O and the force F do work and from our definition of work

$$W_c = \int_0^{2\pi} M_O \, d\theta - \int_{x=0}^{x=h} F \, dx \qquad \text{(iv)}$$

where x is the penetration into the sheet. The negative sign is because x is in the opposite direction from F. (Note that the punch does positive work on the sheet since F and x are, in this case, in the same direction).

By substituting for F from eqn (ii) and integrating we obtain

$$W_c = 2\pi M_O - \frac{F_0 h}{2}. \qquad \text{(v)}$$

If the cycle is to repeat, $\Delta U = 0$, so that from eqns (iii) and (v)

$$M_O = \frac{F_0 h}{4\pi}. \qquad \text{(vi)}$$

This torque is independent of the moment of inertia of the flywheel, and with the assumptions made, is also independent of the geometry of the mechanism. The torque required therefore depends only on the peak force required to pierce the material and the sheet thickness.

Figure 5.51 shows that when the press stamps out a lamination the angular velocity of the flywheel falls from a maximum value ω_{max} at the beginning of stamping to a minimum value ω_{min} at the end of the operation.

We will now apply the First Law of Thermodynamics over the interval during which punching occurs. As the punch touches the strip the position of the crank is θ_1 and the position of the punch is $x = 0$. Let $\theta_2 = \pi$ and $x = h$ be the corresponding positions when the punching operation is complete.

From eqn (iii)

$$W_{12} = U_2 - U_1, \qquad \text{(vii)}$$

where W_{12} is the work done on the system between positions 1 and 2, and U_1, U_2 are the kinetic energies of the system at positions 1 and 2. Again we have assumed that $Q = 0$.

Thus if I_O is the moment of inertia of the flywheel and motor about an axis through O eqn (vii) gives

$$W_{12} = \tfrac{1}{2} I_O (\omega_{min}^2 - \omega_{max}^2). \qquad \text{(viii)}$$

If ω_0 is the mean angular velocity of the flywheel we can write

$$\omega_{max} = \omega_0 + \Delta\omega_1,$$

and $\omega_{min} = \omega_0 - \Delta\omega_2,$

where $\Delta\omega_1, \Delta\omega_2$ are the variations in angular velocity about the mean. Substituting these expressions into eqn (viii), and neglecting terms in $\Delta\omega^2$, since $\Delta\omega_1$,

$\Delta\omega_2$ are assumed to be small, we obtain

$$W_{12} = -I_O\omega_0\Delta\omega, \tag{ix}$$

where $\Delta\omega = \Delta\omega_1 + \Delta\omega_2 = \omega_{max} - \omega_{min}$, is the total variation in the angular velocity of the flywheel.

From eqn (iv)

$$W_{12} = \int_{\theta_1}^{\theta_2 = \pi} M_O \, d\theta - \int_{x_1=0}^{x_2=h} F \, dx.$$

Integrating gives

$$W_{12} = M_O(\pi - \theta_1) - \frac{F_0 h}{2}.$$

Substituting the value of M_O from eqn (vi) gives

$$W_{12} = -\frac{F_0 h}{2}\left(\frac{\pi + \theta_1}{2\pi}\right). \tag{x}$$

This equation can be simplified further by noting that $\theta_1 \approx \pi$ for cases where the lamination thickness is much smaller than the crank radius, i.e. $h \ll R$. For example in a typical press we might have $h \approx 1$ mm and $R \approx 100$ mm. Using this approximation in eqn (x) we find that for most practical purposes

$$W_{12} \approx -\frac{F_0 h}{2}. \tag{xi}$$

Thus, during the stamping of the lamination, the work done by the motor torque is small when compared with the work done by the force F.

Substituting for W_{12} in eqn (ix) gives

$$-\frac{F_0 h}{2} = -I_O\omega_0\Delta\omega. \tag{xii}$$

Writing the variation in the angular velocity of the flywheel as a function of the mean speed gives

$$\frac{\Delta\omega}{\omega_0} = \frac{F_0 h}{2I_O\omega_0^2}. \tag{xiii}$$

Since F_0 and h are fixed by the material and the shape of the lamination, and since ω_0 is determined by the speed of production, I_O is the only parameter that can be varied in eqn (xii) in order to achieve a specified value of $\Delta\omega/\omega_0$. Equation (xiii) can therefore be used to determine the moment of inertia of the flywheel.

5.4 Circuit breaker

The crank slider mechanism shown in plan view in Fig. 5.56 lies in the horizontal plane and represents a simple electrical circuit breaker. To make the circuit

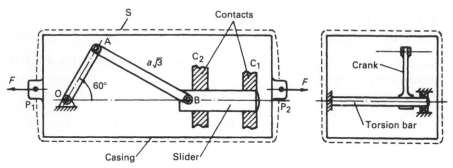

FIG. 5.56

the crank OA is rotated clockwise against a torsion spring at O until the slider B is fully engaged with the stationary contact C_1. The breaker is then latched in this position. When the contact is closed a current flows from the grid powerline into contact C_1 and then back again via the slider B and the fixed contact C_2. The link AB must be constructed from an insulating material.

When a fault develops in the power line, such as would occur during a lightning strike or upon the collapse of a supporting pylon, the circuit must be broken very quickly. This is done by releasing the latch so that the torsion bar can unwind and break the connection at C_1.

In order to achieve a successful break, and to prevent electrical arcing between the contacts, it is necessary that the slider achieves a velocity v_c when the separation between the slider and contact C_1 reaches a specified value λ.

For the purposes of illustration we shall assume that the crank angle θ goes from $\theta = 60°$ when the breaker is in the fully closed position to $\theta = 120°$ when the separation is λ. If we choose OA $= a$ and AB $= a\sqrt{3}$ then from the geometry of Figs. 5.57(a) and (b)

$$\lambda = a. \tag{i}$$

The specification that the velocity of the slider reaches a value v_c at this position is achieved by arranging the torsion bar to have the correct amount of wind up.

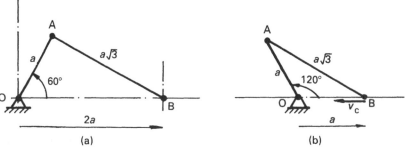

FIG. 5.57

Let us calculate this wind up, θ_0, by applying the First Law of Thermodynamics to the breaker assembly. We will assume, for simplicity, that the crank and the connecting rod are represented by uniform rods of mass m and $m\sqrt{3}$ respectively, and that the slider also has mass m. Friction at the bearings O, A and B and in the slider guideways will be neglected. We shall also assume that the torsion bar has a torsional stiffness k_θ.

Let us draw a system boundary S around the casing containing the breaker assembly, as shown in Fig. 5.56. The forces at the points P_1 and P_2 are the only external forces applied to the system and are provided by the casing support bolts. However, since P_1 and P_2 are stationary, no work will be done on the system by these forces. We shall also assume that no heat is transferred across S. Thus the First Law of Thermodynamics gives

$$W + Q = \Delta U = U_2 - U_1 = 0 \tag{ii}$$

and shows that the internal energy of the breaker assembly is constant during the separation of the contacts.

The first step in our calculation is to derive expressions for the internal energies U_1 and U_2. In the initial position the mechanism is stationary and the internal energy U_1 is determined by the strain energy, V_{s1}, stored in the torsion spring.

If the initial twist in the torsion spring is θ_0, the initial internal energy in the system is

$$U_1 = V_{s1} = \tfrac{1}{2}k_\theta\theta_0^2. \tag{iii}$$

When the separation of the contacts reaches a value λ the slider of the crank-slider mechanism will be moving with velocity v_c to the left.

The internal energy of the system, U_2, will be determined by the residual strain energy stored in the torsion spring and the kinetic energies acquired by the crank, the connecting rod and the slider. For this position the twist in the torsion spring is reduced from θ_0 to

$$\theta_2 = (\theta_0 - \pi/3), \tag{iv}$$

and the strain energy V_{s2} becomes

$$V_{s2} = \tfrac{1}{2}k_\theta(\theta_0 - \pi/3)^2$$

$$= \tfrac{1}{2}k_\theta\left(\theta_0^2 - \frac{2\pi}{3}\theta_0 + \frac{\pi^2}{9}\right). \tag{v}$$

To calculate the kinetic energy of the assembly we must first describe the motion of each component of the mechanism. From the kinematics of the crank-slider mechanism (engineering application 1.6) we know that the crank moves with pure rotation, the connecting rod with general plane motion and the slider with pure translation.

Since the crank moves with pure rotation about an axis through O its kinetic energy is given by

$$T_{OA} = \tfrac{1}{2}I_O\omega_{OA}^2 \tag{vi}$$

where I_O is the moment of inertia of the crank about its axis of rotation and ω_{OA} is its angular velocity. When regarded as a thin rod rotating about an end the moment of inertia of the crank is

$$I_O = \frac{ma^2}{3}. \tag{vii}$$

The kinetic energy of the connecting rod, which moves with general plane motion is given by

$$T_{AB} = \tfrac{1}{2}(m\sqrt{3})v_G^2 + \tfrac{1}{2}I_G\omega_{AB}^2 \tag{viii}$$

where v_G is the velocity of the centre of mass G, ω_{AB} is the angular velocity of the connecting rod, and I_G is the amount of inertia about an axis through its centre of mass.

Given that the connecting rod AB may be considered as a thin uniform rod of mass $m\sqrt{3}$ and length $a\sqrt{3}$ its moment of inertia about an axis through G is

$$I_G = \frac{m\sqrt{3}}{12}(a\sqrt{3})^2$$

$$= \frac{\sqrt{3}}{4}\,ma^2. \tag{ix}$$

The slider moves with pure translation with a velocity v_c and therefore has kinetic energy

$$T_S = \tfrac{1}{2}mv_c^2. \tag{x}$$

Before the kinetic energy of the complete mechanism can be calculated the values of ω_{OA}, ω_{AB} and v_G in eqns (vi) and (viii) must be found in terms of v_c. This requires a *full* kinematic analysis of the mechanism in the position shown in Fig. 5.57(b). This kinematic analysis will be based upon the analytical procedure described in engineering application 1.6.

Let us suppose that the crank-slider mechanism is at some general position, as shown by the line diagram of Fig. 5.58. The angular displacements of the crank and the connecting rod are measured from a datum line given by the axis *OX* and are θ and ϕ, as shown. The displacement x_s of the slider at B is measured along the direction of the X axis from an origin at O.

From the geometry of the mechanism the coordinates of point B on the slider can be written in terms of θ and ϕ as

$$x_s = a\cos\theta - a\sqrt{3}\cos\phi,$$

$$y_s = a\sin\theta - a\sqrt{3}\sin\phi = 0. \tag{xi}$$

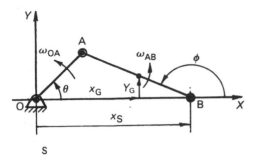

FIG. 5.58

Similarly, the coordinates of the centre of mass, G, of the connecting rod can be written as

$$x_G = a \cos \theta - \frac{a\sqrt{3}}{2} \cos \phi,$$

and $$y_G = \frac{a\sqrt{3}}{2} \sin \phi. \qquad \text{(xii)}$$

If eqns (xi) and (xii) are each differentiated with respect to time we obtain

$$\dot{x}_s = -(a\dot{\theta} \sin \theta - a\sqrt{3} \, \dot{\phi} \sin \phi),$$
$$0 = a\dot{\theta} \cos \theta - a\sqrt{3} \, \dot{\phi} \cos \phi, \qquad \text{(xiii)}$$

and

$$\dot{x}_G = -\left(a\dot{\theta} \sin \theta - \frac{a\sqrt{3}}{2} \dot{\phi} \sin \phi \right),$$

$$\dot{y}_G = \frac{a\sqrt{3}}{2} \dot{\phi} \cos \phi. \qquad \text{(xiv)}$$

Since $\dot{\theta}$ and $\dot{\phi}$ are, by definition, the angular velocities of the crank and the connecting rod and $\dot{x}_s$ is the velocity of the slider, we can set

$$\omega_{OA} = \dot{\theta}$$
$$\omega_{AB} = \dot{\phi} \qquad \text{(xv)}$$

and $$-v_c = \dot{x}_s.$$

(Note: the negative sign in eqn (xv) occurs because the slider velocity $\dot{x}_s$ is defined to be positive towards the right).

When the mechanism is in the final position shown in Fig. 5.57(b) we have $\theta = 120°$ and $\phi = 150°$. Substituting these values into eqn (xiii) gives

$$v_c \doteq a\frac{\sqrt{3}}{2} \omega_{OA} - \frac{a\sqrt{3}}{2} \omega_{AB},$$

$$\qquad \text{(xvi)}$$

and $$0 = \frac{a}{2} \omega_{OA} - \frac{3}{2} a\omega_{AB},$$

from which

$$\omega_{OA} = \sqrt{3}\,\frac{v_c}{a},$$

and $$\omega_{AB} = \frac{1}{\sqrt{3}}\frac{v_c}{a}.$$

(xvii)

The velocity, v_G, of the centre of mass of the connecting rod has components $\dot{x}_G$ along OX and $\dot{y}_G$ along OY which are given by eqns (xiv). When we set

$$\dot{\theta} = \frac{\sqrt{3}\,v_c}{a}, \qquad \dot{\phi} = \frac{1}{\sqrt{3}}\frac{v_c}{a}, \qquad \theta = 120° \quad \text{and} \quad \phi = 150°$$

in eqn (xiv) we obtain

$$\dot{x}_G = -\frac{5}{4}v_c,$$

and $$\dot{y}_G = -\frac{\sqrt{3}\,v_c}{4}.$$

Now $v_G^2 = \dot{x}_G^2 + \dot{y}_G^2$,

so that $v_G^2 = \dfrac{7}{4}v_c^2.$

(xviii)

The kinetic energies of the crank and the connecting rod can now be found by an appropriate substitution of eqns (xvii) and (xviii) into eqns (vi) and (viii). After a little algebra we can show that

$$T_{OA} = \frac{mv_c^2}{2}$$

(xix)

and $$T_{AB} = \frac{11\sqrt{3}}{12}mv_c^2.$$

(xx)

Thus the total kinetic energy T_2 of the system when the slider has moved a distance λ is given by

$$T_2 = T_s + T_{OA} + T_{AB}$$

$$= \left(1 + \frac{11\sqrt{3}}{12}\right)mv_c^2.$$

(xxi)

When the separation of the contact reaches a value λ the internal energy U_2 of the system consists of the kinetic energy of the circuit breaker mechanism plus the remaining strain energy in the torsion bar. Thus

$$U_2 = V_{s2} + T_2$$

(xxii)

where V_{s2} and T_2 are given by eqns (v) and (xxi).

By substituting eqns (iii) and (xxii) into eqn (ii)

$$V_{s2} + T_2 - V_{s1} = 0. \tag{xxiii}$$

Substituting the appropriate values for V and T gives

$$\frac{1}{2}k_\theta\left(\theta_0^2 - \frac{2\pi}{3}\theta_0 + \frac{\pi^2}{9}\right) + \left(1 + \frac{11\sqrt{3}}{12}\right)mv_c^2 = \frac{1}{2}k_\theta\theta_0^2,$$

from which

$$\frac{2}{3}\pi k_\theta\theta_0 = 2\left(1 + \frac{11\sqrt{3}}{12}\right)mv_c^2 + \frac{k_\theta\pi^2}{9}. \tag{xxiv}$$

Equation (xxiv) shows that the wind up in the torsion spring must have a value

$$\theta_0 = \frac{3}{\pi k_\theta}\left(1 + \frac{11\sqrt{3}}{12}\right)mv_c^2 + \frac{\pi}{6} \tag{xxv}$$

to ensure that the slider reaches the design velocity v_c when the contact separation is $\lambda = a$.

It should be remembered that the foregoing analysis has been developed on the assumption that friction is negligible. The value of θ_0 calculated from eqn (xxv) should therefore be treated as a *lower* estimate.

5.5 Disc brake

Figure 5.59 shows a flywheel test rig which is used to assess the performance of disc brakes. The flywheel is first driven up to an angular velocity ω_0 and then allowed to rotate freely in its bearings. Immediately after the flywheel starts to rotate freely the brake is operated by forcing the brake pads against the surface of the disc with a total force F. The friction forces acting on the disc are shown in the free body diagrams of Fig. 5.60 and produce a torque

$$M_O = \mu RF \tag{i}$$

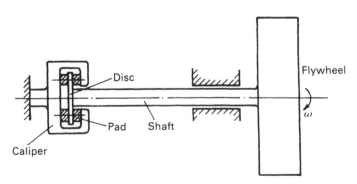

FIG. 5.59

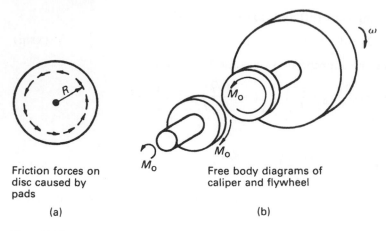

Friction forces on
disc caused by
pads

M_O

(a)

Free body diagrams of
caliper and flywheel

(b)

FIG. 5.60

about the axis of rotation, where R is the mean radius at which the pads are positioned on the disc and μ is the coefficient of friction between the pads and the disc. (See engineering application 3.5 eqn (ix)).

This torque reduces the angular velocity ω of the flywheel. Since the flywheel moves with pure rotation its equation of motion is given from eqn (4.30) by

$$-M_O = I_O\dot{\omega} \qquad\qquad (ii)$$

where I_O is the moment of inertia of the flywheel assembly about its axis of rotation. Equation (ii) can be integrated with respect to time to give

$$\omega = A - \frac{M_O}{I_O}t$$

where A is a constant of integration. When $t = 0$ we know that the flywheel rotates with angular velocity ω_0. Thus

$$A = \omega_0$$

and

$$\omega = \omega_0 - \frac{M_O}{I_O}t. \qquad\qquad (iii)$$

The friction torque M_O causes the angular velocity of the flywheel to reduce uniformly with time. The flywheel will therefore come to rest when

$$t = t_R = \frac{I_O\omega_0}{M_O}. \qquad\qquad (iv)$$

The time t_R is referred to as the 'run down time' and allows eqn (iii) to be rewritten

$$\omega = \omega_0 \left(1 - \frac{t}{t_R} \right) \quad \text{for } t \leq t_R$$

and $\omega = 0$ for $t \geq t_R$. (v)

As the flywheel runs down its kinetic energy falls and we know from experience that the brake will start to heat up. Let us apply the First Law of Thermodynamics, i.e.

$$W + Q = \Delta U, \tag{vi}$$

to the flywheel and the brake and show how the temperature of the brake can be calculated during the run down period.

Figure 5.61 shows a system boundary S drawn around the system of Fig. 5.59.

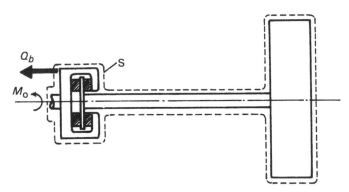

Fig. 5.61

From the free body diagram of the caliper, Fig. 5.60(b), we can see that a torque M_O must be applied to the system boundary at the position where it cuts the caliper. However, since the caliper is *stationary* no work can be done on the system by M_O. When the temperature of the brake assembly becomes greater than that of the surroundings, heat Q_b will be transferred across the system boundary, as shown.

Hence, eqn (vi) becomes

$$-Q_b = \Delta U = U_2 - U_1. \tag{vii}$$

The minus sign on the left-hand side of eqn (vii) indicates that the heat Q_b is *leaving* the system.

Initially the internal energy of the system, U_1, is determined by the kinetic energy of the flywheel and is given by

$$U_1 = \tfrac{1}{2} I_O \omega_0^2. \tag{viii}$$

During run down the internal energy will be determined by the kinetic energy stored in the flywheel and the thermal energy stored in the system. If we now assume, for simplicity, that the thermal energy is due to the temperature rise in the brake and that no heat is conducted along the shaft to the flywheel then

$$U_2 = \tfrac{1}{2}I_O\omega^2 + mC_p\Delta T_B \qquad\qquad (vix)$$

where m and C_p are the mass and specific heat respectively of the brake and ΔT_B is the rise in the temperature of the brake. C_p has units of J/kg°K and ΔT_B is in °K.

Substituting eqns (viii) and (vix) into eqn (vii) gives

$$-Q_b = mC_p\Delta T_B + \tfrac{1}{2}I_O\omega^2 - \tfrac{1}{2}I_O\omega_0^2. \qquad\qquad (x)$$

It can be seen from this equation that if braking occurs instantaneously, so that no heat is lost to the surroundings, i.e. $Q_b = 0$, then an *upper limit* for the rise in temperature ΔT_B can be obtained by setting $\omega = 0$. This maximum rise in temperature is

$$\Delta T_B^* = \frac{\tfrac{1}{2}I_O\omega^2}{mC_p}. \qquad\qquad (xi)$$

In this case the initial kinetic energy of the flywheel has all been stored as thermal energy in the brake. For other circumstances $Q_b > 0$, and eqn (x) gives

$$\Delta T_B < \Delta T_B^*.$$

The *rate* at which heat is lost by the system depends upon the temperature difference ΔT_B, and the value of the heat transfer coefficient h between the brake and its surroundings.

Thus

$$\dot{Q}_b = h\Delta T_B. \qquad\qquad (xii)$$

If we now differentiate eqn (x) with respect to time we obtain

$$-\dot{Q}_b = mC_p\Delta\dot{T}_B + I_O\dot{\omega}\omega. \qquad\qquad (xiii)$$

Hence, by using eqns (ii), (iv) and (xi), eqns (xii) and (xiii) can be combined to give

$$\tau\Delta\dot{T}_B + \Delta T_B = \Delta T_B^*\frac{2\tau}{t_R}\left(1 - \frac{t}{t_R}\right), \qquad\qquad (xiv)$$

where $\tau = mC_p/h$ has the units of time and is called the thermal time constant of the brake.

The thermal time constant τ is an important parameter and determines the rate at which the brake will lose heat when hot. Small values of τ are desirable since this ensures that the brake will lose heat rapidly and not become overheated.

The solution to the first order differential eqn (xiv) is

$$\Delta T_B = B\,e^{-t/\tau} + \frac{\Delta T_B^* 2\tau}{t_R}\left(1 + \frac{\tau}{t_R} - \frac{t}{t_R}\right)$$

where B is an arbitrary constant. (The reader should verify that this is a solution by substituting it back into eqn (xiv)).

When braking starts, the brake and the surroundings are at the same temperature, i.e.

$$\Delta T_B = 0 \quad \text{at } t = 0.$$

This condition gives

$$B = -\Delta T_B^* \frac{2\tau}{t_R}(1 + \tau/t_R)$$

so that $\left(\dfrac{\Delta T_B}{\Delta T_B^*}\right) = \dfrac{2\tau}{t_R}\{(1 + \tau/t_R)(1 - e^{-t/\tau}) - t/t_R\}.$ \hfill (xv)

Equation (xv) gives the temperature rise in the brake as a function of time and the upper limit ΔT_B^*.

Equation (xv) is best interpreted by plotting $\Delta T_B/\Delta T_B^*$ as a function of t/t_R for various values of τ/t_R. These curves are shown in Fig. 5.62 and are plotted for the run down period $0 \leqslant t/t_R \leqslant 1$.

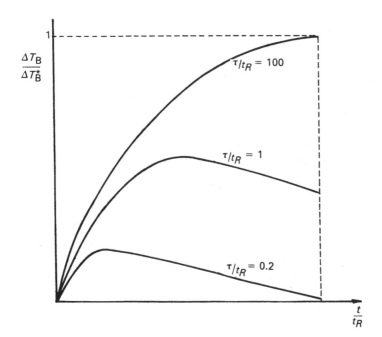

FIG. 5.62

If braking occurs quickly so that the run down period is much less than the thermal time constant, such as illustrated by the curve corresponding to $\tau/t_R = 100$, very little cooling occurs and the brake temperature approaches its upper limit of ΔT_B^*.

When the run down period and the thermal time constant are of similar order, i.e. $\tau/t_R \approx 1$, the temperature of the brake rises to a maximum before run down is complete and thereafter falls.

It can be seen from the slopes of the curves shown in Fig. 5.62 that the rate $d\Delta T_B/dt$ at which the temperature of the brake increases, decreases with time. This effect can be predicted from eqn (xiii) which can be rewritten as

$$\frac{d\Delta T_B}{dt} = -\frac{\dot{Q}_b}{mC_p} - \frac{d}{dt}\frac{(\frac{1}{2}I_O\omega^2)}{mC_p}. \tag{xvi}$$

The second term on the right-hand side of eqn (xvi), including the negative sign, represents the rate at which kinetic energy is lost in the system. From eqn (ii)

$$M_O = -I_O\dot{\omega}.$$

Now

$$M_O\omega = -I_O\dot{\omega}\omega = -\frac{d}{dt}(\frac{1}{2}I_O\omega^2),$$

so that the rate of loss of kinetic energy is proportional to the angular velocity of the system.

Initially, near the start of braking, ω is large so that the rate of loss of kinetic energy is also large. $\dot{Q}_b$ is, however, small since the temperature of the caliper is the same as that of the surroundings. Thus the initial kinetic energy lost is converted into thermal internal energy in the caliper and the temperature rises rapidly. As the temperature rises, $\dot{Q}_b$ increases, but the rate of loss of kinetic energy reduces because the speed falls. The rate of rise of temperature will therefore decrease. It is possible for the heat lost to the surroundings to become greater than the loss in kinetic energy, so that $d\Delta T_B/dt$ becomes negative and the temperature of the caliper starts to fall before the system comes to rest.

Exercises

1   Fig. 5.63 shows a ram of mass m supported at the points A and B by two parallel wires O_1A and O_2B each of length l. If the ram is released from rest when O_1A makes an angle $\theta = \beta$ with the vertical as shown, determine the velocity of the ram at the lowest point of its motion.

2   Two springs of stiffness k_1 and k_2 respectively are connected together in
      (i)   parallel, and
      (ii) series, as shown in Fig. 5.64.

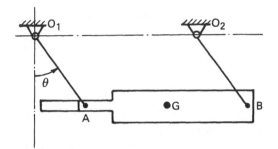

FIG. 5.63

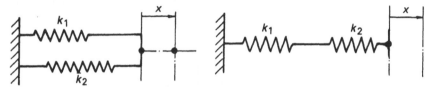

FIG. 5.64

For each configuration determine the strain energy stored in *each* spring when the combined spring is deflected an amount *x*. Also find the stiffness of a single equivalent spring which could replace the two springs in each case.

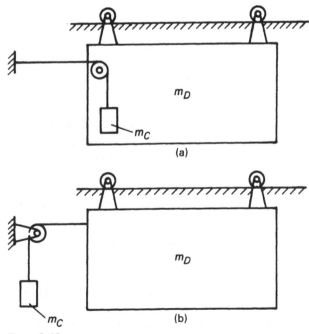

(a)

(b)

FIG. 5.65

3   A garage door of mass m_D is suspended on wheels from a horizontal rail as shown in Fig. 5.65. A self-closing mechanism consisting of a mass m_C supported on a cable and pulley arrangement is to be fitted to the door. Two possible arrangements are shown. Which is the more effective and why?

4   Fig. 5.66 shows a constant force F applied to the end of a rod OA. If the rod is supported at O by means of a pin joint, determine the work done by the force as the rod is displaced from $\theta = 0$ to $\theta = \alpha$. Assume the direction of the force is always vertical as shown.

   If $\theta = \omega t$, determine the power developed by F at the instant where $\theta = \alpha$.

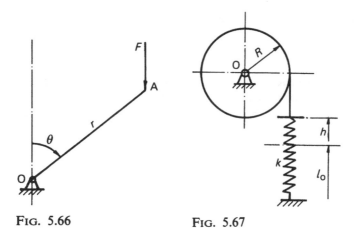

FIG. 5.66   FIG. 5.67

5   A spring of stiffness k and unstretched length l_0 is attached to a disc of mass m and radius R by means of a light inextensible string. The disc is fixed at its centre O by a frictionless pin joint and is rotated such that the spring becomes *extended* by an amount h, as shown in Fig. 5.67.

   If the disc is released from rest in this position, determine the angular velocity of the disc at the instant the spring returns to its unstretched length.

6   A thin uniform rod OA of mass m and length l can turn in bearings at O and is constrained by a torsion spring which exerts a restoring torque $k\theta$ on the rod about O when its deflection is θ as shown in Fig. 5.68. Friction is negligible. If the rod is slightly disturbed from rest in a vertically upwards position, find (i) its angular velocity, and its angular acceleration as a function of θ, and (ii) the component bearing reactions when $\theta = \pi$, assuming that the bar can fall through the vertically downwards position.

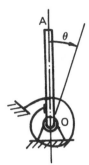

Fig. 5.68

7 A uniform bar AB of mass 4 kg and length 1 m can turn freely in a vertical plane on a hinge at its mid-point O. A small body P of mass 4 kg is to be firmly attached to the bar at a distance p from O. If the bar is released from rest in the horizontal position, determine the value of p which will make the angular velocity of the bar, as it passes through the vertical position, as large as possible.

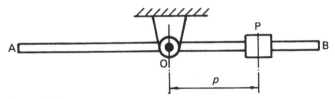

Fig. 5.69

8 The belt and pulley system shown in Fig. 5.70 is driven by a constant torque M_2 applied to pulley 2. Derive an expression for the angular velocity of pulley 1 in terms of its angular displacement θ_1. The moments of inertia of pulleys 1 and 2 about their axes O_1 and O_2 are I_1 and I_2 respectively.

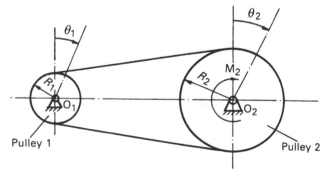

Fig. 5.70

9 The end A of a thin uniform rod AB of mass m and length a moves in a straight line with velocity v as the rod rotates with angular velocity $\dot{\theta}$. Derive an expression for the kinetic energy of the rod in a general position θ, as shown in Fig. 5.71. The moment of inertia of the rod about its centre of mass G is $(ma^2/12)$.

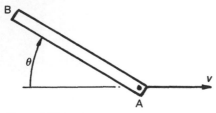

FIG. 5.71

10 The uniform rod AB shown in Fig. 5.72 has mass m and moves in the horizontal plane OXY. If at the instant shown, point A has a velocity v_1 along OX and point B a velocity along the direction of OY, determine the kinetic energy of the rod.

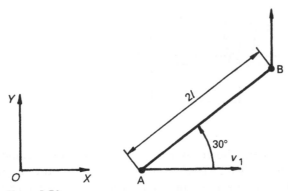

FIG. 5.72

11 To a uniform ring of mass m, radius r and centre C is attached a small body A also of mass m. The combined body can roll without slip on a horizontal plane and is released from rest with A in the topmost position as shown in Fig. 5.73. Using an energy method, find the angular velocity of the ring when it has rolled to a position in which A is in contact with the plane.

12 Fig. 5.74 shows two identical pulleys of mass m_P and radius R connected together by a light inextensible wire to form a simple lifting system. Pulley P_1 is fixed at its centre O_1 by means of a frictionless bearing. A small load of mass m_L is connected to pulley P_2 at its centre O_2. If this system is released from rest with O_1 and O_2 in line, determine the velocity of mass m_L when O_2 has dropped through a distance x as shown.

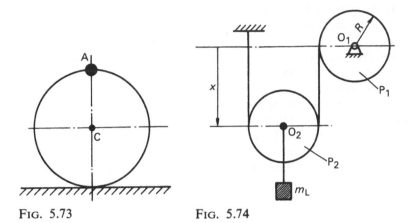

FIG. 5.73   FIG. 5.74

The moment of inertia of each pulley about an axis through its centre is I_O.

13 A body rolls without slip on a plane inclined at 60° to the horizontal as shown in Fig. 5.75. The body consists of two cylinders of diameter 2 m and 1.5 m, the larger cylinder being in contact with the plane. A rope is wrapped around the smaller cylinder and passes over a smooth peg P. The free end of the rope carries a mass m of 120 kg as shown. The body has a mass of 300 kg and a moment of inertia about an axis through its centre of mass of 200 kg m².

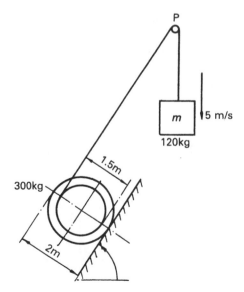

FIG. 5.75

If the mass m has a downward velocity of 5 m/s how far will it have moved when its velocity has reduced to 3 m/s downwards?

14 Show that the kinetic energy of a thin cylinder rolling without slip on a horizontal plane is $mr^2\omega^2$, where m is its mass, r its radius and ω its angular velocity.

Such a cylinder rolling on a horizontal plane, Fig. 5.76, wraps around its circumference a thin flexible cable of mass ρ per unit length. At $t = 0$, when the cylinder just begins to pick up the cable, its angular velocity is known to be ω_0. One revolution later, its angular velocity is ω_1. If a constant horizontal force P is applied to the centre of the cylinder C during the interval, use an energy method to derive an expression for ω_1. Deduce the value of P which makes $\omega_1 = \omega_0$.

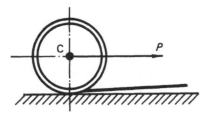

Fig. 5.76

15 A uniform link of mass m and length l is constrained to move as shown in Fig. 5.77. End B is constrained to move in a long vertical slot and end A to move in a horizontal slot. A spring of stiffness k lies in the horizontal slot and is compressed as A moves to the left.

The link is held at an angle θ to the horizontal as shown, and in this position end A of the link is just in contact with the free end of the uncompressed spring. If the link is now released obtain an expression for the velocity of end B of the link as it passes through the horizontal position.

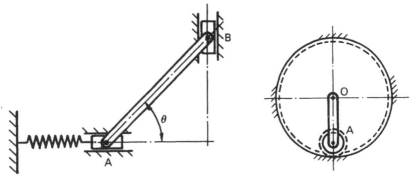

Fig. 5.77   Fig. 5.78

The moment of inertia of the link about an axis through its centre of mass is I and friction is negligible.

16 A geared mechanism (Fig. 5.78) consists of a fixed internally toothed wheel of centre O and pitch circle radius 80 mm, a uniform arm OA which can turn freely about O, and a planet wheel of centre A and pitch circle radius 15 mm which can turn freely on a pin attached to OA at A. The plane of the mechanism is vertical.

If OA is released from rest in a horizontal position, find the angular velocity of OA in the vertical position shown in Fig. 5.78. Friction at O and A may be neglected.

The constants of the system are as follows: mass of rod OA, 1 kg; moment of inertia of rod OA about an axis through O, 1.5×10^3 kg m^2; mass of planet wheel, 0.3 kg; moment of inertia of planet wheel about an axis through A, 0.03×10^{-3} kg m^2.

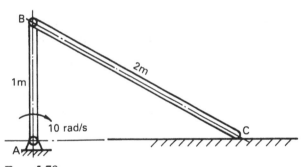

FIG. 5.79

17 The mechanism shown in Fig. 5.79 consists of two uniform rods AB and BC, which have masses 10 kg and 20 kg respectively. If AB rotates clockwise with an angular velocity $\dot{\theta} = 10$ rad/s, determine the kinetic energy of the system at the instant shown.

6

Impulse and momentum

Basic Theory

6.1 Linear momentum

Particle

Let us suppose that the absolute velocity of the particle P shown in Fig. 6.1 is v. If the particle has mass m then the vector quantity defined by

$$p = mv \qquad (6.1)$$

is called the *linear momentum* of the particle. We can see from this definition that the linear momentum vector is in the direction of v and has a magnitude equal to mv.

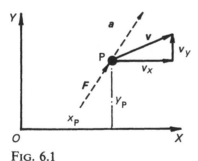

FIG. 6.1

If p is now resolved into components p_x, p_y along the reference directions OX and OY then

$$p_x = mv_x = m\dot{x}_p,$$

and $\qquad p_y = mv_y = m\dot{y}_p.$ $\qquad (6.2)$

Moreover, if p is differentiated with respect to time we obtain

$$\frac{\mathrm{d}p}{\mathrm{d}t} = \frac{m\mathrm{d}v}{\mathrm{d}t} = ma,$$

i.e. the rate of change of the linear momentum of the particle is proportional to its acceleration a. We know from Newton's Second Law of Motion that an external force, $F = ma$ must be applied to the particle in order to produce this acceleration.

Thus $\quad F = \dfrac{\mathrm{d}p}{\mathrm{d}t}.$ \hfill (6.3)

When there is no external force applied, $F = \mathrm{d}p/\mathrm{d}t = 0$, which shows that the linear momentum of the particle is constant during the motion. This represents another way of saying that when no forces act on the particle it continues to move with uniform velocity in a straight line.

Impulse of a force

Let us now consider the force F which is applied to the particle and obtain the integral $\mathscr{I}$ of F over a time interval from t_1 to t_2, so that

$$\mathscr{I} = \int_{t_1}^{t_2} F \, \mathrm{d}t.$$ \hfill (6.4)

The vector quantity $\mathscr{I}$ is called the *impulse* of the force. By substituting eqn (6.3) into eqn (6.4) and integrating we obtain

$$\mathscr{I} = \int_{t_1}^{t_2} \frac{\mathrm{d}p}{\mathrm{d}t} \, \mathrm{d}t = \int_{t_1}^{t_2} \mathrm{d}p = (p_2 - p_1)$$

so that $\quad \mathscr{I} = m(v_2 - v_1).$ \hfill (6.5)

This vector relationship is presented graphically in Fig. 6.2 which shows the change in the linear momentum, $\Delta p = p_2 - p_1$, of the particle. This is equal to the applied impulse $\mathscr{I}$.

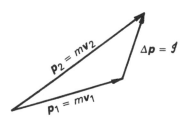

$\Delta p = \mathscr{I}$

$p_2 = mv_2$

$p_1 = mv_1$

FIG. 6.2

Example 6.1   Figure 6.3 shows two blocks, assumed to be particles, resting on a frictionless horizontal surface. Particle P_1 is moving with velocity v_0 and collides with the stationary particle P_2. Let us determine the subsequent motion of the particles. We shall assume that the impact is perfectly elastic and that no energy is lost during the impact.

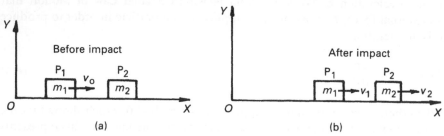

(a)   (b)

FIG. 6.3

The duration $(t_2 - t_1)$ of the contact and the precise variation of the contact force F_c during the contact period cannot be determined very easily—it depends upon the shapes of the contacting surfaces and their respective elasticities. The contact force only has a value during the impact and if measured would be of the form shown in Fig. 6.4.

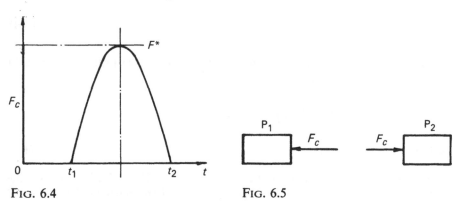

FIG. 6.4   FIG. 6.5

At time t_1 contact occurs, and the force rises rapidly from zero to a maximum value F^*. Thereafter the force reduces and is zero again at time t_2, when the particles separate. At time t_1 particle P_1 has a velocity v_0 and P_2 is at rest. When separation occurs at time t_2 we will assume that the particles P_1 and P_2 move with velocities v_1 and v_2 respectively, as shown in Fig. 6.3(b).

We can calculate the values of v_1 and v_2 by considering the motion of each particle during the impact. When impact occurs the particles will touch and the free body diagrams of P_1 and P_2 will be as shown in Fig. 6.5. The weight of the particles and the forces on them from the ground are not included.

Using the free body diagram of Fig. 6.5 the equation of motion of particle P_2 in the direction of the contact force F_c is, from eqn (6.3)

$$F_c = \frac{dp_2}{dt}. \tag{i}$$

If we now integrate this equation over the period during which the particles are in contact then

$$\int_{t_1}^{t_2} F_c \, dt = (p_2)_{t_2} - (p_2)_{t_1},$$

or $\mathcal{S} = m_2(v_2 - 0) = m_2 v_2.$ (ii)

Likewise, using the free body diagram of Fig. 6.5 for particle P_1,

$$-F_c = \frac{dp_1}{dt}. \tag{iii}$$

The integration of this equation over the time interval t_1 to t_2 yields

$$-\int_{t_1}^{t_2} F_c \, dt = (p_1)_{t_2} - (p_1)_{t_1},$$

i.e. $-\mathcal{S} = m_1 v_1 - m_1 v_0.$ (iv)

The addition of eqns (ii) and (iv) then gives

$$m_1 v_1 + m_2 v_2 = m_1 v_0, \tag{v}$$

which shows that the linear momentum of the system is constant and is not affected by the impact.

If we now draw a system boundary around both particles and assume that the internal energy before and after impact is determined by the kinetic energies of the particles then, from the First Law of Thermodynamics, with Q and W zero,

$$U_2 - U_1 = (\tfrac{1}{2}m_1 v_1^2 + \tfrac{1}{2}m_2 v_2^2) - \tfrac{1}{2}m_1 v_0^2 = 0. \tag{vi}$$

By eliminating v_2 from eqn (vi) using eqn (v) we obtain the quadratic equation

$$m_1(m_1 + m_2)v_1^2 - 2m_1^2 v_0 v_1 + m_1(m_1 - m_2)v_0^2 = 0. \tag{vii}$$

Equation (vii) has two solutions,

$$v_1 = v_0, \tag{viii}$$

and $v_1 = \dfrac{(m_1 - m_2)}{(m_1 + m_2)} v_0.$ (ix)

Substituting $v_1 = v_0$ into eqn (v) gives

$$v_2 = 0. \tag{x}$$

This result is valid and represents the state of the system *before* impact occurs. When we substitute eqn (ix) into eqn (v) we obtain

$$v_2 = \frac{2m_1}{(m_1 + m_2)} v_0. \tag{xi}$$

Equations (ix) and (xi) therefore determine the velocity of each particle after impact. We should note that $v_2 > v_1$, which proves that separation does occur after impact.

Since eqn (xi) shows that $v_2 > 0$ we conclude that the particle P_2 must always move in the direction of v_0 after impact. This is not always the case for particle P_1. When $m_1 > m_2$ eqn (ix) shows that $v_1 > 0$ and that P_1 will continue to move in the same direction. However when $m_1 < m_2$ we obtain $v_1 < 0$, which shows that P_1 has reversed its direction of motion. In the limit when $m_1 \ll m_2$ we find that $v_1 \to -v_0$ and $v_2 \to 0$, i.e. the larger particle P_2 remains stationary after impact and P_1 bounces off without any loss in the magnitude of its velocity.

When the particles have the same mass, i.e. $m_1 = m_2$, we find that $v_1 = 0$ and $v_2 = v_0$, i.e. particle P_1 is brought to rest and P_2 moves off with velocity v_0.

Suppose we now reconsider this example and assume that the particles adhere together during the impact and thereafter move as a single particle of mass $(m_1 + m_2)$ with a velocity v_c as shown in Fig. 6.6.

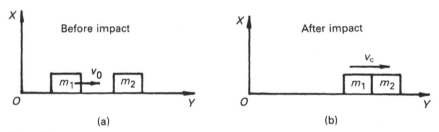

(a)   (b)

FIG. 6.6

If we now set $v_1 = v_2 = v_c$ in eqn (v) we find

$$(m_1 + m_2)v_c = m_1 v_0, \tag{xii}$$

and that

$$v_c = \frac{m_1}{(m_1 + m_2)} v_0. \tag{xiii}$$

The internal energy of the system after impact will be determined by the kinetic energy T_c of the combined mass and is given by

$$T_c = \tfrac{1}{2}(m_1 + m_2) v_c^2. \tag{xiv}$$

Substitution of eqn (xiii) into eqn (xiv) yields

$$T_c = \frac{1}{2} \frac{m_1^2}{(m_1 + m_2)} v_0^2. \tag{xv}$$

Thus the kinetic energy of the system has fallen by an amount

$$\Delta T = \frac{1}{2} m_1 v_0^2 - \frac{1}{2} \frac{m_1^2 v_0^2}{(m_1 + m_2)} = \frac{1}{2} \frac{m_1 m_2}{(m_1 + m_2)} v_0^2 \qquad \text{(xvi)}$$

due to the impact.

Coefficient of restitution

Let us examine the case of normal impact between two spherical particles P_1 and P_2 of mass m_1 and m_2 respectively. We shall assume that the velocities of the particles before and after impact are given by v_1, v_2 and v_1', v_2' as shown in Figs. 6.7(a) and 6.7(b). If we assume that the particles are smooth (i.e. no

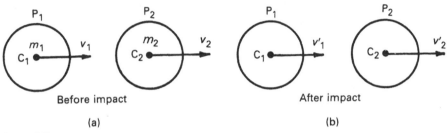

Before impact   After impact

(a)   (b)

FIG. 6.7

friction) then the line of action of the contact force between the particles at impact will be along the common normal at the point where they touch. For spherical particles this line will be along the line joining their centres C_1 and C_2.

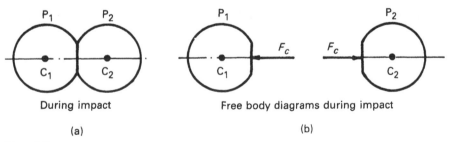

During impact   Free body diagrams during impact

(a)   (b)

FIG. 6.8

When the particles collide, their contacting surfaces will become compressed, and a contact force F_c will act on each particle as shown in Figs. 6.8(a) and (b).

After making contact the centres of the particles will initially continue to approach each other and the contact force will increase, as shown in Fig. 6.9, until a maximum value F^* is reached.

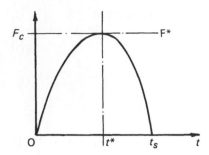

FIG. 6.9

This maximum occurs a time t^* after the initial contact is made. At t^* the particles will be moving with the same velocity v^*. The impulse of F_c for this period of approach is given by

$$\mathscr{I}_a = \int_0^{t^*} F_c \, dt. \tag{6.6}$$

Thereafter the contacting surfaces start to recover and the contact force decreases as the particles, though still in contact, begin to move apart. The contact force reduces to zero at time t_s, which is the instant of separation.

Hence $\mathscr{I}_s = \int_{t^*}^{ts} F_c \, dt \tag{6.7}$

defines the impulse of F_c over the period of recovery.

Let us now consider the impulses applied to each particle during the impact and consider separately the periods of approach and separation.

The impulses applied to particles P_1 and P_2 during the approach are given by

$$-\int_0^{t^*} F_c \, dt = m_1(v^* - v_1) \tag{6.8}$$

on P_1

and $\int_0^{t^*} F_c \, dt = m_2(v^* - v_2) \tag{6.9}$

on P_2.

Eliminating v^* from these equations gives

$$\int_0^{t^*} F_c \, dt = \frac{m_1 m_2}{(m_1 + m_2)} (v_1 - v_2). \tag{6.10}$$

Similarly, during separation, the impulse applied to P_1 is

$$-\int_{t^*}^{t_s} F_c \, dt = m_1(v_1' - v^*),$$ (6.11)

and the impulse applied to P_2 is

$$\int_{t^*}^{t_s} F_c \, dt = m_2(v_2' - v^*).$$ (6.12)

Thus $$\mathcal{I}_s = \frac{m_1 m_2}{(m_1 + m_2)}(v_2' - v_1').$$ (6.13)

Equations (6.10) and (6.13) show that the impulses $\mathcal{I}_a$ and $\mathcal{I}_s$ are in the ratio

$$\frac{\mathcal{I}_s}{\mathcal{I}_a} = \frac{\int_{t^*}^{t_s} F_c \, dt}{\int_0^{t^*} F_c \, dt} = \frac{(v_2' - v_1')}{(v_1 - v_2)}$$

$$= \frac{\text{relative velocity of separation along } C_1 C_2}{\text{relative velocity of approach along } C_1 C_2}$$ (6.14)

This ratio depends only on the velocities of the particles before and after impact and is called the coefficient of restitution e

i.e. $$e = \frac{(v_2' - v_1')}{(v_1 - v_2)}.$$ (6.15)

If we now add eqns (6.8) and (6.11) the total impulse applied to P_1 is

$$-\int_0^{t_s} F_c \, dt = m_1(v_1' - v_1).$$ (6.16)

Likewise eqn (6.9) and (6.12) give the total impulse applied to P_2 as

$$\int_0^{t_s} F_c \, dt = m_2(v_2' - v_2).$$ (6.17)

Equations (6.16) and (6.17) show, as expected, that the linear momentum of the system is unchanged as a result of the impact,

i.e. $$m_1 v_1 + m_2 v_2 = m_1 v_1' + m_2 v_2'.$$ (6.18)

We shall now use eqns (6.15) and (6.18) to determine the change, ΔT, in the kinetic energy of this system of particles due to the impact. The kinetic energies before and after impact are given by

$$T_B = \tfrac{1}{2}m_1 v_1^2 + \tfrac{1}{2}m_2 v_2^2$$

and $$T_A = \tfrac{1}{2}m_1 v_1'^2 + \tfrac{1}{2}m_2 v_2'^2.$$ (6.19)

We may therefore write

$$\Delta T = \tfrac{1}{2}m_1(v_1'^2 - v_1^2) + \tfrac{1}{2}m_2(v_2'^2 - v_2^2)$$

$$= \frac{1}{2(m_1 + m_2)}[m_1(m_1 + m_2)(v_1'^2 - v_1^2) + m_2(m_1 + m_2)(v_2'^2 - v_2^2)]$$

$$= \frac{1}{2(m_1 + m_2)}[-m_1 m_2\{(v_1 - v_2)^2 - (v_2' - v_1')^2\}$$

$$+ (m_1 v_1' + m_2 v_2')^2 - (m_1 v_1 + m_2 v_2)^2]$$

$$= \frac{1}{2(m_1 + m_2)}\left[-m_1 m_2(v_1 - v_2)^2\left\{1 - \left(\frac{v_2' - v_1'}{v_1 - v_2}\right)^2\right\}\right.$$

$$+ (m_1 v_1' + m_2 v_2' - m_1 v_1 - m_2 v_2)(m_1 v_1' + m_2 v_2' + m_1 v_1 + m_2 v_2).$$

Substituting for v_1' and v_2' from eqns (6.15) and (6.18) gives

$$\Delta T = -\frac{m_1 m_2}{2(m_1 + m_2)}(v_1 - v_2)^2(1 - e^2). \qquad (6.20)$$

The minus sign in eqn (6.20) shows that kinetic energy will be lost from the system provided $e < 1$.

Since some of the strain energy of deformation will not be recoverable as kinetic energy, due to internal friction or hysteresis in the material of the spheres, $\Delta T < 0$.

Thus

$$e = \frac{\int_{t^*}^{t_s} F_c \, dt}{\int_0^{t^*} F_c \, dt} \leqslant 1. \qquad (6.21)$$

The impulse of F_c for the period of approach must be greater than the impulse of F_c for the period of separation, as indicated by the shape of Fig. 6.9. If $e = 1$ eqn (6.20) shows that no kinetic energy is lost at impact and the collision is said to be elastic. This also gives $\int_{t^*}^{t_s} F_c \, dt = \int_0^{t^*} F_c \, dt$, which suggests that the function defining the contact force F_c could be symmetric about t^*.

The case $e = 0$ corresponds to maximum loss of kinetic energy and eqn (6.15) shows that $v_1' = v_2'$, i.e. the particles remain in contact after impact.

Using evidence obtained by experiments on colliding spheres Newton suggested that e was a constant for a given pair of materials and proposed that eqn (6.15) should be regarded as an equation of state. This experimental law has been found to give reasonable predictions for particles. Typically e ranges from 0.2 for soft materials such as lead to 0.9 for hard materials such as glass.

If eqns (6.15) and (6.18) are now used to calculate the velocities of the particles after impact we find that

$$v_1' = v_1 - \frac{m_2(1 + e)}{(m_1 + m_2)}(v_1 - v_2) \qquad (6.22)$$

and $\qquad v_2' = v_2 + \dfrac{m_1(1+e)}{(m_1+m_2)}(v_1 - v_2)$ \hfill (6.23)

For positive values of v_2 it can be seen that $v_2' > 0$. Also $v_1' > 0$, provided that

$$(m_1 - m_2 e)v_1 + m_2(1+e)v_2 > 0.$$

If particle P_2 is stationary, as in example 6.1 then

$$v_1' = v_1 \left(\frac{m_1 - m_2 e}{m_1 + m_2} \right). \hfill (6.24)$$

Thus provided that $m_1 > em_2$ the particle will continue to move in the same direction.

The special case where $v_1' = 0$ occurs when $m_1 = em_2$. If $e = 1$, then for v_1' to be zero $m_1 = m_2$, which agrees with the result obtained for an elastic impact.

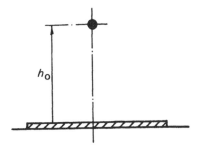

Fig. 6.10

Example 6.2 Figure 6.10 shows a steel ball being dropped from a height h_0 onto a massive horizontal steel plate. Let us calculate the maximum height h_m reached by the ball after m impacts with the plate. The coefficient of restitition between the ball and the plate is e.

Let us suppose that after the nth impact with the plate the ball reaches a height h_n as shown in Fig. 6.11. The velocity of the ball just before it again

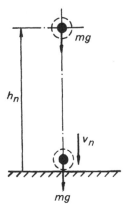

Fig. 6.11

impacts with the plate can be found by applying the First Law of Thermodynamics to the particle and treating the force mg due to gravity as an external force. The change in the internal energy of the particle will be due to the change in its kinetic energy. Thus

$$W = mgh_n = \Delta U = \tfrac{1}{2}mv_n^2. \tag{i}$$

Since the plate is stationary and massive (i.e. $m_2 \to \infty$) eqn (6.24) gives the velocity of the ball after impact as

$$v_n' = -ev_n. \tag{ii}$$

The negative sign indicates that the ball has reversed its direction and is moving upwards, as shown in Fig. 6.12.

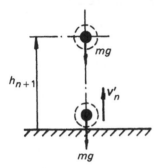

FIG. 6.12

If we now reapply the First Law of Thermodynamics to the particle the height reached is given by the solution of

$$W = \Delta U,$$

so that $\quad -mgh_{n+1} = 0 - \tfrac{1}{2}me^2v_n^2 = -e^2mgh_n.$

Thus $\quad h_{n+1} = e^2h_n.$ (iii)

This is a first order difference equation and has a solution of the form $h_n = A\alpha^n$, where A and α are constants. Thus from eqn (iii)

$$A\alpha^{n+1} = e^2A\alpha^n.$$

Hence $\quad \alpha = e^2$

and $\quad h_n = Ae^{2n}.$

If $h_n = h_0$ when $n = 0$ then

$$h_0 = Ae^0 = A$$

so that $\quad h_n = h_0e^{2n}.$ (iv)

After the mth impact the maximum height reached is $h_m = h_0e^{2m}.$

Closed systems of particles

We shall suppose the system shown in Fig. 6.13, contained within the boundary S, is formed from a collection of particles. Although the system boundary is free to move, the particles are not permitted to leave the system by crossing the boundary. A typical particle P_i is assumed to have mass m_i and velocity v_i as shown. The components of the linear momentum of this particle are given by eqn (6.2) as

$$p_{x_i} = m_i \dot{x}_i,$$

and $\qquad p_{y_i} = m_i \dot{y}_i.$ $\hfill (6.25)$

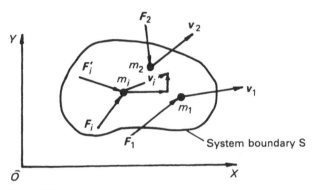

FIG. 6.13

The linear momentum p of the collection of particles is now obtained by summing the momentum of each of the particles. Hence the component of p along OX is

$$p_x = \sum m_i \dot{x}_i. \hfill (6.26)$$

Since the location, along OX, of the centre of mass of the assembly is given by

$$m x_G = \sum m_i x_i,$$

where $m = \sum m_i$ is the total mass, eqn (6.26) can be written

$$p_x = m \dot{x}_G. \hfill (6.27)$$

A similar analysis gives the component p_y as

$$p_y = m \dot{y}_G. \hfill (6.28)$$

Equations (6.27) and (6.28) show that the linear momentum of a system which is composed of a collection of particles is given by

$$p = m v_G \hfill (6.29)$$

where m is the total mass of the system and v_G is the velocity of its centre of mass.

Since a rigid body may be regarded as a closed system of particles rigidly connected together eqn (6.29) must also define the linear momentum of the body, and will apply to all types of motion of the body, i.e. it applies to pure translation, pure rotation and general plane motion.

If an external force F_i, and an internal force F'_i, such as the contact force from an adjacent particle, are applied to P_i, as shown in Fig. 6.13, then from Newton's Second Law of Motion and eqn (6.3)

$$F_i + F'_i = \frac{dp_i}{dt}.$$

If we now add the forces acting on all of the particles in the system, any forces which are internal to the system cancel out in pairs (see page 158) and do not contribute to the total force F applied to the system. Hence

$$F = \sum F_i = \frac{d}{dt} \sum p_i = \frac{dp}{dt}. \tag{6.30}$$

Substitution of eqn (6.29) into eqn (6.30) gives

$$F = m \frac{dv_G}{dt} = ma_G, \tag{6.31}$$

i.e. the total *external* force applied to the system is equal to the mass of the system multiplied by the acceleration of its centre of mass. For the case of a rigid body the force F represents the total *external* force applied to the body and eqn (6.31) is in agreement with the corresponding equations of motion derived in Chapter 4.

If both sides of eqn (6.30) are now integrated with respect to time over the interval t_1 to t_2, we obtain

$$\mathscr{I} = \int_{t_1}^{t_2} F \, dt = \sum p_{i_2} - \sum p_{i_1} = (p_2 - p_1). \tag{6.32}$$

In some cases it may be convenient to consider the motion of the centre of mass G of the system of particles. Equation (6.32) may then be written using eqn (6.29) as

$$\mathscr{I} = m[(v_G)_2 - (v_G)_1]. \tag{6.33}$$

The left hand side of this equation is the resultant linear impulse $\mathscr{I}$ of the externally applied forces and is equal to the change in the linear momentum of the whole system. It should be understood that eqns (6.31), (6.32) and (6.33) also apply to a system of connected rigid bodies since each rigid body in the system can be regarded as a collection of particles. For this case we must be careful to remember that G refers to the centre of mass of the *whole* assembly.

If the resultant external force acting on the system is zero we have $\mathscr{I} = 0$ and therefore the velocity v_G of the centre of mass of the system remains constant. Since impacts between particles only give rise to forces which are internal to the system, the value of v_G will not be affected by these collisions.

Suppose we now return to example 6.1 and consider the two particles P_1 and P_2 of Fig. 6.3 to form a system. Since no external forces are applied to this system along the OX direction, eqn (6.32) gives

$$\mathscr{I}_x = (\textstyle\sum p_{x_i})_2 - (\textstyle\sum p_{x_i})_1 = (m_1v_1 + m_2v_2) - m_1v_0 = 0.$$

which gives the same result as eqn (v) of example 6.1.

Before we can apply eqns (6.30) and (6.31) to problems involving impacts within either a collection of particles or between rigid bodies it is necessary to define absolute and relative *angular* momentum. We shall show in what follows that these quantities allow us to discuss the changes in angular motion caused by impulsive loads.

6.2 Angular momentum

Absolute angular momentum of a particle

The linear momentum of the particle P shown in Fig. 6.14 is

$$p = mv.$$

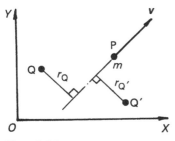

FIG. 6.14

Let us now consider a *stationary* point Q in the OXY plane which is not instantaneously coincident with P. We now define the *absolute* angular momentum of the particle about the point Q as the moment of the linear momentum vector p about the point Q. For the particle shown in Fig. 6.14 let the perpendicular distance from Q to the line of action of p be r_Q. The absolute angular momentum, H_Q, of the particle P about the point Q is thus given by

$$H_Q = r_Q p = r_Q mv. \tag{6.34}$$

Although the absolute angular momentum can be shown to be a vector quantity, for two-dimensional problems it may be treated as a scalar provided that we associate with it a sign to indicate whether the moment is in the clockwise or the anticlockwise direction. It is usual to take the anticlockwise direction as positive. (This is the same as the convention used to define the direction of the moment of a force).

For the arrangement shown in Fig. 6.14 the absolute angular momentum about Q is anticlockwise and is therefore positive,

i.e. $H_Q = + r_Q m v.$ (6.35)

If we now consider the point Q', which lies to the right of the velocity vector v, the absolute angular momentum about Q' is clockwise and is therefore negative,

i.e. $H_{Q'} = -r_{Q'} m v$ (6.36)

An alternative expression for the absolute angular momentum H_Q can be derived by considering the moments of the components p_x and p_y about the point Q. If the positions of Q and P are x_Q, y_Q and x_P, y_P respectively, as shown in Fig. 6.15, then the absolute angular momentum about the point Q is given by

$$H_Q = (x_P - x_Q)p_y - (y_P - y_Q)p_x$$
$$= m[\dot{y}_P(x_P - x_Q) - \dot{x}_P(y_P - y_Q)].$$ (6.37)

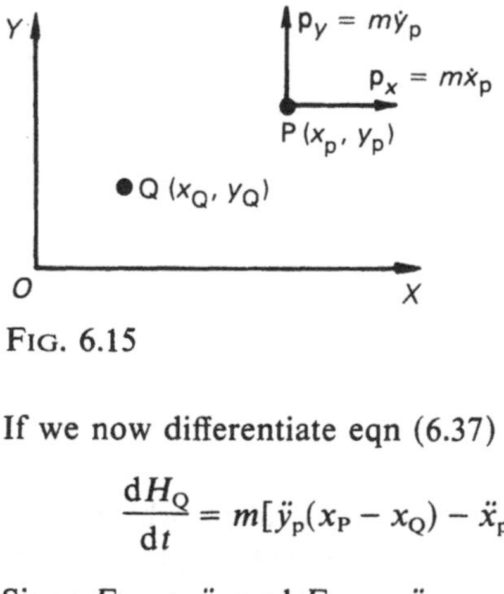

FIG. 6.15

If we now differentiate eqn (6.37) with respect to time we obtain

$$\frac{dH_Q}{dt} = m[\ddot{y}_P(x_P - x_Q) - \ddot{x}_P(y_P - y_Q)].$$ (6.38)

Since $F_x = m\ddot{x}_p$ and $F_y = m\ddot{y}_p$ represent the components of the external force F applied to the particle, eqn (6.38) may be rewritten

$$\frac{dH_Q}{dt} = (x_P - x_Q)F_y - (y_P - y_Q)F_x.$$

The right hand side of this equation is the moment M_Q of the applied force F about the point Q.

Thus $\dfrac{dH_Q}{dt} = M_Q.$ (6.39)

The rate of change of the absolute angular momentum about any point Q is equal to the moment of the applied force about Q. When eqn (6.39) is integrated with respect to time over the interval t_1 to t_2 we have

$$(H_Q)_2 - (H_Q)_1 = \int_{t_1}^{t_2} M_Q \, dt. \tag{6.40}$$

The quantity

$$\mathscr{I}_Q = \int_{t_1}^{t_2} M_Q \, dt \tag{6.41}$$

is called the *angular impulse* of the applied force about the point Q and is equal to the change in the absolute angular momentum of the particle about Q.

It should be understood that, in general, the angular impulse $\mathscr{I}_Q$ about Q is *not* equal to the moment of the linear impulse $\mathscr{I}$ about Q.

Since x_P and y_P are functions of time

$$\int_{t_1}^{t_2} (x_P - x_Q) F_y \, dt \neq (x_P - x_Q) \int_{t_1}^{t_2} F_y \, dt = (x_P - x_Q) \mathscr{I}_y$$

and $$\int_{t_1}^{t_2} (y_P - y_Q) F_x \, dt \neq (y_P - y_Q) \int_{t_1}^{t_2} F_y \, dt = (y_P - y_Q) \mathscr{I}_x.$$

Thus $\mathscr{I}_Q \neq (x_P - x_Q)\mathscr{I}_y - (y_P - y_Q)\mathscr{I}_x.$

Systems of particles

Let us suppose we have a closed system of particles contained within a boundary S as shown in Fig. 6.16(a). The total absolute angular momentum of the whole system about the fixed point Q may be obtained directly from eqns (6.34) and (6.37) as the sum of the individual absolute angular momenta

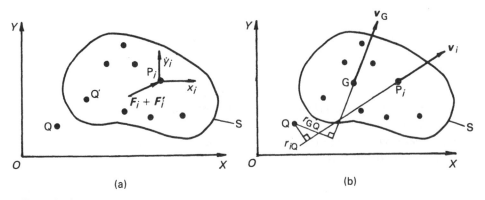

(a)   (b)

FIG. 6.16

about Q and is given by

$$H_Q = \Sigma\, H_{Q_i} = \Sigma\, r_{Q_i} p_i = \Sigma\, m_i[\dot{y}_i(x_i - x_Q) - \dot{x}_i(y_i - y_Q)]. \tag{6.42}$$

The quantities written with the suffix i relate to the typical particle P_i, as shown in Fig. 6.16(a).

Let us now determine the absolute angular momentum of the system about a different point Q'.

Thus
$$\begin{aligned} H_{Q'} &= \Sigma\, m_i[\dot{y}_i(x_i - x_{Q'}) - \dot{x}_i(y_i - y_{Q'})] \\ &= \Sigma\, m_i[\dot{y}_i\{(x_i - x_Q) + (x_Q - x_{Q'})\} \\ &\quad - \dot{x}_i\{(y_i - y_Q) + (y_Q - y_{Q'})\}]. \end{aligned} \tag{6.43}$$

Using eqn (6.42) and noting that $(x_Q - x_{Q'})$ and $(y_Q - y_{Q'})$ are independent of the subscript i we may write

$$H_{Q'} = H_Q + (x_Q - x_{Q'})\,\Sigma\, m_i\dot{y}_i - (y_Q - y_{Q'})\,\Sigma\, m_i\dot{x}_i. \tag{6.44}$$

Since $\Sigma\, m_i = m$, $\Sigma\, m_i x_i = mx_G$ and $\Sigma\, m_i y_i = my_G$ define the total mass and the position of the centre of mass G of the system, eqn (6.44) can be rewritten as

$$H_{Q'} = H_Q + (x_Q - x_{Q'})m\dot{y}_G - (y_Q - y_{Q'})m\dot{x}_G. \tag{6.45}$$

If we now make Q' instantaneously coincident with G, rearrangement of eqn (6.45) gives

$$H_Q = H_G + (x_G - x_Q)m\dot{y}_G - (y_G - y_Q)m\dot{x}_G \tag{6.46}$$

or
$$H_Q = H_G + (x_G - x_Q)p_y - (y_G - y_Q)p_x. \tag{6.47}$$

Equations (6.46) and (6.47) show that the absolute angular momentum of the system about the point Q is equal to the absolute angular momentum of the system about a point coincident with G plus a term which is equal to the moment, $r_{GQ}p$, of the linear momentum vector, $p = mv_G$, about Q, (see Fig. 6.16(b))

i.e.
$$H_Q = H_G + r_{GQ}p. \tag{6.48}$$

These equations also show that if $p = 0$, i.e. $\dot{x}_G = \dot{y}_G = 0$, then

$$H_Q = H_G \tag{6.49}$$

for all points Q. *The value of the absolute angular momentum of a system about a fixed point Q is independent of the position of Q if the centre of mass of the system is stationary.*

Consider now the rate of change of H_Q with respect to time. Differentiation of eqn (6.42) with respect to time gives

$$\frac{dH_Q}{dt} = \Sigma\, m_i[\ddot{y}_i(x_i - x_Q) - \ddot{x}_i(y_i - y_Q)]. \tag{6.50}$$

If F_{xi}, F_{yi} and F'_{xi}, F'_{yi} are the components of the external force F_i and the internal force F'_i applied to particle P_i then, from Newton's Second Law of Motion,

$$F_{xi} + F'_{xi} = m_i \ddot{x}_i,$$

and $\qquad F_{yi} + F'_{yi} = m_i \ddot{y}_i.$

Substituting these values into eqn (6.50) gives

$$\frac{\mathrm{d}H_Q}{\mathrm{d}t} = \sum \{F_{yi}(x_i - x_Q) - F_{xi}(y_i - y_Q)\}, \qquad (6.51)$$

since the summations containing terms in F'_{x_i} and F'_{y_i} are zero.

The right hand side of this equation can be seen to be the total moment M_Q of the *external* forces about the point Q.

Equation (6.51) therefore shows that the rate of change of the absolute angular momentum of the system, about an arbitrary point Q, is equal to the total moment about Q of all the external forces applied to the system, i.e.

$$\frac{\mathrm{d}H_Q}{\mathrm{d}t} = M_Q. \qquad (6.52)$$

Since we can regard a rigid body as a closed system of particles, eqn (6.52) is also true for a rigid body. Equally well, a collection of rigid bodies, either connected or unconnected, may be considered as a system of particles. Equation (6.52) therefore also applies to these systems.

If the total moment about the point Q is zero we have

$$\frac{\mathrm{d}H_Q}{\mathrm{d}t} = 0, \qquad (6.53)$$

which shows that the absolute angular momentum of the system about Q is constant.

Further, if *no* external forces are applied to the system then from eqn (6.51)

$$\frac{\mathrm{d}H_Q}{\mathrm{d}t} = 0 \quad \text{for *all* points Q.}$$

If eqn (6.52) is now integrated with respect to time over the interval t_1 to t_2, the angular impulse of the externally applied forces about Q is given by

$$\mathscr{J}_Q = \int_{t_1}^{t_2} M_Q \, \mathrm{d}t = (H_Q)_2 - (H_Q)_1, \qquad (6.54)$$

and is equal to the change in the absolute angular momentum of the system about Q. This shows that the result given by eqn (6.40) for a single particle also applies to a system of particles.

Alternative formulations of equations of motion

If eqns (6.30) and (6.52) are now considered together we have

$$\frac{\mathrm{d}p_x}{\mathrm{d}t} = F_x,$$

$$\frac{\mathrm{d}p_y}{\mathrm{d}t} = F_y,$$

and $\dfrac{\mathrm{d}H_A}{\mathrm{d}t} = M_A,$ (6.55)

where A is a stationary point.

These equations can be regarded as the three scalar equations of motion of the system. For example if the system is a single rigid body eqns (6.55) would determine its translational and rotational motion once F_x, F_y and M_A are specified.

However, it may be shown that these equations are not the only way of expressing the equations of motion of the system. Although we shall not prove it here eqns (6.55) can be replaced by

$$\frac{\mathrm{d}H_A}{\mathrm{d}t} = M_A,$$

$$\frac{\mathrm{d}H_B}{\mathrm{d}t} = M_B,$$

and $\dfrac{\mathrm{d}H_C}{\mathrm{d}t} = M_C,$ (6.56)

where H_A, H_B, H_C and M_A, M_B, M_C are the absolute angular momenta and the moments applied to the system about three fixed points A, B and C which do not lie on the same straight line.

The solutions of eqns (6.55) and (6.56) will yield identical results.

A third formulation of the equations of motion is also possible and this is given by

$$\frac{\mathrm{d}H_A}{\mathrm{d}t} = M_A,$$

$$\frac{\mathrm{d}H_B}{\mathrm{d}t} = M_B,$$

and $\dfrac{\mathrm{d}p_\xi}{\mathrm{d}t} = F_\xi,$ (6.57) where p_ξ and F_ξ are the components of the linear momentum vector p and the externally applied force F along a direction which is *not perpendicular* to the line joining the points A and B.

If we now return to section in Chapter 3 on statics we can see that eqns (6.55), (6.56) and (6.57) are related to the conditions of equilibrium given by eqns (3.20), (3.28), (3.34).

Which formulation of the equations of motion of a system is best suited to solve a particular problem will, of course, be determined by the nature of the problem, and no hard and fast rules are possible. Nevertheless, the reader must understand that eqns (6.55), (6.56) and (6.57) are equivalent and should be ready to use them in appropriate situations. If there is any doubt about which formulation is most suitable in a particular case we would recommend the use of eqns (6.55).

It should be fully understood that once a formulation has been chosen and then applied to a system the other two formulations *cannot* be used to generate additional equations of motion. If this is done the additional equations of motion will always reduce to give $0 = 0$.

Example 6.3   Figure 6.17(a) shows two *light* pulleys, of radius a and b, which are rigidly connected together and rotate as a single body about the fixed point O. The pulley axis through O is horizontal. Two masses m_1 and m_2 are carried on strings which are attached to the pulleys as shown. Let us find the angular acceleration of the pulley assuming that the bearings at O have negligible friction.

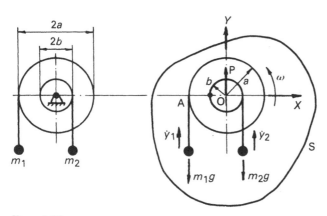

FIG. 6.17

Let us consider the pulley and the masses to be a closed system as shown in Fig. 6.17(b). The external forces applied to the system are the bearing force P at O and the gravitational forces m_1g and m_2g. If the pulley rotates with angular velocity ω, then from kinematic considerations, the velocities of masses

m_1 and m_2 are

$$\dot{y}_1 = -a\omega$$

and $\dot{y}_2 = b\omega,$ (i)

as shown.

Using eqn (6.42) the absolute angular momentum of the system about the point O is

$$H_O = -a(m_1\dot{y}_1) + b(m_2\dot{y}_2)$$

$$= (m_1a^2 + m_2b^2)\omega.$$ (ii)

By taking the moment of the external forces applied to the system about the point O the unknown force **P** is eliminated from the moment equation, and from Fig. 6.17(b)

↺ $M_O = m_1ga - m_2gb.$ (iii)

Since $\dfrac{\mathrm{d}H_O}{\mathrm{d}t} = M_O,$ (iv)

eqns (ii) and (iii) give

$$(a^2m_1 + b^2m_2)\dot{\omega} = m_1ga - m_2gb.$$ (v)

Thus the angular acceleration of the pulley is

$$\dot{\omega} = \frac{(m_1a - m_2b)g}{(a^2m_1 + b^2m_2)}.$$ (vi)

The reader should verify that this is the case by repeating this problem using the equation of motion of each of the masses.

The force P acting at the bearing O can be found by considering the linear momentum p_y of the system along OY.

From Fig. 6.17(b) and eqn (6.25) we have

$$p_y = (m_2\dot{y}_2 + m_1\dot{y}_1)$$

$$= (m_2b - m_1a)\omega.$$ (vii)

The total force acting on the system along OY is

$$F_y = P - (m_1g + m_2g).$$ (viii)

Since $\dfrac{\mathrm{d}p_y}{\mathrm{d}t} = F_y$

eqns (vii) and (viii) give

$$(m_2b - m_1a)\dot{\omega} = P - (m_1g + m_2g).$$ (ix)

Substituting for $\dot{\omega}$ using eqn (vi) yields

$$P = g\frac{(a+b)^2 m_1 m_2}{(a^2 m_1 + b^2 m_2)}. \tag{x}$$

The result could also have been derived considering the absolute angular momentum about some convenient point, e.g. a stationary point instantaneously coincident with point A on the pulley.

From Fig. 6.17(b)

$$Ⓐ \quad H_A = (a+b)m_2\dot{y}_2 = (a+b)m_2 b\omega, \tag{xi}$$

and

$$M_A = aP - (a+b)m_2 g. \tag{xii}$$

Since $\dfrac{\mathrm{d}H_A}{\mathrm{d}t} = M_A$

eqns (xi) and (xii) give

$$(a+b)m_2 b\dot{\omega} = aP - (a+b)m_2 g. \tag{xiii}$$

Substituting for $\dot{\omega}$ using eqn (vi) then yields

$$P = \frac{(a+b)^2 m_1 m_2 g}{(a^2 m_1 + b^2 m_2)}, \tag{xiv}$$

which agrees with eqn (x).

Relative angular momentum of a particle

In many engineering applications it has been found useful to determine the angular momentum of a particle about a point Q which is itself *moving*, such as is shown in Fig. 6.18. The particle P in Fig. 6.18(a) is shown moving with velocity v_p with respect to a stationary (inertial) reference frame *OXY*.

A second reference frame $QX'Y'$ moves, *in pure translation*, with respect to *OXY* with a velocity v_Q. To an observer moving with this frame the particle would appear to move with a velocity

$$v_{PQ} = (v_P - v_Q), \tag{6.57}$$

and would have a *relative* linear momentum p_r given by

$$p_r = mv_{PQ}. \tag{6.58}$$

If this observer was now asked to calculate the moment h_Q of the relative linear momentum about Q he would write

$$h_Q = r_Q p_r = r_Q m v_{PQ}, \tag{6.59}$$

where r_Q is the perpendicular distance from Q to the line of the vector v_{PQ}, as shown in Fig. 6.18(b).

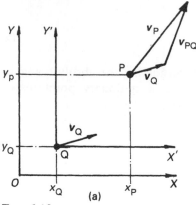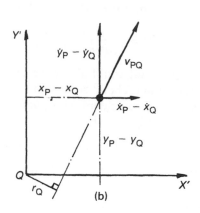

FIG. 6.18

The quantity h_Q is called the *relative* angular momentum of the particle about the moving point Q and is taken to be positive when the moment of the relative linear momentum p_r about Q is anticlockwise.

If v_{PQ} is given in component form h_Q can be written in terms of the coordinates of P and Q as

$$h_Q = m[(\dot{y}_P - \dot{y}_Q)(x_P - x_Q) - (\dot{x}_P - \dot{x}_Q)(y_P - y_Q)]. \tag{6.60}$$

Closed systems of particles

Figure 6.19 shows a closed system of particles contained within a boundary S. We shall now calculate the total relative angular momentum of the system

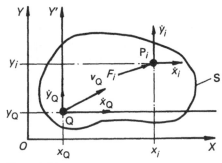

FIG. 6.19

about the moving point Q. This total is obtained by summing the relative angular momenta of each of the particles so that

$$h_Q = \sum h_{Q_i} = \sum r_{Qi} p_{ri}$$
$$= \sum m_i\{(\dot{y}_i - \dot{y}_Q)(x_i - x_Q) - (\dot{x}_i - \dot{x}_Q)(y_i - y_Q)\}. \tag{6.61}$$

This can be expressed as

$$h_Q = \sum m_i \{\dot{y}_i(x_i - x_Q) - \dot{x}_i(y_1 - y_Q)\} - \dot{y}_Q \sum m_i(x_i - x_Q)$$
$$+ \dot{x}_Q \sum m_i(y_i - y_Q). \tag{6.62}$$

Equation (6.42) shows that the first term in eqn (6.62) is the absolute angular momentum H_Q of the system about a fixed point which is *instantaneously coincident* with the moving point Q.

Since the summations $m = \sum m_i$, $m x_G = \sum m_i x_i$ and $m y_G = \sum m_i y_i$ define the total mass and the position of the centre of mass G of the system, eqn (6.62) can now be rewritten as

$$h_Q = H_Q - m\dot{y}_Q(x_G - x_Q) + m\dot{x}_Q(y_G - y_Q). \tag{6.63}$$

If Q is a stationary point, so that $\dot{x}_Q = \dot{y}_Q = 0$, $h_Q = H_Q$, as would be expected.

Now let us substitute for H_Q using eqn (6.46). Equation (6.63) then becomes

$$h_Q = H_G + m(\dot{y}_G - \dot{y}_Q)(x_G - x_Q) - m(\dot{x}_G - \dot{x}_Q)(y_G - y_Q). \tag{6.64}$$

If Q is chosen to be instantaneously coincident with G

$$h_G = H_G. \tag{6.65}$$

Since the absolute and relative angular momenta about the centre of mass G of the system are equal, eqn (6.64) may be written as

$$h_Q = h_G + m(\dot{y}_G - \dot{y}_Q)(x_G - x_Q) - m(\dot{x}_G - \dot{x}_Q)(y_G - y_Q). \tag{6.66}$$

Thus the relative angular momentum about Q is equal to the relative angular momentum about G plus a term which is equal to the moment of the relative linear momentum vector, $p_r = m(v_G - v_Q)$, about Q.

By differentiating eqn (6.61) with respect to time we obtain

$$\frac{dh_Q}{dt} = \sum m_i\{(\dot{y}_i - \dot{y}_Q)(x_i - x_Q) - (\dot{x}_i - \dot{x}_Q)(y_i - y_Q)\}. \tag{6.67}$$

Expanding the terms on the right hand side of eqn (6.67) gives

$$\frac{dh_Q}{dt} = \sum \{(m_i\ddot{y}_i)(x_i - x_Q) - (m_i\ddot{x}_i)(y_i - y_Q)\}$$
$$+ \ddot{x}_Q \sum m_i(y_i - y_Q) - \ddot{y}_Q \sum m_i(x_i - x_Q). \tag{6.68}$$

If F_{x_i}, F_{y_i} and F'_{x_i}, F'_{y_i} are the components of the external and internal forces applied to P_i we have

$$F_{x_i} + F'_{x_i} = m_i\ddot{x}_i \quad \text{and} \quad F_{y_i} + F'_{y_i} = m_i\ddot{y}_i.$$

Since the moments of the internal forces about the point Q cancel in pairs their sum is zero. The first term on the right hand side of eqn (6.68) therefore represents the total moment, M_Q, of the *external* forces about the point Q,

i.e. $M_Q = \sum \{F_{y_i}(x_i - x_Q) - F_{x_i}(y_i - y_Q)\}.$ (6.69)

Hence $\dfrac{dh_Q}{dt} + \ddot{y}_Q \sum m_i(x_i - x_Q) - \ddot{x}_Q \sum m_i(y_i - y_Q) = M_Q.$ (6.70)

Again the summations can be replaced so that

$\dfrac{dh_Q}{dt} + m[\ddot{y}_Q(x_G - x_Q) - \ddot{x}_Q(y_G - y_Q)] = M_Q.$ (6.71)

In this general case h_Q need not be constant if $M_Q = 0$. Equation (6.71) also shows that the angular impulse about the moving point Q, i.e. $\mathcal{J}_Q = \int_{t_1}^{t_2} M_Q \, dt$, is *not* equal to the change in the relative angular momentum about Q. Since eqn (6.52) gives

$\dfrac{dH_{Q'}}{dt} = M_{Q'} = M_Q,$

where Q' is a *stationary point instantaneously coincident with* Q, i.e. $(x_Q, y_Q) \equiv (x_{Q'}, y_{Q'})$, eqn (6.71) can be written

$\dfrac{dh_Q}{dt} + m[\ddot{y}_Q(x_G - x_Q) - \ddot{x}_Q(y_G - y_Q)] = \dfrac{dH_Q}{dt} = M_Q.$ (6.72)

Now if we choose the moving point Q and the stationary point Q' to be instantaneously coincident with G, eqn (6.72) becomes

$\dfrac{dh_G}{dt} = \dfrac{dH_G}{dt} = M_G$ (6.73)

even though Q, Q' and G may have different velocities and accelerations. Equation (6.73) can also be applied to a system in which we choose Q to move with the centre of mass of the system of particles and to be instantaneously coincident with the fixed point Q'.

Hence, if there is no *external* moment about G, i.e. $M_G = 0$, the relative angular momentum about G remains constant.

The angular impulse about G,

$\mathcal{J}_G = \int_{t_1}^{t_2} M_G \, dt = (h_G)_2 - (h_G)_1.$ (6.74)

Thus the angular impulse about G is equal to the change in relative angular momentum about G whether G be stationary, moving with constant velocity, or accelerating.

All of these results, such as eqns (6.71) to (6.74) apply to a closed system of particles. No constraints have been imposed on the motion of the particles relative to one another so that the results will apply to systems containing rigid bodies, whether connected or unconnected.

Angular momentum of a rigid body

In the great majority of applications we do not have to use eqn (6.42) to calculate the absolute angular momentum of a rigid body about a fixed point. The relative angular momentum of the body about a point fixed in the body is more easily determined and eqn (6.63) can be applied to calculate the corresponding value for the absolute angular momentum. We shall therefore only develop an expression for the relative angular momentum of a rigid body about a point which is fixed to the body.

Relative angular momentum of a rigid body about a point Q fixed in the body

Before eqns (6.71) and (6.74) can be used to solve problems involving the motion of rigid bodies it is necessary to derive an expression for the relative angular momentum h_Q of a body about a point Q fixed in the body. Figure 6.20 shows a rigid body moving in the OXY plane with general plane motion. The point Q is assumed to be fixed in the body and the reference frame $QX'Y'$ is taken to move in pure translation relative to axes OXY with velocity v_Q. Since the body is rigid there is a kinematic relationship between the velocity of P_i and the velocity of Q (see Chapter 1, p 26). The magnitude of the relative velocity of the point P_i in the body with respect to point Q is given by

$$v_{P_iQ} = r_i\dot{\theta}, \tag{6.75}$$

where r_i is the distance QP_i and $\dot{\theta}$ is the angular velocity of the body. The direction of v_{P_iQ} is perpendicular to the line QP_i as shown in Fig. 6.20.

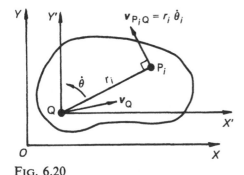

FIG. 6.20

If we now consider the body to be made up of particles, such as the particle of mass m_i at the point P_i, then the relative angular momentum of the body about the point Q is, from eqn (6.59),

$$h_Q = \sum r_i(m_i v_{P_iQ}). \tag{6.76}$$

Substituting for v_{P_iQ} from eqn (6.75),

$$h_Q = \sum m_i r_i^2 \dot{\theta}.$$

Since $\dot{\theta}$ is independent of the subscript i it may be taken outside the summation, so that

$$h_Q = (\sum m_i r_i^2)\dot{\theta}. \tag{6.77}$$

By definition, the summation $\sum m_i r_i^2$ is the moment of inertia I_Q of the body about an axis passing through Q, so that

$$h_Q = I_Q\dot{\theta}. \tag{6.78}$$

The relative angular momentum of the body about the point Q fixed in the body is therefore determined by the moment of inertia of the body about an axis passing through Q and the angular velocity of the body.

We should note that if the body moves with pure translation then $\dot{\theta} = 0$ and $h_Q = 0$, for *all* points Q in the body. Otherwise, i.e. for pure rotation and general plane motion, the relative angular momentum of the body about any point fixed in the body is given by eqn (6.78).

If eqn (6.78) is substituted into eqn (6.71) we obtain

$$I_Q\ddot{\theta} + m[\ddot{y}_Q(x_G - x_Q) - \ddot{x}_Q(y_G - y_Q)] = M_Q \tag{6.79}$$

which is the same as eqn (4.39).

Furthermore, if Q and G are coincident then from eqns (6.65) and (6.78)

$$h_G = H_G = I_G\dot{\theta}.$$

This result can be used to determine the absolute angular momentum of a rigid body about a stationary point Q. Using eqn (6.63) we can write

$$H_Q = I_G\dot{\theta} + m\dot{y}_G(x_G - x_Q) - m\dot{x}_G(y_G - y_Q) = I_G\dot{\theta} + r_{GQ}(mv_G) \tag{6.80}$$

The absolute angular momentum of the body is equal to $I_G\dot{\theta}$ plus the moment of the linear momentum vector, $p = mv_G$, about Q.

In eqns (6.55), (6.56) and (6.57) we showed that the equations of motion of a system could be written in terms of the rates of change of the linear momentum and the absolute angular momentum and that three different, but equivalent, formulations were possible. It is possible to replace the eqns (6.55), (6.56) and (6.57) with equations involving relative angular momentum. If this substitution is made the equations of motion of the system can be written as:

(i) $\dfrac{dp_x}{dt} = F_x,$

$\dfrac{dp_y}{dt} = F_x, \tag{6.81}$

and $\dfrac{dh_A}{dt} + m[\ddot{y}_A(x_G - x_A) - \ddot{x}_A(y_G - y_A)] = M_A,$

(ii) $\dfrac{\mathrm{d}h_\mathrm{A}}{\mathrm{d}t} + m[\ddot{y}_\mathrm{A}(x_\mathrm{G} - x_\mathrm{A}) - \ddot{x}_\mathrm{A}(y_\mathrm{G} - y_\mathrm{A})] = M_\mathrm{A},$

$\dfrac{\mathrm{d}h_\mathrm{B}}{\mathrm{d}t} + m[\ddot{y}_\mathrm{B}(x_\mathrm{G} - x_\mathrm{B}) - \ddot{x}_\mathrm{B}(y_\mathrm{G} - y_\mathrm{B})] = M_\mathrm{B},$

and $\dfrac{\mathrm{d}h_\mathrm{C}}{\mathrm{d}t} + m[\ddot{y}_\mathrm{C}(x_\mathrm{G} - x_\mathrm{C}) - \ddot{x}_\mathrm{C}(y_\mathrm{G} - y_\mathrm{C})] = M_\mathrm{C},$ (6.82)

where A, B, C are moving points which do not, instantaneously, lie on the same straight line,

(iii) $\dfrac{\mathrm{d}h_\mathrm{A}}{\mathrm{d}t} + m[\ddot{y}_\mathrm{A}(x_\mathrm{G} - x_\mathrm{A}) - \ddot{x}_\mathrm{A}(y_\mathrm{G} - y_\mathrm{A})] = M_\mathrm{A},$

$\dfrac{\mathrm{d}h_\mathrm{B}}{\mathrm{d}t} + m[\ddot{y}_\mathrm{B}(x_\mathrm{G} - x_\mathrm{B}) - x_\mathrm{B}(y_\mathrm{G} - y_\mathrm{B})] = M_\mathrm{B},$

and $\dfrac{\mathrm{d}p_\xi}{\mathrm{d}t} = F_\xi$ (6.83)

where p_ξ and F_ξ are the components of the linear momentum vector p and the externally applied force F along a direction which is not instantaneously perpendicular to the line joining A and B.

In eqns (6.81), (6.82) and (6.83) it is sometimes convenient to choose the point A to be at the centre of mass G of the system.

Example 6.4 Figure 6.21 shows a block A of mass m_b moving in a smooth horizontal guideway. The position of the block at any instant is given by the coordinate x_A as shown. A thin uniform rod AB of length $2a$ and mass m_r is connected to the block by a frictionless pin joint at A.

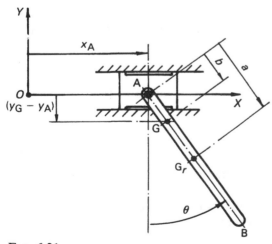

FIG. 6.21

The angular displacement of the rod is θ and is measured from the downward vertical, as shown. Let us calculate the initial values of the linear acceleration $\ddot{x}_A$, and the angular acceleration $\ddot{\theta}$, when the rod is released from rest at a position given by $\theta = 45°$.

Let us treat the block and the rod as a system as in Fig. 6.22, which shows the external forces acting on the system. The forces R_1 and R_2 are the contact forces between the block and the guideway and G is the position of the centre of mass of the assembly, which is a distance

$$b = \frac{m_r a}{m_r + m_b} \tag{i}$$

from A.

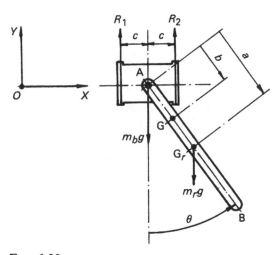

FIG. 6.22

We shall now calculate the relative angular momentum of the system about the moving point A.

Now $h_A = h_{Ar} + h_{Ab}$ \hfill (ii)

where h_{Ar} and h_{Ab} are the relative angular momenta of the rod and block respectively about A.

Equation (6.78) gives

$$h_{Ar} = I_A \dot{\theta} \tag{iii}$$

where $I_A = 4m_r a^2/3$ is the moment of inertia of the rod about an axis through A.

Since the block is in pure translation its relative angular momentum about its centre of mass, which is at A, is zero.

i.e. $h_{Ab} = 0.$ \hfill (iv)

Thus $h_A = I_A \dot{\theta}$.   (v)

Equation (6.71) gives the rate of change of h_A with respect to time as

$$\frac{dh_A}{dt} + (m_r + m_b)[\ddot{y}_A(x_G - x_A) - \ddot{x}_A(y_G - y_A)] = M_A. \qquad \text{(vi)}$$

In this equation M_A represents the moments of the external forces applied to the system about point A and, from Fig. 6.22,

Ⓐ $M_A = -m_r ga \sin\theta + R_2 c - R_1 c.$ (vii)

From Fig. 6.21 we also have

$$\ddot{y}_A = 0, \qquad \text{(viii)}$$

since the block is constrained to move along a horizontal guideway, and

$$(y_G - y_A) = -b \cos\theta$$
$$= -\frac{m_r a}{(m_r + m_b)} \cos\theta. \qquad \text{(ix)}$$

Substitution of eqns (v), (vii), (viii) and (ix) into eqn (vi) yields

$$I_A \ddot{\theta} + m_r a \ddot{x}_A \cos\theta = -m_r ga \sin\theta + (R_2 - R_1)c. \qquad \text{(x)}$$

If we now consider the block alone, as shown in Fig. 6.23, then from eqn (iv)

$$h_{Ab} = 0.$$

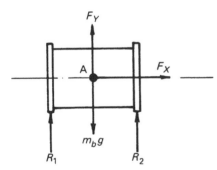

Fig. 6.23

Hence $\dfrac{dh_{Ab}}{dt} = M_{Ab} = 0,$

where M_{Ab} is the moment about A of the external forces applied to the block. The forces F_x, F_y on the block at the pin act through A as does the weight of the block. Thus

Ⓐ $M_{Ab} = R_2 c - R_1 c = 0,$

so that $(R_2 - R_1) = 0.$ (xi)

Equation (x) therefore reduces to

$$I_A \ddot{\theta} + m_r a \ddot{x}_A \cos \theta = -m_r g a \sin \theta. \tag{xii}$$

In order to calculate $\ddot{x}_A$ and $\ddot{\theta}$ we need another equation. Let us obtain this equation by considering the linear momentum of the system along OX. Now

$$p_x = p_{xb} + p_{xr} \tag{xiii}$$

where $p_{xb} = m_b \dot{x}_A$ \hfill (xiv)

is the linear momentum of the block along OX and

$$p_{xr} = m_r \dot{x}_{G_r} \tag{xv}$$

is the linear momentum of the rod along OX. The velocity of the centre of mass G_r of the rod is found from the vector equation

$$\boldsymbol{v}_{G_r} = \boldsymbol{v}_A + \boldsymbol{v}_{G,A} \tag{xvi}$$

where $v_{G,A} = a\dot{\theta}$ as shown in Fig. 6.24. The component of v_{G_r} along OX is thus found from eqn (xvi) and Fig. 6.24 as

$$\dot{x}_{G_r} = \dot{x}_A + a\dot{\theta} \cos \theta. \tag{xvii}$$

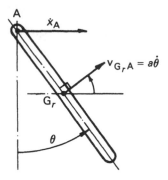

Fig. 6.24

Substitution of eqns (xiv), (xv) and (xvii) into eqn (xiii) gives

$$p_x = m_b \dot{x}_A + m_r(\dot{x}_A + a\dot{\theta} \cos \theta)$$
$$= (m_r + m_b)\dot{x}_A + m_r a\dot{\theta} \cos \theta. \tag{xviii}$$

Now eqn (6.30) gives

$$F_x = \frac{dp_x}{dt}, \tag{xix}$$

where F_x is the total external force applied to the system along OX. From Fig. 6.22 we see that

$$F_x = 0. \tag{xx}$$

Hence eqns (xviii) and (xix) give

$$\frac{d}{dt}[(m_r + m_b)\dot{x}_A + m_r a\dot{\theta} \cos \theta] = 0,$$

i.e.

$$(m_r + m_b)\ddot{x}_A + m_r a\ddot{\theta} \cos \theta - m_r a\dot{\theta}^2 \sin \theta = 0. \qquad \text{(xxi)}$$

Equations (xii) and (xxi) enable $\ddot{x}_A$ and $\ddot{\theta}$ to be calculated once θ and $\dot{\theta}$ have been specified.

When the system is released from rest we know that $\theta = 45°$ and $\dot{\theta} = 0$. The initial values of $\ddot{x}_A$ and $\ddot{\theta}$ are therefore found, from eqns (xii) and (xxi) as the solution to

$$I_A\ddot{\theta} + \frac{m_r a}{\sqrt{2}}\ddot{x}_A = -\frac{m_r ga}{\sqrt{2}} \qquad \text{(xxii)}$$

and

$$\frac{m_r a\ddot{\theta}}{\sqrt{2}} + (m_r + m_b)\ddot{x}_A = 0. \qquad \text{(xxiii)}$$

Taking $I_A = 4m_r a^2/3$ these equations give

$$\ddot{x}_A = \frac{3m_r g}{(5m_r + 8m_b)} \qquad \text{(xxiv)}$$

and

$$\ddot{\theta} = -\frac{6(m_r + m_b)g}{\sqrt{2}(5m_r + 8m_b)a}. \qquad \text{(xxv)}$$

The reader should verify that the contact forces R_1, R_2 can be calculated by either considering the linear momentum of the system along OY, *or* by considering the relative angular momentum of the system about G, together with eqn (xi).

Whichever approach is used the initial values of R_1 and R_2 can be shown to be given by

$$R_1 = R_2 = R = \frac{(m_r + m_b)(m_r + 4m_b)g}{(5m_r + 8m_b)}.$$

6.3   Impacts involving rigid bodies

In Section 6.1 the equations of motion of a particle were integrated with respect to time and it was shown that the impulse vector, $\mathcal{I} = \int F \, dt$, was equal to the change in the linear momentum of the particle to which F was applied. By determining the impulses applied to a pair of particles during a collision it was shown, by an example, how momentum principles could be used to determine the motion of the particles after the impact.

The same method can also be applied to problems which involve impacts between rigid bodies. For these cases we must write down the equations of

motion of each of the bodies during the impact and then, by integration of
these with respect to time, determine the changes in the linear momentum and
relative angular momentum about their respective centres of mass.

Since a single rigid body can be regarded as a closed system of particles
then the eqns (6.30) and (6.71),

$$F = \frac{dp}{dt}$$

and $\quad \dfrac{dh_Q}{dt} + m[\ddot{y}_Q(x_G - x_Q) - \ddot{x}_Q(y_G - y_Q)] = M_Q,$

also apply to a rigid body.

The substitution of the linear momentum $p = mv_G$ from eqn (6.29) into eqn
(6.30) gives

$$F = \frac{d(mv_G)}{dt} = ma_G \tag{6.83}$$

which can be seen to agree with the equations of motion, eqn (4.9), (4.23) and
(4.24) derived for the case of pure translation, pure rotation and general plane
motion.

When eqn (6.83) is integrated with respect to time the linear impulse of the
external forces applied to the body is given by

$$\mathcal{I} = \int_{t_1}^{t_2} F\, dt = m((v_G)_2 - (v_G)_1), \tag{6.84}$$

or, in component form,

$$\mathcal{I}_x = m((\dot{x}_G)_2 - (\dot{x}_G)_1)$$
$$\mathcal{I}_y = m((\dot{y}_G)_2 - (\dot{y}_G)_1), \tag{6.85}$$

no matter what the motion of the body.

For the special case of pure translation we should remember that all points
in the body have the same velocity, so that $v_G = v_i$ for all points.

We shall now consider the angular momentum of a rigid body and its
relationship to the equations for rotational motion by examining separately
the cases of pure translation, pure rotation and general plane motion.

Pure translation

If the point Q in eqn (6.71) is chosen to be coincident with the centre of mass
G of the body then

$$\frac{dh_G}{dt} = M_G \tag{6.86}$$

where the relative angular momentum h_G of a rigid body about G is given by eqn (6.78) as

$$h_G = I_G \dot{\theta}.$$

Since $\dot{\theta} = 0$ for pure translation, $h_G = 0$ and eqn (6.86) reduces to

$$M_G = 0. \tag{6.87}$$

This equation agrees with eqn (4.19).

Integration of eqn (6.86) with respect to time also shows that the angular impulse of the external forces about G is zero. Thus for a body moving with pure translation

$$\mathcal{J}_G = \int_{t_1}^{t_2} M_G \, \mathrm{d}t = (h_G)_2 - (h_G)_1 = 0. \tag{6.88}$$

Example 6.5 The slider shown in Fig. 6.25 is constrained to move vertically in a frictionless guideway. At the bottom of the guideway a stop of mass m_s is supported on a simple coil spring of stiffness k, and is positioned to prevent the slider from leaving the guideway. Suppose that the slider is released from rest a vertical distance X_0 above the equilibrium position of the stop. Let us determine the velocity of the slider and stop assembly at the instant just after the slider strikes the stop and then discuss the subsequent motion of the slider. It is assumed that the slider has a mass m and that its centre of mass G_1 is located relative to the corner contact points ABCD as shown. The centre of mass G_2 of the stop is offset an amount Δ from the centre line of the slider.

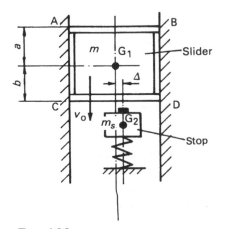

FIG. 6.25

The first step in the solution of this problem is to determine the velocity v_0 of the slider just before impact with the stop. This is done by drawing a system boundary around the slider and considering the work, $W = mgX_0$, done on the system by the gravitational force mg during the vertical displacement X_0.

The forces R_1 and R_2 on the slider from the guideway at A and C do no work since the displacement X_0 is perpendicular to their lines of action. The slider is released from rest as shown in Fig. 6.26 and is assumed to reach a velocity v_0 just before impact.

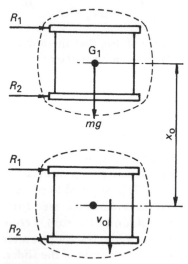

FIG. 6.26

Applying the First Law of Thermodynamics to the slider

$$W = \Delta U = U_2 - U_1, \qquad (i)$$

since no heat is transferred across the boundary. The internal energy of the system is due solely to its kinetic energy. Initially the slider is at rest so that $U_1 = 0$.

Hence $mgX_0 = \frac{1}{2}mv_0^2$, $\qquad (ii)$

so that the velocity of the slider as it contacts the stop,

$$v_0 = \sqrt{2gX_0}. \qquad (iii)$$

During the impact period we shall assume that the stop is not displaced from its equilibrium position and that free body diagrams of the slider and the stop are of the form shown in Fig. 6.27. The forces R_1 and R_2 represent the contact forces between the slider and the guideways and P is the contact force between the slider and the stop.

Since the stop has not yet moved away from its equilibrium position the spring force on the stop will be $m_s g$ as shown. Let us suppose that at the end of the impact period the slider and the stop move with a common velocity v_1 and do not separate.

During impact the equations of motion for the slider and the stop can be determined using the free body diagrams.

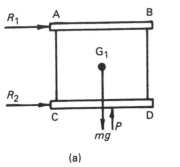

(a)   (b)

FIG. 6.27

For the slider

$$\rightarrow R_1 + R_2 = 0 \tag{iv}$$

$$\downarrow mg - P = m\frac{dv}{dt}, \tag{v}$$

$$\overset{\frown}{G_1} \Delta P + bR_2 - aR_1 = 0, \tag{vi}$$

and for the stop

$$\downarrow P = m_s\frac{dv}{dt}. \tag{vii}$$

Let us now suppose that the period of the impact is δt. During this time the forces P, R_1 and R_2 will increase rapidly from zero and will reach maximum values which we will assume are very much greater than other system forces such as the weight mg.

We will now integrate eqns (iv) to (vii) over the interval δt and note that the velocities of the slider and stop change from v_0 and 0 at the beginning of the impact, to v_1 at the end of the impact.

Thus

$$\int_0^{\delta t} R_1\, dt + \int_0^{\delta t} R_2\, dt = 0, \tag{viii}$$

$$mg\, \delta t - \int_0^{\delta t} P\, dt = mv_1 - mv_0, \tag{ix}$$

$$\Delta \int_0^{\delta t} P\, dt + b\int_0^{\delta t} R_2\, dt - a\int_0^{\delta t} R_1\, dt = 0, \tag{x}$$

and $$\int_0^{\delta t} P\, dt = m_s v_1. \tag{xi}$$

To solve these equations we now make the assumption that the duration of the impact is small so that $\delta t \rightarrow 0$. We will also assume that the forces P, R_1

and R_2 will instantaneously become very large at impact so that their impulses, given by

$$\mathscr{I}_P = \int_0^{\delta t} P\, dt, \qquad \mathscr{I}_{R_1} = \int_0^{\delta t} R_1 dt \quad \text{and} \quad \mathscr{I}_{R_2} = \int_0^{\delta t} R_2\, dt$$

are finite as $\delta t \to 0$.

Hence eqns (viii) to (xi) can be rewritten as

$$\mathscr{I}_{R_1} + \mathscr{I}_{R_2} = 0, \tag{xii}$$

$$-\mathscr{I}_P = m(v_1 - v_0), \tag{xiii}$$

$$\Delta\mathscr{I}_P + b\mathscr{I}_{R_2} - a\mathscr{I}_{R_1} = 0, \tag{xiv}$$

and $\qquad \mathscr{I}_P = m_s v_1. \tag{xv}$

It can be seen, by comparing eqn (ix) with eqn (xiii) that the forces which have a finite magnitude such as the weight mg can be neglected during the impact. The large contact forces produced by the impact therefore determine the initial motion of the combined system.

The solution to the above equations yields the values

$$v_1 = \frac{m}{(m + m_s)}\, v_0, \tag{xvi}$$

$$\mathscr{I}_P = \frac{mm_s}{m + m_s}\, v_0, \tag{xvii}$$

and $\qquad \mathscr{I}_{R_1} = -\mathscr{I}_{R_2} = \dfrac{\Delta}{(a + b)}\dfrac{mm_s}{(m + m_s)}\, v_0. \tag{xviii}$

An alternative approach is to consider the two bodies together as a system of particles. The total external force applied to the system in the direction of the motion during the impact is mg so that, from eqn (6.30),

$$\frac{dp}{dt} = mg$$

where p is the linear momentum of the system along the guideway.

Integration of this equation over the interval δt gives

$$p_2 - p_1 = mg\,\delta t. \tag{xix}$$

If we assume that the impact period is short, $\delta t \to 0$ and $p_2 = p_1$ so that the linear momentum is constant during the impact.

Thus $\qquad (m + m_s)v_1 = mv_0,$

so that $\quad v_1 = \dfrac{mv_0}{m + m_s} \tag{xx}$

which agrees with eqn (xvi).

Thus we can obtain the velocity after impact from momentum considerations alone by considering the two bodies as a system.

The impulses applied to the slider by the guideway are given by eqn (xviii). They are equal and opposite and are applied to the slider corners A and D as shown in Fig. 6.28. Due to the large contact forces associated with these impulses one might expect to observe local bruising of material at the slider corners. It can be seen from eqn (xviii) that these impulses can be reduced to zero if the centres of mass G_1 and G_2 of the slider and the stop are aligned, i.e. if $\Delta = 0$. In this case the impulse $\mathscr{I}_P$ would have a line of action passing through G_1.

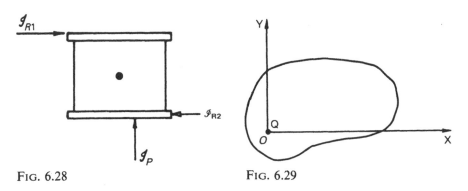

FIG. 6.28   FIG. 6.29

Immediately after the impact the kinetic energy of the slider and stop assembly can be calculated using eqn (xvi) and is given by

$$T = \frac{1}{2}(m + m_s)v_1^2$$

$$= \frac{1}{2}\frac{mv_0^2}{(1 + m_s/m)}. \tag{xxi}$$

There is no change in the strain energy stored in the spring during the impact, since it has been assumed that no displacement of m_s has occured. Equation (xxi) shows that there has been a loss of kinetic energy,

$$T_{\text{Loss}} = \frac{1}{2}mv_0^2 - \frac{1}{2}\frac{mv_0^2}{(1 + m_s/m)} = \frac{1}{2}\frac{mm_sv_0^2}{(m_s + m)}, \tag{xxii}$$

which occurs because the masses remain together after collision.

Pure rotation

When a body moves with pure rotation about an axis passing through a fixed point O, as shown in Fig. 6.29 then it is convenient to choose the point Q in eqn (6.71) to be coincident with O.

For this case $\ddot{x}_O = \ddot{y}_O = 0$ and eqn (6.71) reduces to

$$\frac{dh_O}{dt} = M_O. \qquad (6.89)$$

Since eqns (6.63) and (6.78) give $h_O = H_O = I_O\dot{\theta}$, eqn (6.89) can be rewritten as

$$\frac{dH_O}{dt} = \frac{dh_O}{dt} = I_O\ddot{\theta} = M_O \qquad (6.90)$$

which agrees with eqn (4.29) for pure rotation.

The angular impulse of the external forces about O is given by integrating eqn (6.90) with respect to time,

i.e. $$\mathcal{I}_O = \int_{t_1}^{t_2} M_O \, dt = (H_O)_2 - (H_O)_1 = I_O(\dot{\theta}_2 - \dot{\theta}_1). \qquad (6.91)$$

For a body moving with pure rotation about a fixed point O the change in the angular momentum about O is equal to the angular impulse of the external forces about O. This applies to both relative and absolute angular momentum since in this case they are equal.

Example 6.6   Figure 6.30 shows a thin uniform rod of mass m and length $2a$ supported at its end by means of a frictionless pin joint at O. The rod is initially at rest and hangs freely in the vertical position due to gravity.

A bullet of mass m_b is fired at the rod with a velocity v_0 and strikes the rod at an angle α to the horizontal and at a point A distant b from O. If the bullet remains embedded in the rod, let us find the angular velocity of the rod after impact, and the angle through which the rod rotates before coming to rest.

A force with components P and Q is applied to the rod at A by the bullet during the impact, as shown in the free body diagram of Fig. 6.31. The forces

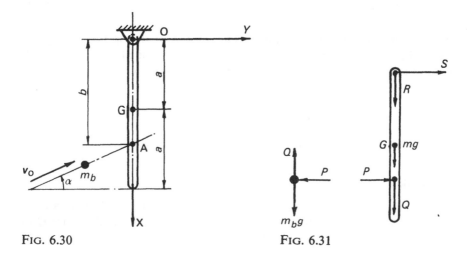

FIG. 6.30   FIG. 6.31

R and S represent the forces applied to the rod by the bearing at O. Over the period δt of the impact these forces are assumed to reach maximum values which are very much greater than the weights mg and $m_b g$. It is also assumed that the angular velocity of the rod increases rapidly from zero to a value $\dot{\theta}_0$ and that the corresponding rotation of the rod during the time δt is negligibly small.

If we now apply eqn (6.91) to the rod, then from the free body diagram of Fig. 6.31, the angular impulse,

$$\mathcal{J}_O = b \int_0^{\delta t} P \, dt = I_O(\dot{\theta}(\delta t) - \dot{\theta}(0)) = I_O \dot{\theta}_0 \qquad \text{(i)}$$

where I_O is the moment of inertia of the rod about O.

Similarly the linear impulse eqns (6.85) for the rod are

$$\mathcal{J}_x = \int_0^{\delta t} R \, dt + \int_0^{\delta t} Q \, dt + mg \, \delta t = 0 \qquad \text{(ii)}$$

along OX and

$$\mathcal{J}_y = \int_0^{\delta t} P \, dt + \int_0^{\delta t} S \, dt = ma\dot{\theta}_0 \qquad \text{(iii)}$$

along OY.

Immediately before impact occurs the velocity of the bullet is v_0 in the direction shown in Fig. 6.30. After impact the velocity of the bullet is the same as that of point A, which is $v_A = b\dot{\theta}_0$ in the direction of OY.

The impulse relationship for the bullet, using eqn (6.4) of section 6.1 and the free body diagram of Fig. 6.31, may be written in component form as

$$-\int_0^{\delta t} Q \, dt + m_b g \, \delta t = m_b v_0 \sin \alpha \qquad \text{(iv)}$$

along OX and

$$-\int_0^{\delta t} P \, dt = m_b(b\dot{\theta}_0 - v_0 \cos \alpha), \qquad \text{(v)}$$

along OY.

Since we shall now let $\delta t \to 0$, the impulse due to the forces mg and $m_b g$ will be very small and can be neglected. The impulses due to the forces P, Q, R and S are finite since it is assumed that these forces become very large during the impact.

If the impulse $\int_0^{\delta t} P \, dt$ is eliminated from eqn (i) to (v) we obtain

$$(I_O + m_b b^2)\dot{\theta}_0 - m_b b v_0 \cos \alpha = 0, \qquad \text{(vi)}$$

from which the angular velocity $\dot{\theta}_0$ of the rod after impact can be obtained.

Since $(I_O + m_b b^2)$ is the moment of inertia of the rod and the bullet about O the first term $(I_O + m_b b^2)\dot{\theta}_0$ in eqn (vi) is the angular momentum about O of the combined system just after impact has occurred. The second term in eqn (vi) is the angular momentum of the bullet about O before it makes contact with the rod.

Equation (vi) shows that, during the impact, the relative angular momentum of the system about O is constant,

i.e. $(h_O)_0 = (h_O)_{\delta t}.$ (vii)

This condition may have been realised in the very beginning of the problem by considering the bullet and the rod together as a closed system, as shown in Fig. 6.32. Since there are no external angular impulses applied to this system about the point O, i.e. $M_O = 0$, eqns (6.53) and (6.63) show that the relative and the absolute angular momentum of this system about O must be constant.

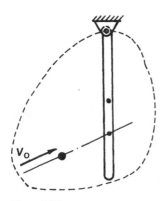

FIG. 6.32

The angular velocity of the assembly and the impulses at O and A are thus given by

$$\dot{\theta}_0 = \frac{m_b b v_0 \cos \alpha}{(I_O + m_b b^2)},$$ (viii)

$$\int_0^{\delta t} P \, dt = \frac{m_b I_O v_0 \cos \alpha}{(I_O + m_b b^2)},$$ (ix)

$$\int_0^{\delta t} Q \, dt = -m_b v_0 \sin \alpha,$$ (x)

$$\int_0^{\delta t} S \, dt = -\frac{(I_O - mab)}{(I_O + m_b b^2)} m_b v_0 \cos \alpha,$$ (xi)

and $$\int_0^{\delta t} R \, dt = m_b v_0 \sin \alpha.$$ (xii)

From eqn (xii) it can be seen that the vertical impulse at the bearing O is determined by the vertical component of bullet's linear momentum before it strikes the rod. The horizontal component of this impulse is given by eqn (xi) and is a proportion of the horizontal component of the bullet's linear momentum. It is interesting to see that this impulse can be reduced to zero if the position of point A, where the bullet strikes the rod, is determined by

$$b = \frac{I_O}{ma}. \tag{xiii}$$

This point is called the centre of percussion of the rod (see example 4.5).

The maximum angle θ_m through which the rod now rotates can be determined by drawing a boundary around the bullet and rod as shown in Fig. 6.33 and applying the First Law of Thermodynamics.

Work is done on the system by the external forces mg and $m_b g$, and the only internal energy of the system is its kinetic energy.

Now $\quad W = \Delta U = U_2 - U_1.$

From Fig. 6.33 the work done by the forces mg and $m_b g$ during the rotation θ_m is negative and is given by

$$W = -mga(1 - \cos \theta_m) - m_b g b (1 - \cos \theta_m). \tag{xiv}$$

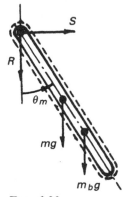

Fig. 6.33

Just after impact the kinetic energy of the rod and bullet assembly is given by

$$T_1 = \tfrac{1}{2}(I_O + m_b b^2)\dot{\theta}_0^2, \tag{xv}$$

where $(I_O + m_b b^2)$ is the moment of inertia about O of the rod and bullet together.

Using eqn (viii) T_1 can be rewritten as

$$T_1 = \frac{1}{2} \frac{m_b^2 b^2 v_0^2 \cos^2 \alpha}{(I_O + m_b b^2)}. \tag{xvi}$$

At the point of maximum swing the angular velocity of the assembly is zero. Hence $T_2 = 0$ so that

$$\Delta U = T_2 - T_1 = -T_1. \tag{xvii}$$

Substituting into the First Law of Thermodynamics gives

$$-(ma + m_b b)g(1 - \cos\theta_m) = -\frac{1}{2}\frac{m_b^2 b^2 v_0^2 \cos^2\alpha}{(I_O + m_b b^2)}. \qquad \text{(xviii)}$$

Since $(1 - \cos\theta_m) = 2\sin^2(\theta_m/2)$, the angle θ_m can be found from the above equation as the solution to

$$\sin\frac{\theta_m}{2} = \frac{m_b b v_0 \cos\alpha}{2\sqrt{g(ma + m_b b)(I_O + m_b b^2)}} \qquad \text{(xix)}$$

Thus if m_b, m and I_O are known the muzzle velocity v_0 of the bullet can be found by measuring θ_m. This is a well known experimental method of measuring muzzle velocity of small guns. Usually the angle α is zero in such experiments.

General plane motion

To derive the rotational equation of motion of a rigid body moving with general plane motion, we choose the point Q in eqn (6.71) to be coincident with the centre of mass G of the body. The absolute and relative angular momentum about G are the same, i.e. $H_G = h_G = I_G \dot\theta$, and eqn (6.71) becomes

$$\frac{dh_G}{dt} = \frac{dH_G}{dt} = I_G \ddot\theta = M_G, \qquad (6.92)$$

which agrees with the equation of motion eqn (4.43).

When eqn (6.92) is integrated with respect to time we obtain the angular impulse of the external moments about G, i.e.

$$\mathscr{J}_G = \int_0^{\delta t} M_G \, dt = (h_G)_2 - (h_G)_1 = (H_G)_2 - (H_G)_1 = I_G(\dot\theta_2 - \dot\theta_1). \qquad (6.93)$$

For a body moving with general plane motion the angular impulse about G is equal to the change in the relative (or absolute) angular momentum of the body about G.

Example 6.7   Figure 6.34 shows a thin rod of mass m resting on a smooth horizontal plane. A particle of mass m_p is fired at the rod with velocity v_0 and is aimed to strike the rod at a point A, distant q from the centre of mass G of the rod as shown. The direction of the velocity vector v_0 is arranged to be perpendicular to the line AG. If the rod and the particle remain in contact after the impact, let us determine the angular velocity of the rod, and the velocity of a point B which is located a distance b from G, as shown.

The first step in the solution of this problem is to recognise that after the impact, the rod and the particle will move as a combined body in general plane motion. When impact occurs a contact force P will be produced at A,

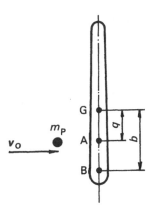

FIG. 6.34

and the free body diagrams of the particle and the rod during the impact period will be as shown in Fig. 6.35.

During impact the particle will have an acceleration $\ddot{x}_p$, as shown. The rod, however, will move with general plane motion and we must therefore describe its motion by considering the acceleration $\ddot{x}_G$ of its centre of mass G, and its angular acceleration $\ddot{\theta}$.

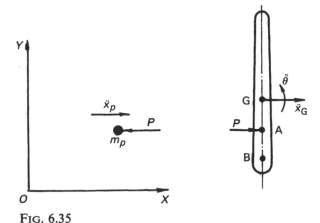

FIG. 6.35

The equation of motion for the particle is therefore

$$-P = m_p\ddot{x}_p. \tag{i}$$

Now for the rod,

$$P = m\ddot{x}_G, \tag{ii}$$

and $$Pq = I_G\ddot{\theta}. \tag{iii}$$

Let us now integrate these equations over the duration δt of the impact and note that the particle has an initial velocity v_0 and that $\dot{x}_G$ and $\dot{\theta}$ are zero just before the impact occurs.

Hence at the end of the impact period we have

$$-\int_0^{\delta t} P \, dt = m_p(\dot{x}_p - v_0),$$ (iv)

$$\int_0^{\delta t} P \, dt = m\dot{x}_G,$$ (v)

and $$q \int_0^{\delta t} P \, dt = I_G \dot{\theta}.$$ (vi)

At the end of the impact the velocity of the particle will be the same as that of point A on the rod so that

$$\dot{x}_p = v_A.$$

Since the rod is in general plane motion the velocity of point A can be related to that of the centre of mass by the vector equation

$$v_A = v_G + v_{AG}.$$ (vii)

All of these vectors are perpendicular to the rod and $v_{AG} = q\dot{\theta}$. Hence

$$\dot{x}_p = \dot{x}_G + q\dot{\theta}.$$ (viii)

Solving eqns (iv), (v), (vi) and (viii) gives the velocity of the centre of mass and the angular velocity of the rod after impact as

$$\dot{x}_G = \frac{m_p I_G v_0}{(m + m_p)I_G + m_p m q^2},$$ (ix)

and $$\dot{\theta} = \frac{m_p m q v_0}{(m + m_p)I_G + m_p m q^2}.$$ (x)

The velocity of the general point B on the rod can now be found using eqn (vii) by writing

$$v_B = v_G + v_{BG}.$$ (xi)

For this case v_B is in the direction of OX so that

$$v_B = \dot{x}_b = \dot{x}_G + b\dot{\theta},$$ (xii)

which then gives

$$\dot{x}_B = \frac{(I_G + mqb)m_p v_0}{(m + m_p)I_G + m_p m q^2}.$$ (xiii)

Alternatively we could treat the particle and rod as a closed system of particles to which no external forces are applied. From eqn (6.73) we see that in this case $dh_G/dt = 0$, so that the relative angular momentum of the whole system about its centre of mass is constant.

From Fig. 6.36 the position of G_c, the centre of mass of the system when the particle is embedded in the rod, is given by

$$(m + m_p)c = m_pq,$$

so that $\quad c = \dfrac{m_pq}{m + m_p}.$ \hfill (xiv)

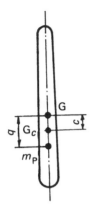

FIG. 6.36

Just before impact occurs the relative angular momentum of the system about G_c is

$$(h_{G_c})_1 = m_pv_0(q - c). \tag{xv}$$

Just after impact has occurred the relative angular momentum about G_c will be

$$(h_{G_c})_2 = I_{G_c}\dot{\theta}. \tag{xvi}$$

Now since no external forces or moments act on the system eqn (6.74) gives

$$(h_{G_c})_1 = (h_{G_c})_2,$$

so that $\quad m_pv_0(q - c) = I_{G_c}\dot{\theta}.$ \hfill (xvii)

Using the parallel axes theorem,

$$I_{G_c} = I_G + mc^2 + m_p(q - c)^2. \tag{xviii}$$

Substituting this value into eqn (xvii) and eliminating c, using eqn (xiv), gives

$$\dot{\theta} = \frac{m_pmqv_0}{(m + m_p)I_G + mm_pq^2}. \tag{xix}$$

This result agrees with eqn (x) which was obtained from the impulse equations for the particle and the rod.

The linear momentum of the system along OX is

$$p_x = \sum m_i \dot{x}_i = m v_{G_c},$$

and since no external forces are applied we have

$$\frac{dp_x}{dt} = 0.$$

Thus $(p_x)_2 - (p_x)_1 = 0,$

where $(p_x)_1 = m_p v_0$ and $(p_x)_2 = (m + m_p)\dot{x}_{G_c}.$

Hence $\dot{x}_{G_c} = \dfrac{m_p v_0}{(m + m_p)}.$ (xx)

If in eqn (xiii) we let $b = c = m_b q/(m + m_b)$, i.e. if B is at G_c, we obtain

$$\dot{x}_B = \dot{x}_{G_c} = \frac{\left(I_G + \dfrac{mm_p}{(m + m_b)}q^2\right)}{(m + m_p)I_G + m_p m q^2} m_p v_0 = \frac{m_p v_0}{(m + m_p)},$$

which agrees with eqn (xx).

The subsequent motion of the system can now be discussed in terms of $\dot{x}_{G_c}$, $\dot{y}_{G_c}$ and $\dot{\theta}$. Since *no* external forces are applied to the combined system after impact $\dot{x}_{G_c}$, $\dot{y}_{G_c}$ and $\dot{\theta}$ will remain constant. Since $\dot{y}_{G_c} = 0$ the point G_c will continue to move in a straight line in the direction of OX and the rod will continue to rotate at a constant angular velocity $\dot{\theta}$ as shown in Fig. 6.37.

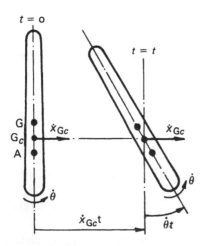

FIG. 6.37

Engineering Applications

When applying the results of this Chapter to an engineering problem we will again make use of the earlier assumptions and idealisations to produce a mathematical model of the total system. It is an essential first step when using momentum methods to define carefully the system under investigation. All external forces applied to the system must be identified and shown on the diagram of the system.

It is important to determine the position and motion of the centre of mass of the system and it is useful to identify any points within the system which are stationary.

6.1 Surveillance satellite

Figure 6.38 shows a surveillance satellite. The satellite is designed to spin about its central axis through its centre of mass G and a radar dish fixed to its exterior, as shown, monitors the surrounding space as the satellite rotates.

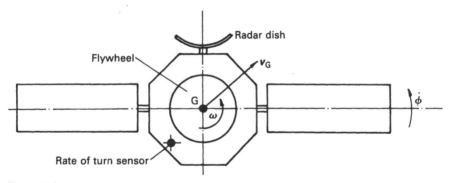

FIG. 6.38

For good surveillance it is important to be able to control the angular velocity $\dot{\phi}$ of the satellite. This is done by varying the angular velocity of a flywheel which is mounted in the body of the satellite, as shown. A rate of turn sensor is mounted on the satellite and is used to measure the angular velocity $\dot{\phi}$. This measurement is read by the satellite computer and compared with the desired angular velocity Ω_0 of the satellite to produce an error signal

$$\varepsilon = (\dot{\phi} - \Omega_0). \tag{i}$$

Let us show how this error signal can be used to drive the flywheel and thereby control the rate at which the satellite rotates.

Since the body of the satellite and the flywheel may be regarded as a closed system of particles, and since no external forces act on this system, then eqn (6.73) shows that the relative angular momentum of the assembly about G is constant,

i.e. $$\frac{dh_G}{dt} = M_G = 0. \qquad\qquad (ii)$$

If the moment of inertia of the satellite body about an axis through G is I_G then for the satellite

$$h_{GS} = I_G\dot{\phi}. \qquad\qquad (iii)$$

Similarly, if the flywheel rotates with an *absolute* angular velocity ω and if I_F is its moment of inertia about its spin axis then

$$h_{GF} = I_F\omega. \qquad\qquad (iv)$$

The total relative angular momentum about G is therefore given by

$$h_G = I_G\dot{\phi} + I_F\omega. \qquad\qquad (v)$$

Substitution of eqn (v) into eqn (ii) yields

$$I_G\ddot{\phi} + I_F\dot{\omega} = 0. \qquad\qquad (vi)$$

Let us now consider the free-body diagram of the flywheel, as shown by Fig. 6.39. The forces R_x and R_y represent the bearing forces at the flywheel support bearings and M_F is the torque applied to the flywheel by its drive motor. The stator of this motor is rigidly fixed to the body of the satellite. Since the flywheel is rotating about an axis through its centre of mass we can write

$$\frac{dh_{GF}}{dt} = I_F\dot{\omega} = M_F \qquad\qquad (vii)$$

Suppose we now arrange for the motor torque M_F to be proportional to the error ε such that

$$M_F = K\varepsilon, \qquad\qquad (viii)$$

where K is a constant.

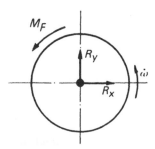

Fig. 6.39

Substitution of eqns (i), (vii) and (viii) into eqn (vi) then yields

$$I_G\ddot{\phi} + K(\dot{\phi} - \Omega_0) = 0$$

or $\tau\ddot{\phi} + \dot{\phi} = \Omega_0,$ (ix)

where $\tau = I_G/K$ has the units of time and is called the spin rate time constant.

Equation (ix) is a first order linear differential equation in $\dot{\phi}$. Since the desired angular velocity of the satellite, Ω_0, is a constant the solution of eqn (ix) is given by

$$\dot{\phi} = \Omega_0 + Ae^{-t/\tau},$$ (x)

where A is an arbitrary constant.

If, when set up, the satellite is not rotating, then $\dot{\phi} = 0$ at time $t = 0$. Thus from eqn (x)

$$A = -\Omega_0.$$

The angular velocity of the satellite thereafter is given by

$$\dot{\phi} = \Omega_0(1 - e^{-t/\tau}).$$ (xi)

The variation of $\dot{\phi}$ with time is shown plotted in Fig. 6.40 for various values of τ. In all cases it is shown that $\dot{\phi}$ approaches the desired value Ω_0 exponentially but at a rate which depends upon the value of the time constant τ. The smaller the value of τ the faster $\dot{\phi}$ approaches the desired value.

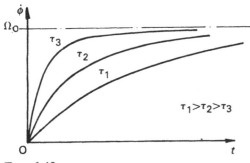

FIG. 6.40

Since the time constant $\tau = I_G/K$, small values can be achieved, for a given satellite, by making the value of K large.

We should note that the foregoing analysis does not assume that the centre of mass of the satellite is stationary. The centre of mass G is free to move in space.

6.2 Robot arm

Figure 6.41 shows a simple two degree of freedom robot arm. The arm CD carries the robot gripper hand at the point C and is driven by means of the

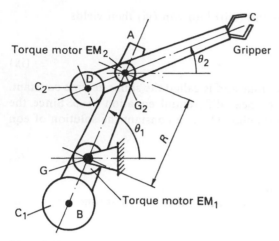

FIG. 6.41

electric torque motor EM_2. The armature and stator of this motor are rigidly connected to arms CD and AB respectively and its output axis passes through the point G_2 as shown. A counterbalance mass C_2 is added to the arm CD at D so that the centre of mass of the arm also lies at G_2.

A second torque motor EM_1 is connected to the arm AB and has its output axis passing through the point G. The stator of this motor is rigidly connected to ground and thereby supports the whole robot arm. A second counterbalance mass C_1 is added to arm AB at B and is chosen so that the centre of mass of the *complete* robot arm *always* lies at G.

We shall assume that the masses of arms AB and CD are m_1 and m_2 respectively and that I_1 and I_2 represent their moments of inertia about axes through G and G_2 respectively. The angular positions of the arms, at any instant, are given by θ_1 and θ_2 as shown, and the distance $GG_2 = R$.

Let us use momentum principles to derive the equations of motion of the arm in terms of the torques generated by the motors EM_1 and EM_2.

We shall now draw the free body diagram of arm CD, Fig. 6.42, and consider the relative angular momentum of the arm about an axis through G_2. The

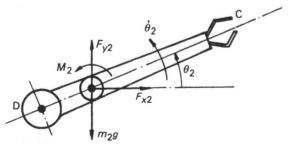

FIG. 6.42

components F_{x_2} and F_{y_2} represent the force applied to CD due to the armature bearings at G_2 and M_2 represents the torque applied to the armature by the stator of EM_2.

Now $h_{G_2} = I_2\dot{\theta}_2.$ (i)

Using eqn (6.73) and Fig. 6.42 we have

$$\frac{dh_{G_2}}{dt} = M_{G_2} = M_2,$$

so that $I_2\ddot{\theta}_2 = M_2.$ (ii)

Let us now draw the free body diagram of the robot arm, as shown in Fig. 6.43, and consider this as a system of connected rigid bodies. The components F_{x_1} and F_{y_1} give the force applied to the armature of EM_1 and M_1 represents the torque applied to the robot by motor EM_1. It should be understood that the torque motor EM_2 does not apply an external torque to the robot arm at G_2. The torques applied to the armature and stator of EM_2 are equal and opposite, so that the torque produced by motor EM_2 is internal to the system.

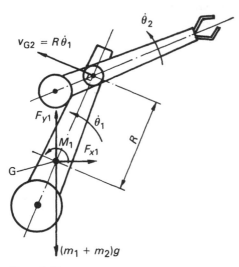

FIG. 6.43

The *absolute* angular momentum of the arm CD about the *fixed point* G is given by eqns (6.48) and (6.65) as

$$H_G^{(CD)} = h_{G_2} + r_{GG_2}m_2v_{G_2},$$ (iii)

where $h_{G_2} = I_2\dot{\theta}_2,$ (iv)

and $r_{GG_2} = R.$ (v)

Since arm AB is in pure rotation about G, Fig. 6.43 shows that

$$v_{G_2} = R\dot{\theta}_1. \tag{vi}$$

Substitution of eqns (iv), (v) and (vi) into eqn (iii) gives the absolute angular momentum of arm CD about an axis through G as

$$H_G^{(CD)} = I_2\dot{\theta}_2 + m_2 R^2 \dot{\theta}_1. \tag{vii}$$

Since G is a fixed point eqns (6.63) and (6.78) give the absolute angular momentum of arm AB about an axis through G as

$$H_G^{(AB)} = I_1\dot{\theta}_1. \tag{viii}$$

Thus the absolute angular momentum of the robot arm about an axis through G is

$$H_G = H_G^{(AB)} + H_G^{(CD)}$$
$$= I_2\dot{\theta}_2 + (I_1 + m_2 R^2)\dot{\theta}_1. \tag{ix}$$

The total *external* moment applied to the robot arm about G is

$$M_G = M_1. \tag{x}$$

We should remember that the weights $m_1 g$ and $m_2 g$ do not contribute to this moment as the centre of mass of the whole arm has been located at G by choosing the two counterbalancing masses at B and D to have appropriate values.

Since $\dfrac{dH_G}{dt} = M_G$

eqns (ix) and (x) give

$$I_2\ddot{\theta}_2 + (I_1 + m_2 R^2)\ddot{\theta}_1 = M_1. \tag{xi}$$

Substituting $I_2\ddot{\theta}_2 = M_2$ from eqn (ii) reduces eqn (xi) to

$$(I_1 + m_2 R^2)\ddot{\theta}_1 = (M_2 - M_1). \tag{xii}$$

Equations (ii) and (xii) determine the angular accelerations of the robot arms in terms of the motor torques M_1 and M_2.

The reader should compare the equations of motion, eqns (ii) and (xii), with those obtained in Application 4.9 for the same robot arm.

6.3   Friction clutch

Figure 6.44 shows a simple power transmission system which connects a motor D_1 to a mixer D_2. The motor and mixer may be considered as rigid bodies which rotate about fixed axes through their centres of mass G_1 and G_2 and have moments of inertia I_1 and I_2 about their axes of rotation. The output from the motor drives a gearbox via a friction clutch, and the mixer is driven

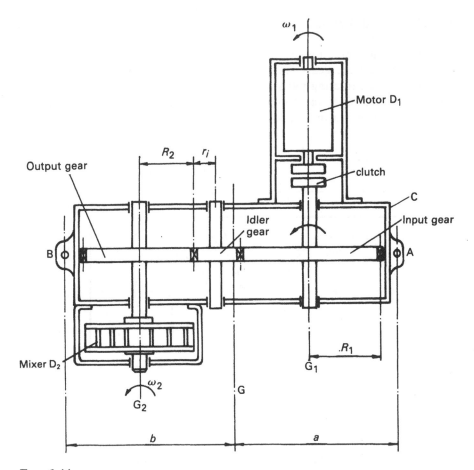

Fig. 6.44

from the gearbox output shaft. The motor, clutch, mixer and gears are supported in bearings in the casing C and form a single unit as shown. The casing
is bolted to ground at the points A and B. The clutch allows the motor to be
disengaged from the gearbox. The input gear is driven via the clutch and drives
the output gear via an idler gear so that the input and output shafts rotate in
the same direction. The dimensions of the gears are as shown on Fig. 6.44.

   Initially the clutch is disengaged and the motor D_1 is assumed to be rotating,
in the anticlockwise direction, with a constant angular velocity ω_0. When the
clutch is engaged the drive is transmitted through the friction plates of the
clutch to the gearbox and the mixer D_2 will start to rotate. The friction torque
M_1 transmitted to the gearbox by the clutch will then accelerate D_2 until its
angular velocity ω_2 reaches a steady value Ω_2. An equal and opposite torque
is applied to D_1 and this decelerates the motor until its angular velocity ω_1 is
reduced to a steady value Ω_1. During slip the angular velocity of the input
gear is ω_1^* and is less than ω_1. We shall also assume that the fixing bolts at

the points A and B on the gearbox apply forces V_1 and V_2 to the gearbox as shown in Fig. 6.45.

Let us obtain expressions for the steady angular velocities Ω_1 and Ω_2. We shall assume that the gears are light (i.e. massless) and that m_1 and m_2 are the masses of the rotating parts of D_1 and D_2.

Let us first consider the complete unit as a system contained within the boundary S, as shown in Fig. 6.45.

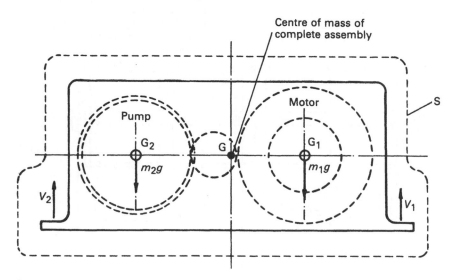

FIG. 6.45

Since D_1 rotates about a *fixed* axis which passes through its centre of mass, eqn (6.49) shows that the absolute angular momentum of D_1 about an arbitrary point Q is independent of the position of Q. Thus if we choose the point Q to be at the centre of mass G *of the rotating parts*, i.e. at a point distant

$$c = (2r + R_1 + R_2)m_2/(m_1 + m_2)$$

to the left of G_1

$$H_Q^{(1)} = H_{G_1}^{(1)} = I_1\omega_1. \tag{i}$$

The same is true for rotor D_2,

i.e. $$H_Q^{(2)} = H_{G_2}^{(2)} = I_2\omega_2. \tag{ii}$$

Hence the total absolute angular momentum of the system about the point G during the period of slip is

$$H_G = I_1\omega_1 + I_2\omega_2. \tag{iii}$$

Now eqn (6.52) gives

$$\frac{dH_G}{dt} = M_G \tag{iv}$$

where $M_G = (aV_1 - bV_2)$ (v)

is the external anticlockwise moment applied to the system about G by the holding down bolts at A and B. By considering the whole unit as a system the torques at the motor and the mixer will be internal to the system and therefore do not contribute to M_G.

The integration of eqn (iv) with respect to time over the period t_s of slip gives

$$\int_0^{t_s} M_G \, dt = (H_G)_{t=t_s} - (H_G)_{t=0}$$

i.e. $$\int_0^{t_s} M_G \, dt = (I_1\Omega_1 + I_2\Omega_2) - I_1\omega_0. \tag{vi}$$

The change in the angular momentum of the system is equal to the angular impulse about the centre of mass of the system applied by the gearbox fixing bolts.

Let us now draw the free body diagrams of D_1 and D_2 as shown in Fig. 6.46.

The forces F_{x_1}, F_{y_1} and F_{x_2}, F_{y_2} are the shear forces at connecting shafts and M_1 and M_2 are the torques which are applied to D_1 by the clutch and to D_2 by its drive shaft.

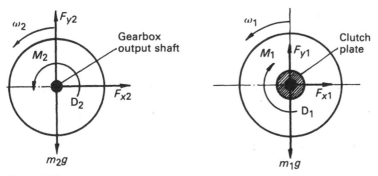

FIG. 6.46

The equations of motion for D_1 and D_2 about their respective axes are, for D_1

$$-M_1 = \frac{dH_{G_1}^{(1)}}{dt} = I_1\dot{\omega}_1, \tag{vii}$$

and for D_2

$$M_2 = \frac{dH_{G_2}^{(2)}}{dt} = I_2\dot{\omega}_2. \tag{viii}$$

If these equations are integrated with respect to time over the period of slip we get

$$-\int_0^{t_s} M_1 \, dt = I_1\Omega_1 - I_1\omega_0 \tag{ix}$$

and

$$\int_0^{t_s} M_2 \, dt = I_2\Omega_2. \tag{x}$$

Suppose we now draw the free body diagram of the gearbox assembly as shown in Fig. 6.47. A system boundary S_b has also been drawn around the

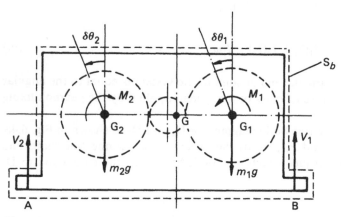

FIG. 6.47

box as shown by the dotted line. Since this system has been assumed to be *massless* and rigid its internal energy U will always be zero. If we assume no heat is transferred from the gearbox, the First Law of Thermodynamics gives the work done on the gearbox as $W = \Delta U = 0$. The work δW done by the torques M_1 and M_2 due to displacements $\delta\theta_1^*$ and $\delta\theta_2$ of G_1 and G_2 is therefore given by

$$\delta W = M_1 \, \delta\theta_1^* - M_2 \, \delta\theta_2 = 0. \tag{xi}$$

The minus sign in eqn (xi) occurs because the torque M_2 on the output gear and the angular displacement $\delta\theta_2$ have opposite directions. The forces at A, B, G_1 and G_2 are fixed and do no work.

If eqn (xi) is divided by δt the limit $\delta t \to 0$ gives

$$M_1\omega_1^* = M_2\omega_2. \tag{xii}$$

The kinematic relationship between the angular velocities ω_1^* and ω_2 can be found by considering the gear train shown in Fig. 6.48. Using the results of Chapter 1 the magnitudes of the velocities of points C_1 and C_2 can be written

$$v_{c_1} = R_1\omega_1^* = r_i\omega_i,$$

and $v_{c_2} = r_i\omega_i = R_2\omega_2$

where $v_{c_1} = v_{c_2}$.

By inspection of Fig. 6.48 it can be seen that ω_1^* and ω_2 are in the same direction.

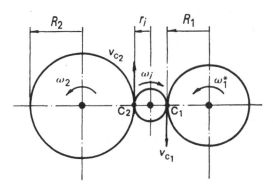

FIG. 6.48

Elimination of v_{c_1} and v_{c_2} from these equations gives the speed ratio of the gearbox as

$$\frac{\omega_1^*}{\omega_2} = \frac{R_2}{R_1}. \tag{xiii}$$

Now, from eqns (xii) and (xiii)

$$\frac{M_1}{M_2} = \frac{R_1}{R_2} = \frac{\Omega_2}{\Omega_1}. \tag{xiv}$$

Hence, by eliminating M_1 and M_2 from eqns (ix) and (x) using eqns (xiv),

$$\left(I_1\Omega_1 + \frac{R_1}{R_2}I_2\Omega_2\right) = \left(I_1 + \left(\frac{R_1}{R_2}\right)^2 I_2\right)\Omega_1 = I_1\omega_0. \tag{xv}$$

Thus the steady angular velocities of the motor and the mixer are

$$\Omega_1 = \frac{I_1\omega_0}{I_1 + \left(\dfrac{R_1}{R_2}\right)^2 I_2} \tag{xvi}$$

and $$\Omega_2 = \frac{(R_1/R_2)I_1\omega_0}{I_1 + \left(\dfrac{R_1}{R_2}\right)^2 I_2}. \tag{xvii}$$

If these expressions are now substituted into eqn (vi) the change in the angular momentum of the system can be found as

$$\Delta H = \int_0^{t_s} M_G \, dt = \frac{I_1 I_2}{I_1 + \left(\dfrac{R_1}{R_2}\right)^2 I_2} \frac{R_1}{R_2}\left(1 - \frac{R_1}{R_2}\right)\omega_0. \qquad \text{(xviii)}$$

Equation (xviii) shows that the angular momentum of the total system is changed when the clutch is engaged and that this change is equal to the angular impulse applied to the gearbox by the fixing bolts at A and B. It can be seen that the angular momentum will remain constant only if $R_1 = R_2$, i.e. for a gear ratio of unity. For this case a $aV_1 = bV_2$ and $\int_0^{t_s} M_G \, dt = 0$.

Exercises

1 A rocket is made up of two stages of mass m_1 and m_2 as shown in Fig. 6.49. Before separation the assembly moves with velocity v. When separation occurs the stages are caused to move apart with relative velocity v_s. Determine the new velocity v^* of the carrier stage m_2.

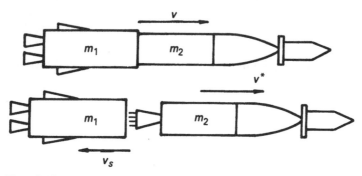

FIG 6.49

2 A ferry barge is secured to a dock by means of an inextensible rope as shown in Fig. 6.50. If a lorry of mass m is driven off the barge such that it leaves the barge with velocity v_0 after time Δt sec, determine the linear impulse applied to the barge by the rope and the average force in the rope due to this impulse.

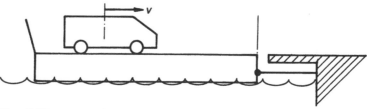

FIG 6.50

3   Fig. 6.51 shows two small, identical, spheres A and B colliding. Sphere A is fired off from a point P with a velocity v_0 such that at impact the line joining the centres O_1 and O_2 makes an angle θ with the direction of v_0.

   If no energy is lost determine the velocities of A and B after the impact.

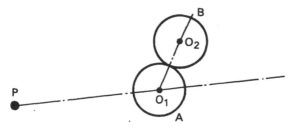

FIG 6.51

4   A block of mass M is dropped from a height h onto a spring scale as shown in Fig. 6.52. If the pan of the spring scale has mass m and the spring has stiffness k, determine the compression of the spring after impact.

   You should assume the block and the pan remain in contact just after impact.

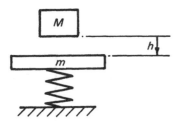

FIG 6.52

5   An empty box of mass 2 kg is projected with an initial velocity of 6 m/s up a smooth inclined plane, as shown in Fig. 6.53. After the box has travelled 2 m up the plane, a component of mass 4 kg drops vertically into it from a height of 1 m. Find how much further the box containing

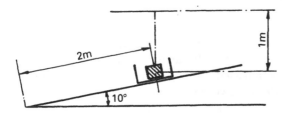

FIG 6.53

the component will travel up the plane. The plane is inclined to the horizontal at 10° and friction may be neglected.

6   The block shown in Fig. 6.54 has mass m and is located in a frictionless guideway. If a spherical particle of mass m_p is fired along the centre line of the guide with velocity v as shown, determine the velocity of the block after impact has occurred. Assume a coefficient of restitution e.

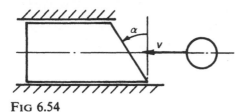

FIG 6.54

7   Two shafts carry discs of moments of inertia I_1 and I_2 about their common axis and can be coupled by a clutch at A (Fig. 6.55). If initially I_1 is rotating with an angular velocity ω while I_2 is at rest, and then the clutch is engaged, find the common angular velocity of the shafts after slipping has ceased.

Find also the loss of energy.

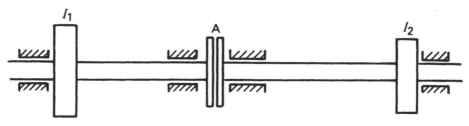

FIG 6.55

8   The satellite shown in Fig. 6.56 changes its angular velocity ω by means of two small thrusters A and B, positioned as shown. If the thrusters each eject particles of mass Δm with velocity v *relative* to the satellite, determine the change, $\Delta\omega$, in the angular velocity. Assume that the moment of inertia of the satellite about an axis through G has an initial value I.

9   A uniform thin rod AB of mass 2 kg is 1 m long, and can rotate freely in a vertical plane about a hinge at A. It carries at its end B a small scoop of mass 1 kg. The bar is released from rest in the horizontal position, $\theta = 90°$, and swings under its own weight as shown in Fig. 6.57 until it reaches the vertical position. The scoop then picks up a small component C of mass 0.5 kg and the assembly swings up to a maximum inclination

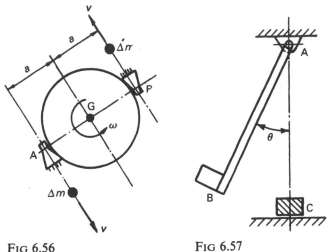

FIG 6.56   FIG 6.57

α on the other side of the vertical through A. Neglecting friction, find the value of α.

10   A particle of mass m_b is allowed to fall a height h into a cup which is located, as shown in Fig. 6.58, on a heavy uniform rod of mass m_r and length $2b$. If the bar is pivoted at its centre of mass O, determine the angular velocity of the bar immediately after impact. Show that this angular velocity will be a maximum if $a = b\sqrt{m_r/3m_b}$.

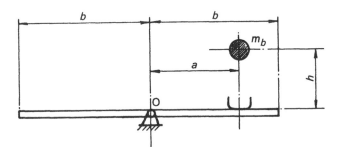

FIG 6.58

11   The disc shown in Fig. 6.59 has three equispaced slots in which are located small identical blocks each of mass m. The blocks are latched into position at a radius a and the disc rotates about a fixed axis through its centre O with angular velocity Ω_0. If the latches are released determine the angular velocity of the disc when the blocks reach the ends of their slots. The moment of inertia of the disc about an axis through O is I_0.

  Does this angular velocity depend upon friction between the blocks and their slots?

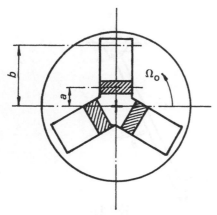

FIG 6.59

12 A disc of radius r slides with pure translation with velocity v_0 on a smooth horizontal surface and encounters an obstruction as shown in Fig. 6.60. Obtain expressions for the velocity of the centre O of the disc after impact with the obstruction if

(a) the disc slips at the contact with the obstruction and the coefficient of friction between the two surfaces is μ, and

(b) there is no slip at the contact point.

Assume that the disc has mass m and moment of inertia I_O about an axis through O.

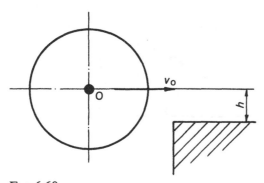

FIG 6.60

13 The disc shown in Fig. 6.61 rolls without slip along a perfectly rough horizontal plane with velocity v_1. Neglecting *slip* and *rebound* determine the velocity of its centre O as it starts to roll up the inclined plane. ,

14 Fig. 6.62 shows two uniform discs of mass m_1 and m_2 connected together by a loose inextensible string. If disc 1 is given an initial angular velocity ω_0 determine the angular velocities of the discs the instant after the string

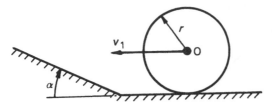

FIG 6.61

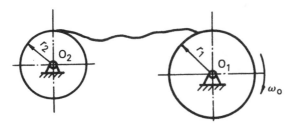

FIG 6.62

becomes taut. Neglect friction. Is the angular momentum of the system conserved?

15 The uniform disc shown in Fig. 6.63 has mass m and radius r and rolls without slip on a horizontal plane.
   (i)   Determine the absolute angular momentum of the disc about the point A.
   (ii) Derive expressions for the relative angular momentum of the disc about the points O, A and B.

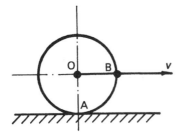

FIG 6.63

16 Fig. 6.64 shows an epicyclic gear unit with a fixed ring gear. Derive an expression for the *absolute* angular momentum of the gear system about the point O. The planet and the sun gears can be considered as uniform discs of radius r and R and mass m_p and m_s respectively. The arm OA can be regarded as a uniform rod of mass m_r, which rotates with angular velocity ω as shown.

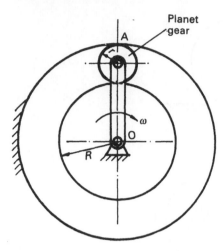

FIG 6.64

Also derive an expression for the *relative* angular momentum of the planet gear about the point A.

17 The bearings O_1 and O_2 of the discs A and B shown in Fig. 6.65 are supported on bearing blocks which are free to move in horizontal guides. Initially disc A is rotating with angular velocity ω_0 and is not touching disc B which is at rest. Determine the steady angular velocities of the discs after they have been forced together with a force P as shown.

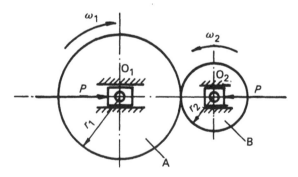

FIG 6.65

18 Figure 6.66 shows a thin uniform rod AB of mass m and length $2l$ held so that end B rests on a smooth horizontal surface. If the rod is released, determine, using momentum principles, the initial angular acceleration of the rod and the initial acceleration of end B.

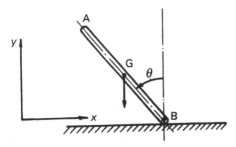

FIG 6.66

19 Figure 6.67 shows a uniform rod AB of mass m and length $2l$ in which end B is in contact with a smooth vertical wall and end A is in contact with a smooth horizontal surface. The rod is free to move in the plane of the figure under the action of gravity. The free body diagram of the rod and the position of its instantaneous centre I.C. are shown in Fig. 6.67.

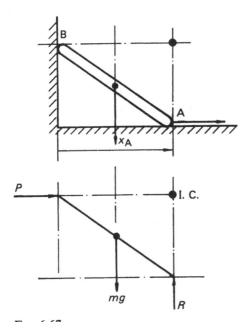

FIG 6.67

By considering the absolute angular momentum of the rod about a stationary point *instantaneously* coincident with I.C. show that the equation of motion of the rod is given by

$$\frac{dH_{IC}}{dt} = \frac{mgx_A}{2}$$

$$\text{where } H_{IC} = 4/3 \frac{ml^2 \dot{x}_A}{\sqrt{4l^2 - x_A^2}}$$

Hint   refer to eqn (6.72) in Chapter 6.

20   Define the angular momentum about a fixed point O of a rigid body moving (i) in pure translation and (ii) in pure rotation about O.

A uniform solid rectangular block ABCD moves along a production line by sliding down an inclined plane until its edge A meets a fixed horizontal surface, as shown in Fig. 6.68. It then turns on the edge A until AD makes contact with the horizontal plane, at which stage the block is in an erect position. There is no slip at A during the rotation.

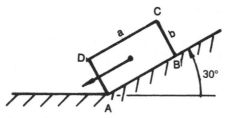

FIG 6.68

If the velocity of the block just before it makes initial contact with the horizontal plane at A is v, find the angular velocity of the block at the start of its turn, and hence determine the minimum value of v that is required to carry the block over to an erect position.

The moment of inertia of a uniform rectangular block about an edge is $m(a^2 + b^2)/3$, where a and b are the lengths of its sides and m is its mass.

21   A rod AB has a mass of 2 kg, a length of 0.5 m, and a moment of inertia of 0.04 kgm^2 about its centre of mass C. It lies flat on a frictionless horizontal plane and is struck by an impulse of magnitude $P = 0.2 \text{ Ns}$,

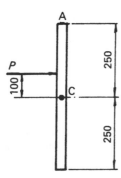

All dimensions in mm
FIG 6.69

as shown in Fig. 6.69. Derive expressions for the subsequent position of C and the angular displacement of the rod.

22 A forge-hammer (Fig. 6.70) consists of a heavy bar OA, with a striking head at A, which can rotate with negligible friction about an axis through its end O. The mass of the hammer is m, its centre of mass is at G where $OG = b$, its length $OA = l$, and its moment of inertia about O is mk^2.

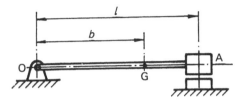

FIG 6.70

When released from rest in a vertical position, the hammer falls freely under gravity until it reaches a horizontal position when it strikes the workpiece and rebounds. It is found experimentally that the *upward* velocity of A just after impact is σ times its *downward* velocity just before impact, where $\sigma < 1$.

Derive expressions for the vertical impulse P delivered to the workpiece and the vertical impulse Q at O during impact, and establish the design condition relating b, l and k which minimises Q.

7

Momentum and mass transfer

Basic Theory

In Chapter 6, Section 6.1, we developed an expression for the linear momentum of a closed system of particles and showed that the rate of change of linear momentum of the system was equal to the sum of the external forces applied to the system, i.e. $F = \mathrm{d}p/\mathrm{d}t$. Since this system was assumed to be closed no mass could, by definition, enter or leave the system across its boundaries. As the system boundary moved the mass moved with it. Although this formulation is very useful for systems built from an assembly of rigid bodies it is not a convenient way of dealing with devices, such as pumps, compressors, turbines and jet engines, which require a flow of fluid into and out of the system.

7.1 Variable mass and fluid flow systems

We shall now show how the results of Section 6.1 can be extended to take into account the transfer of mass across the system boundary. Figure 7.1(a)

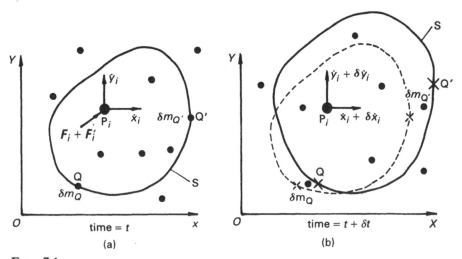

Fig. 7.1

shows a system of particles such as P_i contained within a boundary S at time t. The particle P_i is shown situated in the centre region of the system whilst the particle at the point Q on the boundary is *just inside* the boundary and represents a *small* mass which is just about to leave the system.

Let us assume that this particle has mass δm_Q and is moving with velocity v_Q which has components $\dot{x}_Q$, $\dot{y}_Q$. It is assumed that the velocity of this particle is different from the velocity v_Q^* of the coincident point Q on the system boundary. Q on the boundary therefore has coordinates x_Q^*, y_Q^* which at time t are the same as the coordinates x_Q, y_Q of the coincident particle. The asterisk will be used to denote the displacements and velocities of points on the boundary.

Outside the boundary of the system there are other particles. Let us consider the particle at the point Q'. This particle has mass $\delta m_{Q'}$ and is *just outside* the system boundary. It therefore must be excluded from the system at time t. It is assumed that this mass moves with velocity $v_{Q'}$ which is different from the velocity $v_{Q'}^*$ of the coincident point Q' on the system boundary.

At time t the linear momentum of the system is therefore given by

$$p_x = \sum m_i \dot{x}_i + \sum_s \delta m_Q \dot{x}_Q$$

$$p_y = \sum m_i \dot{y}_i + \sum_s \delta m_Q \dot{y}_Q. \tag{7.1}$$

The first sum represents the momentum of the particles 'well' inside the boundary and the second sum, which is summed around the system boundary, represents the momentum of the particules which are *at* the system boundary and which are *about to leave* the system.

At time $t + \delta t$ we shall assume that the system boundary has moved and the points Q and Q' on the system boundary have now positions given by $x_Q^* + \delta x_Q^*$, $y_Q^* + \delta y_Q^*$ and $x_{Q'}^* + \delta x_{Q'}^*$, $y_{Q'}^* + \delta y_{Q'}^*$ respectively. During this time, the mass δm_Q has left the system and has moved to a new position outside S given by $x_Q + \delta x_Q$, $y_Q + \delta y_Q$, as shown in Fig. 7.1(b). Correspondingly the mass $\delta m_{Q'}$, has entered the system and has moved to a position within the system given by $x_{Q'} + \delta x_{Q'}$, $y_{Q'} + \delta y_{Q'}$.

The linear momentum of the system at time $t + \delta t$ is therefore given by the momentum of those particles which have stayed inside the boundary plus the momentum of those which have entered the system, i.e.

$$p_x + \delta p_x = \sum m_i(\dot{x}_i + \delta \dot{x}_i) + \sum_s \delta m_{Q'}(\dot{x}_{Q'} + \delta \dot{x}_{Q'})$$

$$p_y + \delta p_y = \sum m_i(\dot{y}_i + \delta \dot{y}_i) + \sum_s \delta m_{Q'}(\dot{y}_{Q'} + \delta \dot{y}_{Q'}). \tag{7.2}$$

Subtracting eqns (7.2) and (7.1) gives

$$\delta p_x = \sum m_i \delta \dot{x}_i + \sum_s \delta m_{Q'} \dot{x}_{Q'} - \sum_s \delta m_Q \dot{x}_Q$$

$$\delta p_y = \sum m_i \delta \dot{y}_i + \sum_s \delta m_{Q'} \dot{y}_{Q'} - \sum_s \delta m_Q \dot{y}_Q \tag{7.3}$$

where the terms $\delta m_Q \delta \dot{x}_Q$ etc. are of second order and have been neglected.

The first term in eqn (7.3) represents the change in the linear momentum of the system due to the change in the velocity of the particles contained by the system boundary. The second and third terms are the linear momentum gained and lost by the system due to particles entering and leaving the system during δt. If eqns (7.3) are divided by δt, the limit as $\delta t \to 0$ gives

$$\frac{\mathrm{d}p_x}{\mathrm{d}t} = \sum m_i \ddot{x}_i + \sum_s \dot{m}_{Q'} \dot{x}_{Q'} - \sum_s \dot{m}_Q \dot{x}_Q \tag{7.4}$$

and

$$\frac{\mathrm{d}p_y}{\mathrm{d}t} = \sum m_i \ddot{y}_i + \sum_s \dot{m}_{Q'} \dot{y}_{Q'} - \sum_s \dot{m}_Q \dot{y}_Q, \tag{7.5}$$

where the terms $\dot{m}_Q$ and $\dot{m}_{Q'}$ represent the mass flow rates out of and into the system at the points Q and Q'.

Newton's Second Law of Motion for the particle P_i gives

$$F_{x_i} + F'_{x_i} = m_i \ddot{x}_i$$

and $\hspace{11cm}$ (7.6)

$$F_{y_i} + F'_{y_i} = m_i \ddot{y}_i,$$

where F_{x_i}, F_{y_i} and F'_{x_i}, F'_{y_i} are the external and internal forces applied to m_i. Thus

$$\sum m_i \ddot{x}_i = \sum (F_{x_i} + F'_{x_i})' = F_x$$
$$\hspace{10cm} (7.7)$$
and $\quad \sum m_i \ddot{y}_i = \sum (F_{y_i} + F'_{y_i}) = F_y,$

since the sum of the internal forces $\sum F'_{x_i} = \sum F'_{y_i} = 0$. The forces F_x and F_y represent the components of the *total* external force applied to the system. Hence

$$\frac{\mathrm{d}p_x}{\mathrm{d}t} = F_x + \sum_s \dot{m}_{Q'} \dot{x}_{Q'} - \sum_s \dot{m}_Q \dot{x}_Q \tag{7.8}$$

and $\quad \dfrac{\mathrm{d}p_y}{\mathrm{d}t} = F_y + \sum_s \dot{m}_{Q'} \dot{y}_{Q'} - \sum_s \dot{m}_Q \dot{y}_Q. \tag{7.9}$

The rate at which mass can leave the system at the point Q will depend upon the velocity of the elemental mass δm_Q *relative* to the coincident point Q on the system boundary. This relative velocity is shown in Fig. 7.2(a) and is given by

$$v_{Qr} = v_Q - v_Q^*. \tag{7.10}$$

If δm_Q leaves the system then v_{Qr} must have a direction which points *away* from the system boundary, as shown in Fig. 7.2(a). If we consider the system boundary at Q to be represented by a surface area of A_Q, as shown in Fig. 7.2(b), then δm_Q will only pass across A_Q if v_{Qr} has a component $\hat{v}_{Qr}$ which is normal

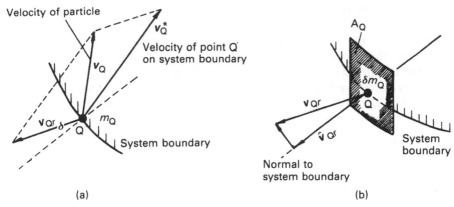

FIG. 7.2

to the boundary surface A_Q and is in the direction shown. If ρ is the density of the fluid at Q then the mass leaving the system at Q in time δt is given by

$$\delta m_Q = \{\rho \hat{v}_{Qr} A_Q\} \, \delta t.$$

Thus $\quad \dot{m}_Q = \rho \hat{v}_{Qr} A_Q.$ \hfill (7.11)

Similarly $\dot{m}_{Q'}$ can be written as

$$\dot{m}_{Q'} = \rho \hat{v}_{Q'r} A_{Q'}. \hfill (7.12)$$

Since $\delta m_{Q'}$ is entering the system the velocity $\hat{v}_{Q'r}$ must have a direction which points into the system.

Thus eqns (7.8) and (7.9) can be rewritten as

$$\frac{dp_x}{dt} = F_x + \sum_s (\rho A_{Q'} \hat{v}_{Q'r}) \dot{x}_{Q'} - \sum_s (\rho A_Q \hat{v}_{Qr}) \dot{x}_Q \hfill (7.13)$$

and $\quad \dfrac{dp_y}{dt} = F_y + \sum_s (\rho A_{Q'} \hat{v}_{Q'r}) \dot{y}_{Q'} - \sum_s (\rho A_Q \hat{v}_{Qr}) \dot{y}_Q. \hfill (7.14)$

Equations (7.13) and (7.14) give the equation of motion of a system whose mass is varying with time.

Example 7.1 Ram jet missile Figure 7.3 shows an anti-tank missile being fired from its launching tube. The solid fuel rocket motor is designed so that it burns fuel at a constant rate $\dot{m}_e$ and expels it from the exhaust nozzle at a

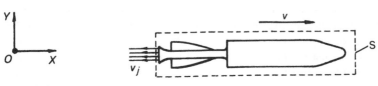

FIG. 7.3

velocity v_j relative to the missile. The positive direction of the vector v_j is chosen to be in the *opposite* direction from that of the velocity v of the missile. If the missile has mass m_0 and is designed to carry mass m_1 of fuel, let us determine the velocity v of the missile at time t after launch. We will assume that the missile travels horizontally during its short flight.

Let us draw a system boundary around the missile as shown by the dotted line in Fig. 7.3. Since fuel is burnt at a constant rate $\dot{m}_e$ the mass of the missile at time t is given by the solution to the differential equation

$$\frac{dm}{dt} = -\dot{m}_e. \tag{i}$$

The minus sign in this equation indicates that mass is being lost from the system. Equation (i) can be integrated directly to give

$$m = (m_0 + m_1) - \dot{m}_e t \quad \text{for } t \leqslant \frac{m_1}{\dot{m}_e}$$

and $\quad m = m_0 \qquad\qquad\qquad \text{for } t > \dfrac{m_1}{\dot{m}_e}, \tag{ii}$

where $(m_0 + m_1)$ is the missile mass at launch, i.e. at $t = 0$.

If the velocity of the missile at time t is assumed to be v then the linear momentum of the system contained within the boundary S is given by

$$p_x = mv, \tag{iii}$$

where m is given by eqn (ii).

The *absolute* velocity of the fuel leaving the system boundary, measured in the direction of the missile motion, is

$$\dot{x}_j = (v - v_j). \tag{iv}$$

For this system eqn (7.9) gives the equation of motion in the direction OX as

$$\frac{dp_x}{dt} = F_x - \dot{m}_e \dot{x}_j, \tag{v}$$

where F_x is the external force applied to the missile and is due to air resistance.

If, for the purposes of illustration, we assume that $F_x = 0$ the substitution of eqns (i), (iii) and (iv) into eqn (v) gives

$$[(m_0 + m_1) - \dot{m}_e t]\frac{dv}{dt} = \dot{m}_e v_j. \tag{vi}$$

The left-hand side of eqn (vi) represents the instantaneous mass of the rocket multiplied by its acceleration. The term on the right-hand side therefore has units of force and may be interpreted as the force exerted on the missile by the jet. Thus the thrust developed developed by the rocket motor is given by $\dot{m}_e v_j$. This shows the importance of ensuring that the nozzle velocity v_j is made as large as possible.

This first order differential equation can be separated and using the condition that $v = 0$ at $t = 0$ gives

$$\int_0^v \frac{dv}{\dot{m}_e v_j} = \int_0^t \frac{dt}{(m_0 + m_1 - \dot{m}_e t)}. \tag{vii}$$

Integrating and rearranging gives

$$v = v_j \ln \frac{m_0 + m_1}{(m_0 + m_1 - \dot{m}_e t)} \quad \text{for } t \leqslant \frac{m_1}{\dot{m}_e}. \tag{viii}$$

The maximum velocity reached by the missile occurs at the instant the fuel runs out, i.e. at $t = m_1/\dot{m}_e$, so that

$$v_{\max} = v_j \ln\left(1 + \frac{m_1}{m_0}\right) \tag{ix}$$

Thus the maximum velocity reached depends upon the ratio m_1/m_0 and v_j. Once the fuel has been used up the velocity remains constant at $v_{\max}$.

It should be appreciated that eqn (viii) represents an upper estimate for v since air resistance has been neglected in the calculation.

7.2 Conditions for steady flow

Let us suppose that we observe a system at time t and note the directions of the momentum vectors at all points in the system and the shape of the system boundary, as shown in Fig. 7.4(a). Suppose we now repeat the observation at

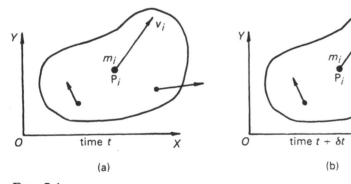

(a) (b)

Fig. 7.4

time δt later and find that the shape of the system boundary has been preserved and that the momentum vectors are unchanged, despite the flow of fluid across the system boundary. This state will occur when the particle passing through the typical point P_i in the system is immediately followed by an identical particle which has the same velocity.

In these circumstances

$$\frac{\mathrm{d}p_x}{\mathrm{d}t} = \frac{\mathrm{d}p_y}{\mathrm{d}t} = 0 \tag{7.15}$$

i.e. the momentum of the system is constant.

For this special case the behaviour of the system is *independent* of time and is said to be 'steady'. Equations (7.8) and (7.9) now give the total external forces applied to the system as

$$F_x = \sum_s \dot{m}_Q \dot{x}_Q - \sum_s \dot{m}_{Q'} \dot{x}_{Q'}$$

$$= \sum_s \rho A_Q \hat{v}_{Q'r} \dot{x}_Q - \sum_s \rho A_{Q'} \hat{v}_{Q'r} \dot{x}_{Q'} \tag{7.16}$$

and $$F_y = \sum_s \dot{m}_Q \dot{y}_Q - \sum_s \dot{m}_{Q'} \dot{x}_{Q'}$$

$$= \sum_s \rho A_Q \hat{v}_{Q'r} \dot{y}_Q - \sum_s \rho A_{Q'} \hat{v}_{Q'r} \dot{y}_{Q'}. \tag{7.17}$$

The forces F_x and F_y are determined by the net rate at which momentum is transferred *out* of the system.

For steady conditions to prevail within the system it is also necessary that the total mass of the system remains constant. Since the mass lost by the system in time δt is given by summing $(\delta m_Q - \delta m_{Q'})$ around the system boundary we require

$$\sum_s (\delta m_Q - \delta m_{Q'}) = 0$$

for the mass content to remain unchanged, so that

$$\sum_s \dot{m}_Q = \sum_s \dot{m}_{Q'}. \tag{7.18}$$

Thus, for steady flow, mass must enter and leave the system at the same rate.

Equations (7.16), (7.17) and (7.18) can be used to calculate the forces applied to a system due to the steady flow of fluid into and out of the system.

Example 7.2 Water jet   Figure 7.5 shows a jet of water emerging with a steady velocity v_0 from a nozzle of cross sectional area A. The water then passes smoothly over the surface of a curved plate without loss of velocity and finally leaves the plate at an angle α with respect to its initial direction. We shall calculate the resultant force applied to the plate by the water.

Let us draw a system boundary around the fluid which is passing over the plate, as shown in Figs. 7.5 and 7.6. Q' and Q represent the regions, each of area A, where the fluid enters and leaves the system boundary. The system boundaries at Q and Q' are stationary and are chosen to be perpendicular to the direction of the fluid flow.

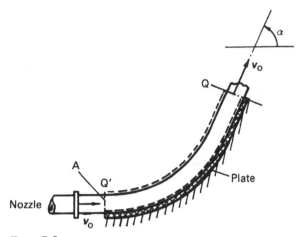

FIG. 7.5

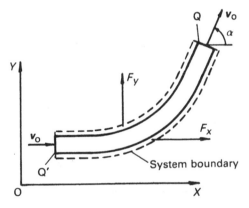

FIG. 7.6

Thus, if the velocity of the fluid is assumed to be uniform across the areas Q and Q', we have

$$\hat{v}_{Qr} = \hat{v}_{Q'r} = v_0. \qquad (i)$$

The mass flow rate into and out of the system is given by

$$\dot{m}_{Q'} = \dot{m}_{Q} = \rho A v_0. \qquad (ii)$$

If the plate applies a force F to the system boundary then the components of F are given by eqns (7.17) and (7.16) as

$$F_x = \sum_s \dot{m}_Q \dot{x}_Q - \sum_s \dot{m}_{Q'} \dot{x}_{Q'}$$

$$(iii)$$

and $$F_y = \sum_s \dot{m}_Q \dot{y}_Q - \sum_s \dot{m}_{Q'} \dot{y}_{Q'}.$$

From Fig. 7.6

$$\dot{x}_Q = v_0 \cos \alpha \qquad \dot{y}_Q = v_0 \sin \alpha$$

and $\dot{x}_{Q'} = v_0, \qquad \dot{y}_{Q'} = 0.$ (iv)

Substituting eqns (iv) into eqns (iii) gives

$$F_x = \rho A v_0 (v_0 \cos \alpha - v_0) = -\rho A v_0^2 (1 - \cos \alpha)$$

and $F_y = \rho A v_0^2 \sin \alpha.$ (v)

The force F_p applied to the plate is, by Newton's Third Law, equal and opposite to F.

The components of F_p are

$$F_{px} = -F_x = \rho A v_0^2 (1 - \cos \alpha)$$

and $F_{py} = -F_y = -\rho A v_0^2 \sin \alpha,$ (vi)

as shown on the vector diagram of Fig. 7.7.

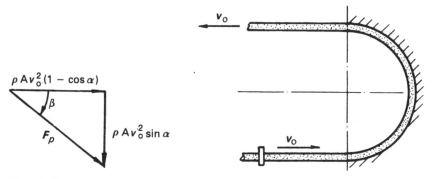

$\rho A v_0^2 (1 - \cos \alpha)$

β

F_p $\rho A v_0^2 \sin \alpha$

Fig. 7.7   Fig. 7.8

The magnitude of F_p is

$$F_p = \rho A v_0^2 [(1 - \cos \alpha)^2 + \sin^2 \alpha]^{1/2}$$
$$= 2\rho A v_0^2 \sin(\alpha/2),$$

and F_p has a line of action which is inclined to the horizontal at an angle given by

$$\tan \beta = \frac{\sin \alpha}{1 - \cos \alpha} = \frac{1}{\tan(\alpha/2)},$$

i.e. $\beta = (\pi - \alpha)/2.$

This force F_p has a maximum value $2\rho A v_0^2$ which occurs when $\alpha = \pi$, i.e. when the flow is turned through 180°, as shown in Fig. 7.8.

Engineering Applications

7.1 Impulse turbine

Let us now show how the steady flow momentum eqns (7.16) and (7.17) can be used to calculate the output of an impulse turbine.

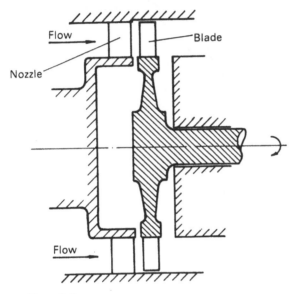

Fig. 7.9

Figure 7.9 shows the construction of such a turbine and shows that the fluid is directed into the moving blades by a set of stationary nozzles fixed in the casing of the machine. The shapes of the nozzle and the blades are shown in Fig. 7.10.

The nozzle in an impulse turbine has a fixed outlet angle β and the blades are designed to have equal inlet and outlet angles α. We shall treat the space between the blades as our system.

If we assume that the blades are thin then our system will have the shape shown in Fig. 7.11.

The regions marked Q' and Q define the inlet and outlet sections of the blade passage and correspond to the region of the system boundary described in Section 7.1, where the fluid enters and leaves the system.

Since the blades are assumed to have a peripheral velocity v_b, and since the system boundary S is constrained to move with the blades, the mass flow rate $\dot{m}$ into and out of the system will be determined, according to eqns (7.11) and

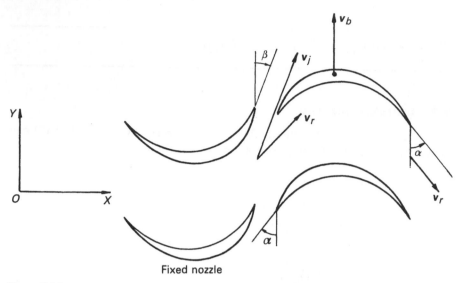

Fixed nozzle

FIG. 7.10

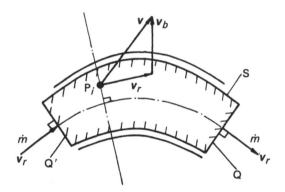

FIG. 7.11

(7.12) by the velocity v_r of the fluid relative to the blades. The system boundary at Q′ and Q is chosen to be perpendicular to the local directions of v_r at entry to and exit from the blades. Furthermore if it is assumed that these sections have equal cross-sectional areas, i.e. $A_Q = A_{Q'}$, then the magnitude of v_r at entrance and exit will remain unchanged. The absolute velocity v of the fluid at any point P_i in the system can be calculated from the vector equation

$$v = v_b + v_r. \qquad \text{(i)}$$

At inlet to the blades the velocity v is equal to the velocity v_j of the fluid leaving the nozzles and can be calculated in terms of v_b and v_r from eqn (i) using the vector diagram shown in Fig. 7.12. The angles α and β shown on

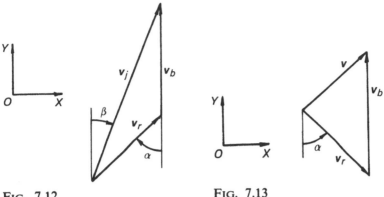

FIG. 7.12   FIG. 7.13

this diagram correspond to the blade inlet and nozzle outlet angles shown in Fig. 7.10.

The components of the *absolute* velocity of fluid *entering* the system are therefore

$$\dot{x}_{Q'} = v_j \sin \beta = v_r \sin \alpha \qquad \text{(ii)}$$

and $\dot{y}_{Q'} = v_j \cos \beta = v_b + v_r \cos \alpha.$ (iii)

At exit from the blades, the *direction* of the relative velocity v_r is different from that at inlet. The absolute velocity v at outlet can be found from eqn (i) using the graphical solution shown in Fig. 7.13.

The components of the *absolute* velocity of the fluid *leaving* the system are therefore

$$\dot{x}_Q = v_r \sin \alpha \qquad \text{(iv)}$$

and $\dot{y}_Q = v_b - v_r \cos \alpha.$ (v)

By substituting eqns (ii), (iii), (iv) and (v) into eqns (7.16) and (7.17) the force applied to the system is given by

$$F_x = \dot{m}(\dot{x}_Q - \dot{x}_{Q'}) = 0 \qquad \text{(vi)}$$

and $F_y = \dot{m}(\dot{y}_Q - \dot{y}_{Q'}) = -2\dot{m}v_r \cos \alpha.$ (vii)

The force F_b which is applied to the blades by the fluid is, by Newton's Third Law, equal and opposite to the force applied to the system,

i.e. $F_{bx} = 0$ (viii)

and $F_{by} = 2\dot{m}v_r \cos \alpha.$ (ix)

Eqns (vi) and (viii) show that no axial force is applied to the blades. Therefore, in this special case for which the inlet and outlet angles of the blades are equal, there is no axial force on the wheel and its supporting bearings.

The force F_{by} is tangential to the turbine wheel and will cause the turbine to rotate and to generate power. For each blade the power produced is given by

$$P_b = F_{yb}v_b$$

$$= 2\dot{m}v_bv_r \cos \alpha. \qquad \text{(x)}$$

The rate at which energy is supplied to the turbine from each nozzle is determined by the kinetic energy of the fluid leaving the nozzle per second

i.e. $\qquad P_{in} = \tfrac{1}{2}\dot{m}v_j^2. \qquad \text{(xi)}$

Thus the efficiency of the blades in converting the kinetic energy contained in the fluid into useful work can be defined as

$$\eta_b = \frac{P_b}{P_{in}} = \frac{4v_bv_r}{v_j^2} \cos \alpha. \qquad \text{(xii)}$$

Using eqn (iii) v_r can be eliminated from eqn (xii) giving

$$\eta_b = 4r(\cos \beta - r), \qquad \text{(xiii)}$$

where $\quad r = v_b/v_j.$

When the efficiency η_b is plotted against the speed ratio r, as shown in Fig. 7.14, it can be seen that the blade efficiency has a maximum value. This

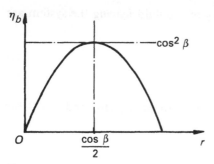

FIG. 7.14

optimum value r_{op} is found by solving $d\eta_b/dr = 0$, and corresponds to

$$r_{op} = \tfrac{1}{2} \cos \beta, \qquad \text{(xiv)}$$

so that

$$(\eta_b)_{\max} = \cos^2 \beta. \qquad \text{(xv)}$$

The condition $\beta = 0$ gives the maximum efficiency, i.e. $\eta_b = 1$. This cannot be realised in practice in an axial flow machine because with $\beta = 0$ the flow would be tangential to the turbine so that the nozzle would interfere with the passage of the blades. Nevertheless, it is advantageous to make β as small as possible. For a given nozzle angle β and jet velocity v_j, the efficiency of the turbine increases as the blade speed is increased until the maximum efficiency at

$$v_b = \frac{v_j \cos \beta}{2} \text{ is reached.}$$

The case $\beta = 0$ corresponds to the condition in a Pelton Wheel turbine.

Exercises

1   Fig. 7.15 shows the thrust reversal system in an aircraft engine. If the mass flow rate through the engine is $\dot{m}$, and is exhausted with a relative velocity v_r as shown, calculate the thrust produced for an aircraft landing with velocity v.

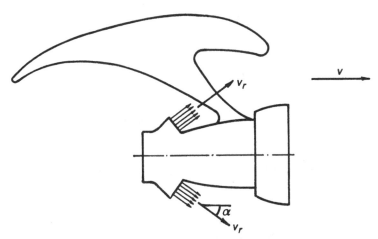

FIG 7.15

2   Fig. 7.16 shows the nozzle of a fire hose supported on its centre line at point A by means of two pinjointed links AB and AC. If the nozzle has inlet and outlet areas A_i and A_o respectively and delivers water at a rate $\dot{m}$, determine the forces at the pin joints B and C. You may assume that the force applied to the nozzle by the hose can be neglected.

3   Fig. 7.17 shows the construction of a water propelled boat. Water enters the propulsion system horizontally through a chute of cross-section A_0 position on the keel. As it passes along the chute its velocity relative to the boat is increased by means of the impeller, I. The water finally leaves the boat horizontally with velocity v_j relative to the boat.

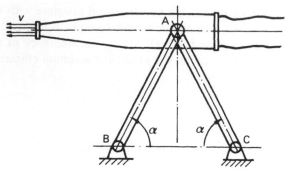

FIG 7.16

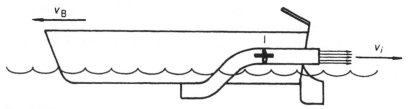

FIG 7.17

If the drag force F on the boat is related to its velocity by

$$F = kv^2$$

calculate the velocity v_j needed to propel the boat at a constant speed v_B.

4   A truck of mass m_0 enters a filling station with an initial velocity v_0. When
the truck reaches the position shown in Fig. 7.18 a sensor causes a hopper

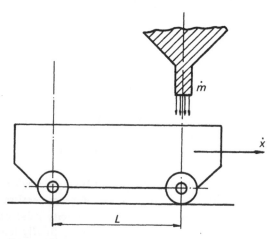

FIG 7.18

to discharge bulk material into the truck at a constant rate $\dot{m}$. After the truck has travelled a further distance L the hopper is closed and the truck is allowed to leave the filling station. Calculate the velocity at which the truck leaves the station.

5 The arrangement shown in Fig. 7.19 is used in an experiment to measure the thrust developed by water flowing from a nozzle of area A and impinging on a flat plate. If the beam CD is in equilibrium with weight mg at end D, derive an expression for the velocity v of the water flowing from the nozzle. Neglect the weights of the beam and plate.

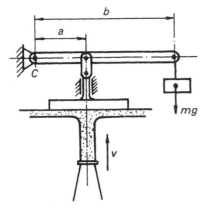

FIG 7.19

6 A jet of water of cross-sectional area A and velocity v_j enters a hole in the block shown in Fig. 7.20. The jet is turned through 90° and exits from the block at right angles to the jet. The block is carried on bearings as shown.

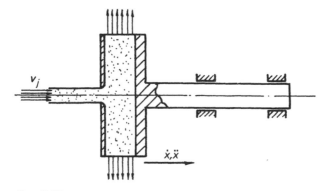

FIG 7.20

Obtain an expression for the acceleration $\ddot{x}$ of the block in terms of its velocity $\dot{x}$ and the velocity of the jet. Assume that the block remains full of water. The mass of the block is m and the mass of the water contained in the cavity is m_w.

Solutions to exercises

Chapter 1

1 $v = at$, $x = \frac{1}{2}at^2 + x_0$, $v = \sqrt{2a(x - x_0)}$

2 $b = \dfrac{8}{\omega}$, $a = \dfrac{3 + 8\omega}{2}$, $\dfrac{m}{s^2}$, $m, \dfrac{1}{s}$.

3 $\dfrac{K}{c}, \dfrac{2.3}{c}$, $1.4\dfrac{K}{c^2}$.

4 82.5 m/s at 76° anticlockwise from OX.

6 (a) 32 m/s²   (b) $\pm\dfrac{16}{\sqrt{3}}$ rad/s²   (c) 40 m/s².

7 $v_0 = \dfrac{v_A + v_B}{2}$, $\omega = \dfrac{v_A - v_B}{2R}$.

8 $a_B = 20.9$ m/s² at 19.2° to the horizontal.

9 178 rev/min. anticlockwise.

10 700 rev/min. anticlockwise   Output would be clockwise.

11 $\dfrac{v}{r}$, 2v.

12 1.74 m/s.

13 $\omega\left(1 + \dfrac{b}{a}\right)$ clockwise.

14 $\omega_2 = \dfrac{\omega_1 r_1}{2r_2}$, $\omega_{arm} = \dfrac{\omega_1}{2}$.

15 $\dfrac{\omega_0}{\omega_i} = \dfrac{(r_D + r_B)(r_C - r_B)}{r_C r_D}$

16 $\dfrac{1000}{11}$ rev/min.

17 $\dot{x} = \varepsilon\omega(\cos\theta - \sin\theta\tan\alpha)$ $\ddot{x} = -\varepsilon\omega^2(\cos\theta\tan\alpha + \sin\theta)$

18 $\dot{\theta} = \dfrac{v}{\sqrt{l^2 - x^2}}$ $\ddot{\theta} = \dfrac{v^2 x}{(l^2 - x^2)^{3/2}}$

19 $\dot{x}_P = -a\dot{\theta}_1\sin\theta_1 - b\dot{\theta}_2\sin\theta_2$
      $\dot{y}_P = a\dot{\theta}_1\cos\theta_1 + b\dot{\theta}_2\cos\theta_2$
      $\ddot{x}_P = -a\ddot{\theta}_1\sin\theta_1 - a\dot{\theta}_1^2\cos\theta_1 - b\ddot{\theta}_2\sin\theta_2 - b\dot{\theta}_2^2\cos\theta_2$
      $\ddot{y}_P = a\ddot{\theta}\cos\theta_1 - a\dot{\theta}_1^2\sin\theta + b\ddot{\theta}_2\cos\theta_2 - b\dot{\theta}_2^2\sin\theta_2.$

20 $\omega_{AB} = 14.4$ rad/s $\circlearrowright$ $\omega_{BC} = 7.2$ rad/s $\circlearrowright$
      $\alpha_{AB} = 115$ rad/s^2 $\circlearrowright$ $\alpha_{BC} = 87.5$ rad/s^2 $\circlearrowleft$
21 $\omega_{O_2B} = 11.5$ rad/s, $\circlearrowright$ $\alpha_{O_2B} = 2230$ rad/s^2 $\circlearrowleft$

22   576 rad/s^2 $\circlearrowright$ 7200 mm/s^2
                             $\rightarrow$

23 $\omega_{BC} = 4.8$ rad/s $\circlearrowright$ $\omega_{CD} = 4.71$ rad/s $\circlearrowleft$
      $\alpha_{BC} = 23$ rad/s^2 $\circlearrowleft$ $\alpha_{CD} = 16$ rad/s^2 $\circlearrowleft$

24   2.24 m/s   137 m/s^2.

Chapter 3

1 $\dfrac{W}{\left(\dfrac{r}{l} + \sqrt{1 - \dfrac{r^2}{l^2}}\right)}$; A, B and C would be collinear

2   10.3 kN,   7.44 kN.

3 $\dfrac{3W}{\sqrt{2}}$,   1.55 W.

4 $\sqrt{2}$ kN (tension),   3/2 kN.
                                   $\leftarrow$

5   8 T.

6   Along DC,   No.

7   $0.707P$ 225° clockwise from AB, $0.707P$ along CB.

8   $\tan \alpha$.

9   350 N,   150 N.

10   16.9 kN.

11   $\cos \theta = \dfrac{R - h}{R}$.

12   8.6 Nm.

13   3111 N,   3111 N,   2786 N.

14   AD 3440 N,   DB 342 N,   DE 1730 N,
     EC 4500 N. (all in compression).

15   17 500 N;   5200 N,   22 500 N.

16   1570 N,   1790 N (tension).

17   $\dfrac{(b - e\mu)M}{\mu r(a + b)}$, $M\dfrac{[(a + \mu e)^2 + \mu^2(a + b)^2]^{1/2}}{r\mu(a + b)}$

18   $\dfrac{2(r_1 + r_2)}{r_1}$

19   $\dfrac{\mu PR}{2 \sin \alpha}$

20   $\dfrac{M_0(a_2 e^{-\mu\beta} - a_1)}{br(1 - e^{-\mu\beta})}$

21   $x = 1.5$ m, $y = 2$ m.

22   0.785 r.

23   0, $3R/8$.

24   9.8 kN,   7.6 kN.

25   20.9 kN,   17.6 kN 48.9° clockwise from OC.

26 12 N/m^2, 48 N/m^2,   0.25 m.

27 200 N,   200 Nm.

28 $2W$, Wa.

29 57.5 kN,   4 m from A.

30 0.207 L.

31 $\dfrac{\omega L^2}{9\sqrt{3}}$, $\dfrac{L}{\sqrt{3}}$

32 0,   300 N,   300 Nm.

Chapter 4

1 $F = P\left(1 - \dfrac{a}{l}\right).$

2 $\tan \theta = \dfrac{a_0}{g + ka_0}$

$T = ma_0[1 + (k + g/a_0)^2]^{1/2}.$

3 $\tan \theta = \dfrac{\omega^2 a}{g}(1 + (b/a)\sin \theta).$

4 $\omega_{\max} = 23.8 \text{ rad/s}$, $\omega_{\min} = 6.6 \text{ rad/s}.$

5 $M_A = \dfrac{mP}{(m^* + m)} \cdot \dfrac{l}{2}.$

6 $T = 1.25 \cos \theta(\sin \theta + 0.3 \cos \theta) \text{ Nm}$

7 $\mu = 0.254$

9 $I_C = \dfrac{68}{96} m^* R^2$

10 $I_X = P\dfrac{R^3}{6}[2 + 3\pi - 6\alpha - 2\cos 2\alpha + 3\sin 2\alpha]$

$I_Y = P\dfrac{R^3}{6}[2 + 3\pi - 6\alpha + 2\cos 2\alpha - 3\sin 2\alpha]$

$I_Z = PR^3[\tfrac{2}{3} + \pi - 2\alpha]$

11 $I_z = \dfrac{m}{12}(b^2 + c^2).$

12 $\theta = 10t - 5\sin 2t.$

13 147.2 N.

15 $a_C = 0.981$ m/s^2

$F = 35.32$ N.

16 $F_N = \dfrac{me\omega^2}{\sqrt{1 + \mu^2}},$ $\qquad F_T = \dfrac{\mu me\omega^2}{\sqrt{1 + \mu^2}},$ $\qquad P = \dfrac{\mu me\omega^3 r}{\sqrt{1 + \mu^2}}.$

17 $\dfrac{3T}{8ma^2}$

18 $\dot{\omega}_1 = 59.7$ rad/s^2

$\omega_s = 80$ rad/s.

19 $F = 6$ N, $\qquad$ 5.16 N.

20 $\ddot{x} = Tr/(2I + mr^2).$

21 $\ddot{\theta} = \dfrac{r_1 mg}{I_O + mr_1^2}$ $\qquad P = \dfrac{mg}{1 + \dfrac{mr_1^2}{I_O}}.$

22 $\tan \alpha = 3\mu.$

23 $\ddot{x} = F \Big/ \left(\dfrac{3m}{4} + M\right).$

24 $\ddot{\theta} = \dfrac{2Pa\sin\theta}{m(k^2 + a^2)}$

25 $T = \left[I_1 + (r_1 + r_2)^2 m + \left(\dfrac{r_1 + r_2}{r_2}\right)^2 I_2\right]\ddot{\theta}_1.$

26 $F = \dfrac{ma_A}{\left(1 + \dfrac{mh^2}{I_G}\right)}$

Chapter 5

1 $v = \sqrt{2gl(1 - \cos \beta)}$

2   (i) $\frac{1}{2}k_1 x^2$, $\frac{1}{2}k_2 x^2$, $(k_1 + k_2)$

   (ii) $\dfrac{1}{2} \dfrac{k_1 x^2}{\left(1 + \dfrac{k_1}{k_2}\right)^2}$, $\dfrac{1}{2} \dfrac{k_2 x^2}{\left(1 + \dfrac{k_2}{k_1}\right)^2}$, $\dfrac{k_1 k_2}{k_1 + k_2}$

3   (b) Requires smaller mass m_c for a given closing velocity.

4 $Fr(1 - \cos \alpha)$,

   $Fr\omega \sin \alpha$.

5 $\dfrac{h}{r} \sqrt{\dfrac{2k}{m}}$

6 $\dot{\theta} = \sqrt{\dfrac{mg(1 - \cos \theta) - k\theta^2}{I_O}}$, $\ddot{\theta} = \dfrac{lmg \sin \theta}{2I_O} - \dfrac{k\theta}{I_O}$,

   $\rightarrow R_x = -\dfrac{mlk\pi}{2I_O}$

   $\uparrow R_y = mg + \dfrac{m^2 g l^2}{I_O} - \dfrac{mlk\pi^2}{2I_O}$.

7 $p = \dfrac{1}{\sqrt{12}} m$

8 $\dot{\theta}_1^2 = \dfrac{2M_2(R_1/R_2)\theta_1}{\left(I_1 + \left(\dfrac{R_1}{R_2}\right)^2 I_2\right)}$

9 $T = \dfrac{m}{2}\left[\dfrac{a^2 \dot{\theta}^2}{3} + v^2 + av\dot{\theta} \sin \theta\right]$

10 $\dfrac{2mv_1^2}{3}$

11 $\sqrt{\dfrac{2g}{r}}$

12 $\left(\dfrac{2(m_L + m_P)gx}{(m_L + m_P) + \dfrac{5I_O}{R_2}}\right)^{1/2}$

13   8.1 m.

14 $\omega_1^2 = \dfrac{2\pi r P + m r^2 \omega_0^2 - 2\pi \rho r^2 g}{m r^2 + 2\pi r^3 \rho}$, $P = \rho r^2 \omega_0^2 + \rho r g$

15 $\dot{\theta}^2 = \dfrac{4[\,mgl \sin\theta - k l^2 (1 - \cos\theta)^2\,]}{(m l^2 + 4I)}$

16   17.5 rad/sec.

17   1166.7 Nm.

Chapter 6

1 $v + \dfrac{m_1 v_s}{m_1 + m_2}$

2 $m v_0$, $\dfrac{m v_0}{\Delta t}$

3 $v_0 \sin\theta$, $v_0 \cos\theta$.

4 $\dfrac{M}{k}\left[\, g + \sqrt{g^2 + \dfrac{2kgh}{M + m}}\,\right]$

5   0.49 m.

6 $v_{\mathrm{B}} = \dfrac{m_p v (1 + e) \cos^2\alpha}{m + m_p \cos^2\alpha}$

7 $\dfrac{I_1 \omega}{(I_1 + I_2)}$, $\dfrac{1}{2}\dfrac{I_1 I_2 \omega^2}{(I_1 + I_2)}$

8 $\Delta\omega = -\dfrac{2\Delta m a v}{I - 2\Delta m a^2}$

9 $\alpha = \cos^{-1}\left(\dfrac{5}{13}\right)$.

10 $\omega = \dfrac{a m_b \sqrt{2gh}}{\dfrac{m_r b^2}{3} + m_b a^2}$

11 $\left(\dfrac{I_O + 3 m a^2}{I_O + 3 m b^2}\right)\Omega_0$,   NO.

12   (i) $v_0 \left[\dfrac{h}{r} - \mu \dfrac{\sqrt{r^2 - h^2}}{r} \right]$

 (ii) $\dfrac{m v_0 h}{I_0 + m r^2}$

13   $\dfrac{v_1}{3}(1 + 2\cos\alpha)$

14   $\omega_1 = \dfrac{\omega_0}{1 + m_2/m_1}$, $\omega_2 = \dfrac{r_1}{r_2}\dfrac{\omega_0}{1 + m_2/m_1}$,   NO.

15   $H_A = \dfrac{3}{2} mrv$ clockwise, $h_0 = \dfrac{mrv}{2}$ clockwise,

 $h_A = \dfrac{3}{2} mrv$ clockwise, $h_B = \dfrac{3}{2} mrv$ clockwise.

16   $H_O = (R - r)\left[\dfrac{2R + r}{2} m_p + R m_s + \dfrac{(R + r)}{3} m_r \right]\omega$ clockwise;

 $h_A = \dfrac{3}{2} m_p r^2 \left(1 + \dfrac{R}{r}\right)\omega$ anticlockwise.

17   $\omega_1 = \dfrac{I_1 \omega_0}{I_1 + I_2 \left(\dfrac{r_1}{r_2}\right)^2}$ $\omega_2 = -\dfrac{(r_1/r_2)I_1 \omega_0}{I_1 + I_2 \left(\dfrac{r_1}{r_2}\right)^2}$

18   $\ddot{\theta} = \dfrac{g \sin\theta}{l\left(\dfrac{1}{3} + \sin^2\theta\right)}$, $\ddot{x}_B = l\ddot{\theta}\cos\theta$.

20   $\omega = \dfrac{3}{2}\left(\dfrac{b}{a^2 + b^2}\right) \cdot v$;

 $v_{\min}^2 = \dfrac{4}{3}\left(\dfrac{a^2 + b^2}{b^2}\right)[(a^2 + b^2)^{1/2} - \dfrac{a}{2} - \dfrac{b\sqrt{3}}{2}]g$

21   $x_C = 0.1t$ $\theta = 0.5t$.

22   $P = m\dfrac{k}{l}(1 + \sigma)\sqrt{2gb}$, $b = k^2/l$ gives $Q = 0$.

 $Q = m(1 + \sigma)\sqrt{2gb}\left(\dfrac{b}{k} - \dfrac{k}{l}\right)$.

Chapter 7

1 $\dot{m}(v + v_r \cos \alpha)$

2 $\dfrac{\dot{m}^2(A_i - A_0)}{2\rho A_0 A_i \cos \alpha}$

3 $v_B\left(1 + \dfrac{k}{\rho A_0}\right)$

4 $v_0 \exp\left(-\dfrac{\dot{m}L}{m_0 v_0}\right)$

5 $v^2 = \dfrac{bmg}{a\rho A}$

6 $\ddot{x} = \dfrac{\rho A v_j(v_j - \dot{x})}{m + m_\omega}$

Index